ISNM
International Series of Numerical Mathematics
Vol. 125

Multivariate Approximation and Splines

Edited by

Günther Nürnberger
Jochen W. Schmidt
Guido Walz

Springer Basel AG

Editors:

Günther Nürnberger
Fakultät für Mathematik und Informatik
Universität Mannheim
D-68131 Mannheim
Germany
e-mail: nuernberger@math.uni-mannheim.de

Jochen W. Schmidt
Institut für Numerische Mathematik
Technische Universität Dresden
D-01062 Dresden
Germany
e-mail: jschmidt@math.tu-dresden.de

Guido Walz
Fakultät für Mathematik und Informatik
Universität Mannheim
D-68131 Mannheim
Germany
e-mail: walz@math.uni-mannheim.de

1991 Mathematics Subject Classification 41A15, 41-06

A CIP catalogue record for this book is available from the Library of Congress, Washington D.C., USA

Deutsche Bibliothek Cataloging-in-Publication Data
Multivariate approximation and splines / ed. by Günther Nürnberger ... – Basel
; Boston ; Berlin : Birkhäuser, 1997
(International series of numerical mathematics ; Vol. 125)
ISBN 978-3-0348-9808-9 ISBN 978-3-0348-8871-4 (eBook)
DOI 10.1007/978-3-0348-8871-4

Originally published by Birkhäuser Verlag in 1997
Softcover reprint of the hardcover 1st edition 1997

Printed on acid-free paper produced of chlorine-free pulp. TCF ∞
Cover design: Heinz Hiltbrunner, Basel

ISBN 978-3-0348-9808-9

9 8 7 6 5 4 3 2 1

Contents

Preface

This book contains the refereed papers which were presented at the international conference on "Multivariate Approximation and Splines" held in Mannheim, Germany, on September 7–10, 1996. Fifty experts from Bulgaria, England, France, Israel, Netherlands, Norway, Poland, Switzerland, Ukraine, USA and Germany participated in the symposium.

It was the aim of the conference to give an overview of recent developments in multivariate approximation with special emphasis on spline methods. The field is characterized by rapidly developing branches such as approximation, data fitting, interpolation, splines, radial basis functions, neural networks, computer aided design methods, subdivision algorithms and wavelets.

The research has applications in areas like industrial production, visualization, pattern recognition, image and signal processing, cognitive systems and modeling in geology, physics, biology and medicine.

In the following, we briefly describe the contents of the papers.

Exact inequalities of Kolmogorov type which estimate the derivatives of multivariate periodic functions are derived in the paper of BABENKO, KOFANOV and PICHUGOV. These inequalities are applied to the approximation of classes of multivariate periodic functions and to the approximation by quasi-polynomials.

BAINOV, DISHLIEV and HRISTOVA investigate initial value problems for nonlinear impulse differential-difference equations which have many applications in simulating real processes. By applying iterative techniques, sequences of lower and upper solutions are constructed which converge to a solution of the initial value problem.

The construction of bivariate biorthogonal cosine wavelets on certain rectangular grids with bell functions not necessarily of tensor product type is the aim of the paper by BITTNER, CHUI and PRESTIN. The biorthogonal system, the frame and the Riesz basis conditions are given explicitly. A main tool are bivariate folding operators.

DAVYDOV, SOMMER and STRAUSS give a survey of recent developments in multivariate interpolation by functions from arbitrary finite-dimensional spaces. A basic result says that almost interpolation sets are characterized by a Schoenberg-Whitney type condition. Of special interest are spaces of generalized splines defined on polyhedral partitions.

In a further paper, these authors describe methods for constructing almost interpolation sets. This is done for spaces with locally independent systems of basis functions. Several examples of such systems, including translates of box splines and finite-element functions, are given.

DELVOS introduces the concept of harmonic Hilbert spaces in the multivariate setting as an extension of periodic Hilbert spaces. Approximation methods for these spaces are studied via Fourier partial integrals and exponential-type interpolation.

The classical results of Runge and Faber show that in general, interpolating polynomials do not converge to the given function. In the univariate and multivariate case, VON GOLITSCHEK investigates further interpolation operators for which the operator norm grows with the dimension of the approximation spaces.

In the paper of KAMONT, the several characterizations of multivariate Besov spaces are given, which only involves values of the functions on dyadic points. The results are used to study the regularity of realizations of random fields such as fractional Brownian motion, fractional Lévi fields and fractional anisotropic Wiener fields.

LENZE constructs three-layer feedforward neural networks which are used for one-sided approximation and interpolation of regular gridded data. The concrete networks are obtained in real-time by using a one-shot learning scheme. An application of this strategy is discussed.

A review of unconstrained minimization of multivariate quadratic splines and of some related problems is given by LI. In this context, the author discusses numerical methods for solving such type of problems and error bounds which are useful for analyzing the convergence of the algorithms.

Multivariate interpolation by translates of a given function, such as radial basis function interpolation, has a strong connection with variational principles. LIGHT describes a general approach to a variational principle, gives some applications and shows how the method is related to results of other authors.

LYCHE and SCHERER derive an upper bound for the condition number of the multivariate Bernstein basis with respect to the uniform norm. It is shown that the upper bound grows like $(s+1)^n$ and that it is independent of s for $n \leq s+1$, where s is the number of variables and n is the total degree of the polynomials.

MASON and VENTURINO show that discrete orthogonality formulae hold for four kinds of Chebyshev polynomials. Each formula yields a general quadrature formula of Clenshaw-Curtis type for integrating weighted functions. For weights of Jacobi type, the Clenshaw-Curtis formula reduces to a Gauss quadrature formula. Aspects of error estimates are discussed.

MULANSKY derives properties of tensor products of convex cones from finite-dimensional spaces. It is shown that cones arising in shape preserving interpolation by tensor products are the intersection of injective cones. Therefore, sufficient conditions for multivariate shape constraints can be derived from the univariate conditions.

In general, the computational cost of solving numerical problems grows exponentially with the number of variables. NOVAK and RITTER study numerical integration and show that by applying non-classical methods, a high number of variables can be compensated by a high degree of smoothness of the underlying functions or by a favorable structure of the problem.

NÜRNBERGER, DAVYDOV, WALZ and ZEILFELDER give a survey of recently developed methods for constructing interpolation points for spaces of splines of arbitrary degree and smoothness on general crosscut partitions. For certain regular type partitions, the approximation order of the corresponding interpolating splines is given.

In the paper of PLONKA , conditions for the orthonormality of scaling vectors in terms of its two-scale symbol and the corresponding transfer operator are investigated. In particular, it is shown that well-known conditions for the two-scale symbol and criteria for the transfer operator are equivalent.

The numerical solution of systems of linear equations with positive definite, double symmetric block-Toeplitz matrices is subject of the paper of POTTS, STEIDL and TASCHE. For such matrices, the authors construct optimal and strong type preconditioners by using the Fejèr and Fourier sum of the generating function of the Wiener class. An estimate of the steps needed for the preconditioned conjugate gradient method is given.

REIMER proves results on the size of Gram determinants which are defined via reproducing kernels. As a consequence, it is shown that there exist interpolation points such that the uniform norm of the corresponding Lagrange functions is small. These results also hold for functions on non-compact sets.

The paper of SCHABACK discusses some problems arising in the error analysis of interpolation by radial basis functions. This is done by applying the general theory to the case of natural cubic splines. By applying an optimality principle for quasi-interpolants which reproduce polynomials, the author obtains improved local error bounds for interpolation by natural cubic splines.

The modeling of high resolution radar imaging and clinical magnetic resonance tomography with the aid of coherent wavelets is the subject of the paper by SCHEMPP. It is shown that the construction of matched filter banks depends on Kepler's spatiotemporal strategy applied to quantum holography and can be described by Fourier analysis of the Heisenberg nilpotent Lie group.

Various versions of sampling theorems are discussed by SCHMEISSER and VOSS. It is shown that there is an equivalence between the sampling of signals and the sampling of entire harmonic functions. In this way, a uniqueness result on entire harmonic functions of exponential type is obtained.

SCHMIDT and WALTHER investigate interpolation of data by biquadratic and biquartic splines on rectangular grids. Conditions are developed under which the first partial derivatives of the interpolating splines satisfy certain restrictions. Moreover, it is shown that the interpolation problem is solvable if additional knots are added to the original grid in a suitable way.

By using generalized Laurent operators, STÖCKLER investigates multivariate affine frames which are generated by multiresolution with a single scaling function.

The relations to the transfer operator are described, a new representation of the lifting scheme is given and the connections to generalized Toeplitz operators are discussed.

In conclusion, the editors would like to thank Deutsche Forschungsgemeinschaft and the University of Mannheim for their support, and Birkhäuser-Verlag for agreeing to publish the proceedings in the ISNM series.

Günther Nürnberger
Mannheim

Jochen W. Schmidt
Dresden

Guido Walz
Mannheim

Summer 1997

Multivariate Inequalities of Kolmogorov Type and Their Applications

V. F. Babenko, V. A. Kofanov, and S. A. Pichugov

Abstract. Connections of inequalities of Kolmogorov type given in the form of inequalities for support functions of convex sets with some problems of analysis will be investigated. Our results will be oriented to applications in multivariate approximation. Exact inequalities which estimate L_2-norms of derivatives of a multivariate periodic function with the help of L_∞-norms of this function and its partial derivatives (of higher order) as well as analogues inequalities for primitives will be proved. Some results on approximation of classes of multivariate periodic functions by other classes which are obtained with the help of inequalities for primitives as well as some results on approximation of function classes by quasipolynomials will be presented.

1. Introduction

Exact inequalities of Kolmogorov type for norms of derivatives of univariate functions

$$\left\|x^{(k)}\right\|_{L_p(G)} \leq K \|x\|^{\alpha}_{L_q(G)} \cdot \left\|x^{(n)}\right\|^{1-\alpha}_{L_r(G)}, \tag{1}$$

where $0 < k < n, 0 < \alpha < 1$, are of great importance for many problems of analysis and, in particular, of approximation theory. Surveys of many known exact inequalities of the form (1) are given in [2], [17]. Many applications of such inequalities are connected with problem of approximation of a function class by another class. This problem plays an important role in the solution of many other problems of approximation of function classes. This was shown in particular in Korneichuk's papers [12], [13], in which the method of intermediate approximation was successfully applied to the solution of some difficult problems of approximation theory. On the other hand the problem of approximation of a function class by another one is closely connected with inequalities of Kolmogorov type . This was shown

Multivariate Approximation and Splines
G. Nürnberger, J. W. Schmidt, and G. Walz (eds.), pp. 1–12.

ISBN 3-7643-5654-5.

in particular in Arestov's [1], Klots' [9] and Ligun's [18] papers. Many interesting applications of these connections to the problems of approximation of function classes by polynomials and splines are presented in the monograph [16].

We will continue in Section 2 investigation of the connections between inequalities of Kolmogorov type and problem of approximation of one function class by another class. Our results in this direction will be oriented to applications in multivariate approximation. We will note also connections of inequalities of Kolmogorov type and problems of approximation of one class by another with inequalities for upper bounds of seminorms and estimations of $K-$ functionals on some function classes.

In Section 3 we will present some new exact inequalities of Kolmogorov type for multivariate periodic functions and in Section 4 we will describe some of their applications. We refer the reader to [4]–[7], [11], [20], [21] for some known exact inequalities of such type.

2. Equivalence Theorems

Let X be a real linear space, θ_X be zero in X, $p(x)$ be some (non-symmetric) norm in X,

$$H_{X,p} := \{x \in X : p(x) \leq 1\},$$

$X'(p)$ be the space of linear bounded (with respect to p) functionals on X, $\langle x, y\rangle$ be the value of a functional $y \in X'(p)$ on an element $x \in X$,

$$p^*(y) := \sup\{\langle x, y\rangle : x \in H_{X,p}\}$$

be (non-symmetric) norm in $X'(p)$. If X is a normed space and $p(x) = \|x\|_X$, then $X'(p) = X^*$ is usual dual space. We will use the following notations. If $M, M_1 \subset X, K \subset X'(p), x \in X, y \in X'(p)$, and ψ is an arbitrary sublinear functional on X, then

$$S_M(y) := \sup\{\langle x, y\rangle : x \in M\}$$

is the support function of the set M,

$$S_K(x) := \sup\{\langle x, y\rangle : y \in K\},$$

$$M^0 := \{y \in X'(p) : S_M(y) \leq 1\},$$

$$E(x, M_1)_{X,p} := \inf\{p(x-u) : u \in M_1\},$$

$$E(M, M_1)_{X,p} := \sup\{E(x, M_1)_{X,p} : x \in M\},$$

$$\psi(M) := \sup\{\psi(x) : x \in M\}.$$

In the normed case we will write $H_X, E(x, M_1)_X$ and $E(M, M_1)_X$ instead of $H_{X,\|\cdot\|}, E(x, M_1)_{X,\|\cdot\|}$ and $E(M, M_1)_{X,\|\cdot\|}$ respectively. If $H_1, \ldots, H_m \subset X$, then for $x \in X$ and any $t = (t_1, \ldots, t_m) \in \mathbb{R}^m_+$ set

$$K_p(X; H_1, \ldots, H_m; x; t) := \inf_{\substack{x_j \in \operatorname{cone} H_j,\\ j=1,\ldots,m}} \left\{ p\left(x - \sum_{j=1}^m x_j\right) + \sum_{j=1}^m t_j S_{H_j^0}(x_j) \right\}.$$

Denote by $\mathcal{F}_m$ the set of all lower semicontinuous, concave functions $\Phi : \mathbb{R}_+^m \to \mathbb{R}_+$. For $\Phi \in \mathcal{F}_m$ set $\overline{\Phi}(t) = -\Phi(t)$, if $t \in \mathbb{R}_+^m$, and $\overline{\Phi}(t) = +\infty$, if $t \notin \mathbb{R}_+^m$. Let $\overline{\Phi}^*$ be the Legendre transform of the function $\overline{\Phi}$, i.e. $\overline{\Phi}^*(s) := \sup_{t\in\mathbb{R}^m}\{\langle t, s\rangle - \overline{\Phi}(t)\}$, $s \in \mathbb{R}^m$. Let finally $\sum_{j=1}^m H_j$ be the algebraic sum of sets $H_1, ..., H_m \subset X$.

Theorem 1. *Let $\Phi \in \mathcal{F}_m$; $H, H_1, ..., H_m \subset X$ be arbitrary convex sets including θ_X. Then the following statements are equivalent:*

1) If $y \in X'(p), y \neq \theta_{X'(p)}$ and $S_{H_j}(y) < +\infty, j = 1, ..., m$, then

$$S_H(y) \le p^*(y) \cdot \Phi\left(\frac{S_{H_1}(y)}{p^*(y)}, ..., \frac{S_{H_m}(y)}{p^*(y)}\right).$$

2) If y is such as in 1) and $N = (N_1, ..., N_m) \in \mathbb{R}_+^m$, then

$$S_H(y) \le \overline{\Phi}^*(-N)\, p^*(y) + \sum_{j=1}^m N_j S_{H_j}(y).$$

3) For any $N \in \mathbb{R}_+^m$

$$E\left(H; \sum_{j=1}^m N_j H_j\right)_{X,p} \le \overline{\Phi}^*(-N).$$

4) If sublinear functional ψ on X is such that the values $\psi(H_{X,p})$, $\psi(H_j), j = 1, ..., m$, are finite, then

$$\psi(H) \le \overline{\Phi}^*(-N) \cdot \psi(H_{X,p}) + \sum_{j=1}^m N_j \psi(H_j).$$

5) If ψ from 4) is such that $\psi(H_{X,p}) \neq 0$, then

$$\psi(H) \le \psi(H_{X,p}) \cdot \Phi\left(\frac{\psi(H_1)}{\psi(H_{X,p})}, ..., \frac{\psi(H_m)}{\psi(H_{X,p})}\right).$$

6) For any $x \in H$ and $t \in \mathbb{R}_+^m$

$$K_p(X; H_1, ..., H_m; x; t) \le \Phi(t).$$

Parts 1) and 2) of this Theorem are abstract versions of inequalities (1) of Kolmogorov type in multiplicative and additive forms respectively; part 3) is an estimation of approximation of one set by another; parts 4) and 5) are abstract

versions of inequalities for upper bounds of seminorms ; part 6) is an estimate on the class H for characteristic of $K-$ functional of m spaces type.
Proof: First of all let us prove that for any $t = (t_1, ..., t_m) \in \mathbb{R}_+^m$

$$\Phi(t) = \inf_{N \in \mathbb{R}_+^m} \left\{ \overline{\Phi}^*(-N) + \sum_{j=1}^{m} N_j t_j \right\}. \tag{2}$$

Really, using the definition of $\overline{\Phi}$, Fenhel-Moreau theorem (see, for example, [19, p.49]) and the definition of the Legendre transform we will have

$$\Phi(t) = -\overline{\Phi}(t) = -\overline{\Phi}^{**}(t) = -\sup_{N \in \mathbb{R}^m} \left\{ \sum_{j=1}^{m} (-N_j) t_j - \overline{\Phi}^*(-N) \right\}. \tag{3}$$

Since $\overline{\Phi}^*(-N) = \infty$ for $N \notin \mathbb{R}_+^m$, the sup in (3) can be taken only over $N \in \mathbb{R}_+^m$. Therefore the relation (2) follows from (3).

It is easy to verify with the help of the relation (2) that 1) $\Leftrightarrow$ 2) and 4) $\Leftrightarrow$ 5).

Let us prove that 2) $\Rightarrow$ 3) $\Rightarrow$ 4) $\Rightarrow$ 2). First we will prove that 2) $\Rightarrow$ 3). Using duality theorem for the best approximation by convex set with respect to non-symmetric norm p (see, for example, [14,§1.3]) we obtain

$$E := E\left(H, \sum_{j=1}^{m} N_j H_j\right)_{X,p} = \sup_{x \in H} \sup_{\substack{y \in X'(p), \\ p^*(y) \le 1}} \left\{ \langle x, y \rangle - \sup_{u \in \sum_{j=1}^{m} N_j H_j} \langle u, y \rangle \right\}$$

$$= \sup_{x \in H} \sup_{\substack{y \in X'(p), \\ p^*(y) \le 1}} \left\{ \langle x, y \rangle - \sum_{j=1}^{m} N_j \sup_{u_j \in H_j} \langle u_j, y \rangle \right\}.$$

Changing the positions of the first and the second supremums and taking into account the definition of the support function we will have

$$E = \sup_{\substack{y \in X'(p), \\ p^*(y) \le 1}} \left\{ S_H(y) - \sum_{j=1}^{m} N_j S_{H_j}(y) \right\}. \tag{4}$$

Now the statement 3) follows from the statement 2) and the relation (4).

Show that 3) $\Rightarrow$ 4). For any $x \in H$ and $\varepsilon > 0$ denote by $u_j^\varepsilon(x)$, $j = 1, ..., m$, such elements from H_j that for the element $u^\varepsilon(x) := \sum_{j=1}^{m} N_j u_j^\varepsilon(x)$ the inequality

$$p\left(x - u^\varepsilon(x)\right) \le E\left(H, \sum_{j=1}^{m} N_j H_j\right)_{X,p} + \varepsilon \tag{5}$$

holds true. Since ψ is positive homogeneous we have for $x \neq \theta_X$

$$\psi(x) = \psi\left(\frac{x}{p(x)}\right) p(x) \leq \psi(H_{X,p})p(x). \tag{6}$$

Using (5) and (6) we will have

$$\psi(x) \leq \psi\left(x - u^{\varepsilon}(x)\right) + \psi\left(u^{\varepsilon}(x)\right) \leq \psi\left(H_{X,p}\right) p\left(x - u^{\varepsilon}(x)\right) + \sum_{j=1}^{m} N_j \psi\left(u^{\varepsilon}(x)\right)$$

$$\leq \psi\left(H_{X,p}\right)\left(E\left(H, \sum_{j=1}^{m} N_j H_j\right)_{X,p} + \varepsilon\right) + \sum_{j=1}^{m} N_j \psi(H_j). \tag{7}$$

Now the statement 4) follows from the statement 3) and (7).

Show that 4) $\Rightarrow$ 2). For fixed $y \in X'(p)$ set $\psi(z) := \langle z, y\rangle$. It is clear that the functional ψ is sublinear. Using the statement 4) and the definition of support function we obtain

$$S_H(y) = \sup_{z \in H} \langle z, y\rangle = \psi(H) \leq \overline{\Phi}^*(-N)\psi(H_{X,p}) + \sum_{j=1}^{m} N_j \psi(H_j)$$

$$= \overline{\Phi}^*(-N)p^*(y) + \sum_{j=1}^{m} N_j S_{H_j}(y).$$

The fact that 4) $\Rightarrow$ 2) is proved.

What is left to show that 3) $\Rightarrow$ 6) $\Rightarrow$ 1).

Let us prove that 3) $\Rightarrow$ 6). Using the definition of K_p- functional, we will have for any $t = (t_1, ..., t_m),\ N = (N_1, ..., N_m) \in \mathbb{R}_+^m$ the following relation

$$K_p(X; H_1, ..., H_m; x, t)$$

$$\leq \inf_{\substack{x_j \in \operatorname{cone} H_j, \\ S_{H_j^0}(x_j) \leq N_j, j=1,...,m}} \left\{ p(x - \sum_{j=1}^{m} x_j) + \sum_{j=1}^{m} t_j N_j \right\}. \tag{8}$$

In view of the definition of H_j^0

$$N_j H_j \subset \{x_j : x_j \in \operatorname{cone} H_j : S_{H_j^0}(x_j) \leq N_j\}, \quad j = 1, ..., m.$$

Therefore it follows from the statement 3) and the relation (8) that

$$K_p(X; H_1, ..., H_m; x, t) \leq \inf_{\substack{x_j \in N_j H_j, \\ j=1,...,m}} \left\{ p(x - \sum_{j=1}^{m} x_j) + \sum_{j=1}^{m} t_j N_j \right\}$$

$$\leq E\left(H, \sum_{j=1}^{m} N_j H_j\right)_{X,p} + \sum_{j=1}^{m} t_j N_j \leq \overline{\Phi}^*(-N) + \sum_{j=1}^{m} t_j N_j. \tag{9}$$

Since the inequality (9) holds true for any $N = (N_1, ..., N_m) \in \mathbb{R}_+^m$, the statement 6) follows from (9) and (2).

Finally let us show that 6) $\Rightarrow$ 1). Let $x \in H$, $y \in X'(p)$ $\left(y \neq \theta_{X'(p)}\right)$, $t = (t_1, ..., t_m), N = (N_1, ..., N_m) \in \mathbb{R}_+^m$ be fixed. Taking into account that $p(x) = \sup\{\langle x, y\rangle : p^*(y) \leq 1\}$, we obtain from statement 6) the following inequality:

$$\Phi(t) \geq K_p(X; H_1, ..., H_m; x, t)$$

$$\geq \frac{1}{p^*(y)} \inf_{\substack{x_j \in \text{cone} H_j, \\ j=1,...,m}} \left\{ \left\langle x - \sum_{j=1}^{m} x_j, y \right\rangle + \sum_{j=1}^{m} t_j S_{H_j^0}(x_j) p^*(y) \right\}. \tag{10}$$

Setting in (10) $t_j = S_{H_j}(y)/p^*(y)$ and using the inequality

$$\langle x_j, y\rangle \leq S_{H_j}(y) \cdot S_{H_j^0}(x_j),\ x_j \in H_j,\ y \in X'(p),$$

we obtain

$$\Phi\left(\frac{S_{H_1}(y)}{p^*(y)}, ..., \frac{S_{H_m}(y)}{p^*(y)}\right)$$

$$\geq \frac{1}{p^*(y)} \inf_{\substack{x_j \in \text{cone} H_j, \\ j=1,...,m}} \left\{ \langle x, y\rangle - \sum_{j=1}^{m} \langle x_j, y\rangle + \sum_{j=1}^{m} S_{H_j}(y) S_{H_j^0}(x_j) \right\} \geq \frac{1}{p^*(y)} \langle x, y\rangle .$$

Since this inequality holds true for any $x \in H$, the statement 1) is proved.

This completes the proof.

3. New Multivariate Inequalities

The elements of the standard basis in $\mathbb{R}^m$ will be denoted by e_j, $j = 1, .., m$. Let $L_q(\mathrm{T}^m)$, $1 \leq q \leq \infty$, $\mathrm{T}^m = (0, 2\pi]^m$, be the spaces of $2\pi-$ periodic in each variable functions $x : \mathbb{R}^m \to \mathbb{R}$ with the corresponding norms $\|x\|_p = \|x\|_{L_p(\mathrm{T}^m)}$ (defining $\|x\|_p = \|x\|_{L_p(\mathrm{T}^m)}$ we assume that T^m is endowed with the normed Lebesgue measure).

For $t = (t_1, ..., t_m) \in \mathbb{R}^m$ and $j = 1, ..., m$ set $t^1 = (t_2, ..., t_m)$, $t^m = (t_1, ..., t_{m-1})$ and $t^j = (t_1, ..., t_{j-1}, t_{j+1}, ..., t_m), 1 < j < m$. Denote by $L_p^j = L_p^j(\mathrm{T}^m)$ the spaces of functions $x \in L_p(\mathrm{T}^m)$ of the form $x(t) = g\left(t^j\right)$, where $g \in L_p\left(\mathrm{T}^{m-1}\right)$. Given set $M \subset \{1, ..., m\}$ let

$$L_p^M = L_p^M(\mathrm{T}^m) := \sum_{j\in M} L_p^j(\mathrm{T}^m).$$

For $j = 1, ..., m$ denote by $L_{p,j} = L_{p,j}(\mathrm{T}^m)$ the set of functions $x \in L_p(\mathrm{T}^m)$ such that $\int\limits_{\mathrm{T}^1} x(t)\,dt_j = 0$ for almost every t^j, and for $M \subset \{1, ..., m\}$ set

$$L_{p,M} = L_{p,M}(\mathrm{T}^m) := \bigcap_{j \in M} L_{p,j}(\mathrm{T}^m).$$

In the case $M = \{1, ..., m\}$ we will write $L_{p,0}$ instead of $L_{p,M}$.

Our approach will be based on concepts of fractional integral and of fractional derivative in sense of Weil. Let us give these concepts in the form useful for us.

Let vector $\gamma = (\gamma_1, ..., \gamma_m) \in \mathbb{R}_+^m$ be given and let $\mathrm{supp}\gamma := \{j : \gamma_j \neq 0\}$. The fractional integral $I_\gamma x$ of function $x \in L_{1,\mathrm{supp}\gamma}$ will be defined as convolution

$$I_\gamma x(t) := (B_\gamma * x)(t) = (2\pi)^{-m} \int\limits_{\mathrm{T}^m} B_\gamma(t - \tau)\, x(\tau)\, d\tau,$$

where for $u = (u_1, ..., u_m)$

$$B_\gamma(u) := \prod_{j:\gamma_j \neq 0} B_{\gamma_j}(u_j) = \prod_{j:\gamma_j \neq 0} 2 \sum_{k=1}^{\infty} k^{-\gamma_j} \cos\left(k u_j - \gamma_j \frac{\pi}{2}\right).$$

It follows from the embedding theorem that if $\gamma \in \mathbb{R}_+^m$ and $p, s \in [1, \infty]$ are such that

$$a)\ \min\{\gamma_j : j = 1, ..., m\} > p^{-1} - s^{-1} \quad \text{or} \quad b)\ p \geq s,$$

then for any $x \in L_{p,\mathrm{supp}\gamma}$ the function $I_\gamma x$ belongs to $L_s(\mathrm{T}^m)$ and the operator $I_\gamma : L_{p,\mathrm{supp}\gamma} \to L_s(\mathrm{T}^m)$ is bounded. Let such γ, p, s, be given. Define the corresponding Sobolev space

$$\mathcal{L}_{p,s}^\gamma = \mathcal{L}_{p,s}^\gamma(\mathrm{T}^m) := \{y : y = I_\gamma x + h : x \in L_{p,\mathrm{supp}\gamma}, h \in L_s^{\mathrm{supp}\gamma}\}$$

and the corresponding Sobolev class

$$\mathcal{W}_{p,s}^\gamma = \mathcal{W}_{p,s}^\gamma(\mathrm{T}^m) := \left\{y \in \mathcal{L}_{p,s}^\gamma : y = I_\gamma x + h; \|x\|_p \leq 1\right\}.$$

Note that the representation $y = I_\gamma x + h$ $(x \in L_{p,\mathrm{supp}\gamma},\ h \in L_s^{\mathrm{supp}\gamma})$ for function $y \in \mathcal{L}_{p,s}^\gamma$ is unique. Therefore for such functions y the fractional derivative $\mathcal{D}^\gamma y$ of order γ is correctly defined by formula

$$\mathcal{D}^\gamma y = \mathcal{D}^\gamma(I_\gamma x + h) = x.$$

For $s,\ t \in \mathbb{R}^m$ set $(t, s) := \sum_{j=1}^m t_j s_j$ and $|s| = \sum_{j=1}^m |s_j|$. Note that if $x \in L_{p,\mathrm{supp}\gamma}(\mathrm{T}^m)$ is defined by its Fourier series

$$x(t) \sim \sum_{k \in \mathbb{Z}^m} \widehat{x}_k e^{i(k,t)},$$

then the Fourier series of $I_\gamma x$ is

$$I_\gamma x(t) \sim \sum_{k:\text{supp}\gamma\subset\text{supp}k} \frac{\widehat{x}_k}{(ik)^\gamma} e^{i(k,t)} \tag{11}$$

(here and below we adopt the convention that $0^0 := 1$ and use the notation $(ik)^\gamma := \prod_{j\in\text{supp}\gamma} |k_j|^{\gamma_j} \exp\left(i\frac{\pi}{2}\gamma_j \text{sgn} k_j\right)$).

For $\alpha, \beta \in \mathbb{R}, \alpha, \beta > 0$, and $n \in \mathbb{N}$ let $\varphi_n(\cdot; \alpha, \beta) := B_n * \varphi_0(\cdot; \alpha, \beta)$, where $\varphi_0(t; \alpha, \beta), t \in \mathbb{R}$, is 2π-periodic function which equals α for $t \in \left[0, \frac{2\pi\beta}{\alpha+\beta}\right)$, and equals $-\beta$ for $t \in \left[\frac{2\pi\beta}{\alpha+\beta}, 2\pi\right)$. Note that $\varphi_n := \varphi_n(\cdot; 1, 1)$ is usual Euler's spline.

In the case $m = 1$ the following inequality is well known and has many applications in approximation theory: if $x \in L^r_{\infty,\infty}(\mathrm{T}^1)$, $r \in \mathbb{N}$, then for any $k \in \mathbb{N}$, $k < r$, and any $p \in [1, \infty]$

$$\left\|x^{(k)}\right\|_{L_p(\mathrm{T}^1)} \leq \frac{\|\varphi_{r-k}\|_{L_p(\mathrm{T}^1)}}{\|\varphi_r\|^{1-k/r}_{L_\infty(\mathrm{T}^1)}} \|x\|^{1-k/r}_{L_\infty(\mathrm{T}^1)} \left\|x^{(r)}\right\|^{k/r}_{L_\infty(\mathrm{T}^1)}. \tag{12}$$

This inequality for $p = \infty$ is a special case of the Kolmogorov's inequality [10, p.277-290]. For $p \in [1, \infty)$ this inequality was proved by Ligun [18].

Note that inequality (12) can be represented as inequality for primitives of function $x \in L_\infty(\mathrm{T}^1)$ which have zero mean value on a period and precisely such form of (12) is useful for many applications.

Non-symmetric generalization of (12) gives us the following inequality which we present as inequality for primitives : if r, k, p are such as in (12), then for any $\alpha, \beta > 0$ and any $x \in L_{\infty,0}(\mathrm{T}^1)$

$$\frac{\|(I_k x)_\pm\|_{L_p(\mathrm{T}^1)}}{\|\varphi_k(\,\cdot\,; \alpha, \beta)_\pm\|_{L_p(\mathrm{T}^1)}} \leq \frac{E_0\left(I_r x\right)^{k/r}_{L_\infty(\mathrm{T}^1)}}{E_0\left(\varphi_r(\,\cdot\,; \alpha, \beta)\right)^{k/r}_{L_\infty(\mathrm{T}^1)}} \left\|\frac{x_+}{\alpha} + \frac{x_-}{\beta}\right\|^{1-k/r}_{L_\infty(\mathrm{T}^1)}, \tag{13}$$

where $x_\pm = \max\{\pm x(t), 0\}$, $E_0(x)_{L_\infty(\mathrm{T}^1)}$ is the best uniform approximation of a function x by constants.

For $p = \infty$ this inequality is a special case of Hörmander's inequality [8]. In the case $p \in [1, \infty)$, $\max\{\alpha, \beta\} = \infty$ the inequality (13) was proved by Ligun (see, for example,[16, p.119]). For $p \in [1, \infty)$ and any $\alpha, \beta > 0$ this inequality was proved by Babenko [3] (see also [15, p.69]).

The following two theorems (Theorems 2, 3) give us multivariate analogs of the inequality (12) for L_2-norms of derivatives and of primitives respectively. In contrast to the case $m = 1$ in multivariate case inequalities for derivatives and primitives are not equivalent and precisely inequalities for primitives are efficient for many applications.

Theorem 2. *Let $\gamma = (\gamma_1, ..., \gamma_m) \in \mathbb{R}^m_+, \gamma_j > 0, \ \ j = 1, ..., m,$ and $r, |\gamma| \in \mathbb{N}, \ |\gamma| < r$. Then for any function $x \in \bigcap_{j=1}^{m} \mathcal{L}^{re_j}_{\infty,\infty}(\mathrm{T}^m)$*

$$\|\mathcal{D}^\gamma x\|_{L_2(\mathrm{T}^m)} \leq \frac{\left\|\varphi_{r-|\gamma|}\right\|_{L_2(\mathrm{T}^1)}}{\|\varphi_r\|^{1-|\gamma|/r}_{L_\infty(\mathrm{T}^1)}} \|x\|^{1-|\gamma|/r}_{L_\infty(\mathrm{T}^m)} \prod_{j=1}^{m} \|\mathcal{D}^{re_j} x\|^{\gamma_j/r}_{L_\infty(\mathrm{T}^m)}. \tag{14}$$

This inequality becomes equality for functions $\psi_{n,r}(t) := \varphi_r((n,t)), \ n \in \mathbb{N}^m$.

In the case $m = 2$ the inequality (14) has been obtained in [4] for some special values of $\gamma \in \mathbb{R}^2_+$.

Theorem 3. *Let $\alpha, \beta > 0; \gamma = (\gamma_1, ..., \gamma_m) \in \mathbb{R}^m_+, \ \prod_{j=1}^m \gamma_j > 0, |\gamma| \in \mathbb{N}; |\gamma| < r_j \in \mathbb{N}, j = 1, ..., m$. Then for any function $x \in L_{\infty,0}(\mathrm{T}^m)$*

$$\frac{\|I_\gamma x\|_{L_2(\mathrm{T}^m)}}{\left\|\varphi_{|\gamma|}(\cdot;\alpha,\beta)\right\|_{L_2(\mathrm{T}^1)}} \leq \prod_{j=1}^{m} \frac{E\left(I_{r_j e_j} x, L^j_\infty\right)^{\frac{\gamma_j}{r_j}}_{L_\infty(\mathrm{T}^m)}}{E_0\left(\varphi_{r_j}(\cdot;\alpha,\beta)\right)^{\frac{\gamma_j}{r_j}}_{L_\infty(\mathrm{T}^1)}} \left\|\frac{x_+}{\alpha} + \frac{x_-}{\beta}\right\|^{1-\sum_{j=1}^{m}\frac{\gamma_j}{r_j}}_{L_\infty(\mathrm{T}^m)}. \tag{15}$$

In particular

$$\frac{\|I_\gamma x\|_{L_2(\mathrm{T}^m)}}{\left\|\varphi_{|\gamma|}\right\|_{L_2(\mathrm{T}^1)}} \leq \prod_{j=1}^{m} \frac{E\left(I_{r_j e_j} x, L^j_\infty\right)^{\gamma_j/r_j}_{L_\infty(\mathrm{T}^m)}}{\left\|\varphi_{r_j}\right\|^{\gamma_j/r_j}_{L_\infty(\mathrm{T}^1)}} \|x\|^{1-\sum_{j=1}^m \frac{\gamma_j}{r_j}}_{L_\infty(\mathrm{T}^m)}. \tag{16}$$

Inequality (15) becomes equality, if $x(t) = \varphi_0((n,t);\alpha,\beta)$, $n \in \mathbb{N}^m$.

We present the proof of Theorem 3. Theorem 2 can be proved similarly.

Proof: Let $x \in L_{\infty,0}(\mathrm{T}^m)$ be given. Using (11) and Parseval's equality we have

$$\|I_\gamma x\|^2_{L_2(\mathrm{T}^m)} = \sum_{k:\mathrm{supp}\gamma\subset\mathrm{supp}k} |\widehat{x}_k|^2 (k^2)^{-\gamma} = \sum_{k:\mathrm{supp}\gamma\subset\mathrm{supp}k} |\widehat{x}_k|^2 (k^2)^{-\sum_{\nu=1}^m |\gamma| e_\nu},$$

where $(k^2)^{-\gamma} = \prod_{\nu=1}^m (k^2_\nu)^{-\gamma_\nu}$ for $k = (k_1, ..., k_m)$, and

$$\|I_{|\gamma|e_j} x\|^2_{L_2(\mathrm{T}^m)} = \sum_{k\in\mathrm{Z}^m} |\widehat{x}_k|^2 (k^2)^{-|\gamma|e_j}.$$

It is not difficult to prove using well known generalized Hölder's inequality that

$$\|I_\gamma x\|_{L_2(\mathrm{T}^m)} \leq \prod_{j=1}^{m} \|I_{|\gamma|e_j} x\|^{\frac{\gamma_j}{|\gamma|}}_{L_2(\mathrm{T}^m)}. \tag{17}$$

Note that inequality (17) becomes equality for any function x of the form $x(t) = \varphi\left(\sum_{j=1}^m t_j\right), \ \varphi \in L_{2,0}(\mathrm{T}^1), \ t = (t_1, ..., t_m) \in \mathbb{R}^m$.

Let us represent $\|I_{|\gamma|e_j}x\|^2_{L_2(\mathrm{T}^m)}$ in the form

$$\|I_{|\gamma|e_j}x\|^2_{L_2(\mathrm{T}^m)} = \frac{1}{(2\pi)^{m-1}} \int\limits_{\mathrm{T}^{m-1}} \left(\frac{1}{2\pi} \int\limits_{\mathrm{T}^1} \left[I_{|\gamma|e_j}x(s)\right]^2 ds_j \right) ds^j.$$

Using the inequality (13) for estimation of the interior integral we will have (for any $g \in L_\infty(\mathrm{T}^{m-1})$)

$$\frac{\|I_{|\gamma|e_j}x\|^2_{L_2(\mathrm{T}^m)}}{\|\varphi_{|\gamma|}(\,\cdot\,;\alpha,\,\beta)\|^2_{L_2(\mathrm{T}^1)}}$$

$$\le (2\pi)^{1-m} \int\limits_{\mathrm{T}^{m-1}} \left(\frac{\|I_{r_je_j}x(s) - g(s^j)\|^{|\gamma|/r_j}_{L_\infty(\mathrm{T}^1)}}{E_0\left(\varphi_{r_j}(\,\cdot\,;\alpha,\,\beta)\right)^{|\gamma|/r_j}_{L_\infty(\mathrm{T}^1)}} \left\| \frac{x_+}{\alpha} + \frac{x_-}{\beta} \right\|^{1-|\gamma|/r_j}_{L_\infty(\mathrm{T}^1)} \right)^2 ds^j$$

$$\le \left(\frac{\|I_{r_je_j}x(s) - g(s^j)\|^{|\gamma|/r_j}_{L_\infty(\mathrm{T}^m)}}{E_0\left(\varphi_{r_j}(\,\cdot\,;\alpha,\,\beta)\right)^{|\gamma|/r_j}_{L_\infty(\mathrm{T}^1)}} \left\| \frac{x_+}{\alpha} + \frac{x_-}{\beta} \right\|^{1-|\gamma|/r_j}_{L_\infty(\mathrm{T}^m)} \right)^2. \tag{18}$$

The inequality (15) follows from (17) and (18) .
Theorem is proved.

4. Some applications

As an application of Theorems 1 and 3 we present the solution of the problem of L_1- approximation of the class $\mathcal{W}^\gamma_{2,1}$ by the class $\sum_{j=1}^m N_j W^{r_je_j}_{1,1}$ and by quasipolynomials.

Theorem 4. *Let $\gamma = (\gamma_1, ..., \gamma_m)$, $N = (N_1, ..., N_m) \in \mathbb{R}^m_+, \gamma_j > 0,\ N_j > 0,\ j = 1, ..., m,\ |\gamma| \in \mathbb{N}$, and $r_j \in \mathbb{N},\ r_j > |\gamma|,\ j = 1, ..., m,$ be given. Then*

$$E\left(W^\gamma_{2,1}, \sum_{j=1}^m N_j W^{r_je_j}_{1,1} \right)_{L_1(\mathrm{T}^m)}$$

$$\le \left(1 - \sum_{j=1}^m \frac{\gamma_j}{r_j}\right) \left(\frac{\|\varphi_{|\gamma|}\|_{L_2(\mathrm{T}^1)}}{\prod\limits_{j=1}^m \|\varphi_{r_j}\|^{\gamma_j/r_j}_{L_\infty(\mathrm{T}^1)}} \prod_{j=1}^m \left(\frac{\gamma_j}{r_jN_j}\right)^{\frac{\gamma_j}{r_j}} \right)^{\frac{1}{1-\sum_{j=1}^m \frac{\gamma_j}{r_j}}}.$$

Given $p \in [1, \infty]$, $n \in \mathbb{Z}^m_+$ denote by $Q_{n,p}$ the set of quasipolynomials of the form

$$u_n(t) = \sum_{j=1}^m \sum_{k=-n_j}^{n_j} f_{k,j}\left(t^j\right) e^{ikt_j},$$

where $f_{k,j}$ are complex-valued functions for which $\mathrm{Re} f_{k,j}$, $\mathrm{Im} f_{k,j} \in L_p(\mathrm{T}^{m-1})$ and $f_{-k,j} = \overline{f_{k,j}}$ (the overline means the complex conjugation). Note that various problems of approximation by quasipolynomials were consider by Bernstein, Brudnyi and other mathematicians. The following result is some multivariate analog of Nikolskii result (see, for example, [16, p. 130] or [14, §4.2]) about L_1-approximation by trigonometric polynomials.

Theorem 5. *Let* $r = (r_1, ..., r_m)$, $n = (n_1, ..., n_m) \in \mathbb{N}^m$ *be given and* $n - 1 := (n_1 - 1, ..., n_m - 1)$. *Then for any* $(N_1, ..., N_m) \in \mathbb{R}_+^m$

$$E\left(\sum_{j=1}^{m} N_j W_{1,1}^{r_j e_j}, Q_{n-1,1}\right)_{L_1(\mathrm{T}^m)} = \sum_{j=1}^{m} N_j \frac{\|\varphi_{r_j}\|_{L_\infty(\mathrm{T}^1)}}{n_j^{r_j}}.$$

Using Theorem 4 ,Theorem 5 and the method of intermediate approximation, we obtain the following Theorem.

Theorem 6. *Let* $m \in \mathbb{N}$, $n = (n_1, ..., n_m) \in \mathbb{N}^m$, $\gamma = (\gamma_1, ..., \gamma_m) \in \mathbb{R}^m$, $\gamma_j > 0$, $j = 1, ..., m$, $|\gamma| \in \mathbb{N}$, $n^\gamma = n_1^{\gamma_1}, ..., n_m^{\gamma_m}$. *Then*

$$E\left(W_{2,1}^{\gamma}, Q_{n-1,1}\right)_{L_1(\mathrm{T}^m)} = \frac{\|\varphi_{|\gamma|}\|_{L_2(\mathrm{T}^1)}}{n^\gamma}.$$

In the case $m = 1$ quasipolynomials become trigonometric polinomials and and Theorem 6 reduces to the result of Taikov (see, for example,[16, p.130] or [14, §4.2]) (for $p = 2$).

References

1. Arestov V. V., On some extremal problems for univariate differentiable functions, Proc. Steklov Inst. Math. **138** (1975), 3–26.
2. Arestov V. V. and Gabushin V. N., The best approximation of unbounded operators by bounded ones, Izv. Vyssh. Uchebn. Zaved. Mat. **11** (1995), 44–66.
3. Babenko V. F., Nonsymmetric extremal problems of approximation theory, Soviet Math. Dokl. **269**, 3 (1983), 521–524.
4. Babenko V. F., Kofanov V. A., and Pichugov S. A., On inequalities of Landau-Hadamard-Kolmogorov type for the L_2- norm of intermediate derivatives, East J. Approx. **2**, 3 (1996), 343–368.
5. Buslaev A. P. and Tikhomirov V. M., On inequalities for derivatives in multivariate case, Mat. Zametki **25**, 1 (1979), 59–74.
6. Chen W. and Ditzian Z., Mixed and directional derivatives, Proc. Amer. Math. Soc. **108**, 1 (1990), 177–185.
7. Ding Dung and Tikhomirov V. M., On inequalities for derivatives in L_2-metric, Vestn. MGU, Ser. Matem., Meh. **2** (1979), 7–11.

8. Hörmander L., New proof and generalization of inequality of Bohr, Math. Scand. **2** (1954), 33–45.
9. Klots B. E., Approximation of differentiable functions by more smooth functions, Mat. Zametki **21**, 1 (1977), 21–32.
10. Kolmogorov A. A., *Selected Works*, Kluwer Academic Publishers, Dordrecht, 1991.
11. Konovalov V. N., Exact inequalities for norms of functions, third partial and second mixed derivatives, Mat. Zametki **23**, 1 (1978), 67–78.
12. Korneichuk N. P., On best uniform approximation on some classes of continuous functions, Soviet Math. Dokl. **140**, 4 (1961), 748–751.
13. Korneichuk N. P., Inequalities for differentiable periodic functions and the best approximation of one function classes by another, Izv. AN USSR, Ser. matem. **36** (1972), 423–434.
14. Korneichuk N. P., *Exact Constants in Approximation Theory*, Cambridge University Press, 1991.
15. Korneichuk N. P., Ligun A. A., and Babenko V. F., *Extremal Properties of Polynomials and Splines*, Nova Science Publishers, New York, 1996.
16. Korneichuk N. P., Ligun A. A., and Doronin V. G., *Approximation with Constraints*, Naukova dumka, Kiev, 1982.
17. Kwong M. K. and Zettl A., *Norm Inequalities for Derivatives and Differences*, Lect. Notes in Math. **1536**, Springer, Berlin , 1992.
18. Ligun A. A., Inequalities for upper bounds of functionals, Analysis Math. **2**, 1 (1976), 11–40.
19. Tikhomirov V. M., Convex analysis, in *Modern problems of mathematics. Fundamental directions*, **14** (1987), Nauka, Moscow, 1987, 5–101.
20. Timofeev V. G., Inequality of Landau type for multivariate functions, Mat. Zametki **37**, 5 (1985), 676–689.
21. Timoshin O. A., Exact inequality between norms of derivatives of second and third order, Dokl. Ross. Acad. Nauk **344**, 1 (1995), 20–22.

Dnepropetrovsk State University
Pr. Gagarina, 72
Dnepropetrovsk, 320625
Ukraine
`babenko@ftf.dsu.dp.ua`

Monotone Iterative Technique for Impulsive Differential-Difference Equations with Variable Impulsive Perturbations

Drumi Bainov, Angel Dishliev, and Snezhana Hristova

Abstract. This paper deals with impulsive differential-difference equations for which the impulses are realized at moments when the integral curve of the initial value problem considered meets fixed curves in the extended phase space of the equation. The corresponding impulsive differential-difference inequalities are considered. Sufficient conditions are given for the absence of the phenomenon "beating" and for the continuation of the solutions of the initial value problems for impulsive differential-difference equations and inequalities. Monotone sequences of lower and upper solutions of the initial value problem for the impulsive differential-difference equations are constructed. Under some natural assumptions it is proved that these sequences are convergent to the solutions of the same problem.

1. Introduction

The impulsive differential equations are widely used in the simulation of real processes studied in physics, chemical technologies, population dynamics, pharmacokinetics, impulse technique, industrial robotics and economics. In these processes (due to external short-time perturbation) a rapid change of the main parameters describing the process is observed. The duration of the changes is negligible and for this reason they are realized momentarily in the form of impulses. In view of the many applications of the impulsive equations it is necessary to develop methods for such equations. Unfortunately, a comparatively small class of impulsive differential-difference equations can be solved analytically. Therefore, it is necessary to establish approximation methods for finding solutions of impulsive equations. The monotone iterative technique of V. Lakshmikantham [7]–[9] is such a

Multivariate Approximation and Splines
G. Nürnberger, J. W. Schmidt, and G. Walz (eds.), pp. 13–28.

ISBN 3-7643-5654-5.

method which can be easily applied in practice. This technique combines the ideas of the method of lower and upper solutions with appropriate monotone methods. Up to now, the monotone iterative technique of V. Lakshmikantham has been applied to impulsive differential-difference equations in the case, when the impulsive moments (the moments when the main parameters change abruptly) are fixed in advance [4]–[6].

In the present paper, we study impulsive differential-difference equations and inequalities for which the impulses are realized when the integral curve of these equations or inequalities meets the given curves in the extended phase space. In these equations, the impulsive moments depend on the initial function.

The phenomenon when the integral curve meets a given impulsive curve more than once is called "beating" of the solution. In the case, when this phenomenon occurs, it is possible that the solution cannot be continued further from a given point. Sufficient conditions are given for the absence of the phenomenon "beating" and for the continuation of the solutions of initial value problems for impulsive differential-difference equations and inequalities. Monotone sequences of lower and upper solutions of the initial value problem are constructed. Sufficient conditions are given such that the constructed sequences converge to the solutions of the initial value problem for impulsive differential-difference equations.

2. Statement of the Problem. Preliminary Notes

The paper deals with the initial value problem for impulsive nonlinear differential-difference equation

$$\frac{dx}{dt} = f(t, x(t), x(t-h)), \qquad t > t_0, \quad t \neq \tau(x(t)), \tag{1}$$

$$x(t+0) = x(t) + I(x(t)), \qquad t = \tau(x(t)), \tag{2}$$

$$x(t) = \varphi(t), \qquad t_0 - h \leq t \leq t_0, \tag{3}$$

where $f: [t_0, T] \times \mathbb{R} \times \mathbb{R} \to \mathbb{R}$; $\tau: \mathbb{R} \to (t_0, T)$; $I: \mathbb{R} \to \mathbb{R}$; $\varphi: [t_0 - h, t_0] \to \mathbb{R}$; t_0, T, $h \in \mathbb{R}$, $t_0 < T$, $h > 0$. In the sequel, we shall use the notation

$$\sigma = \Big\{(t, x): \quad t = \tau(x), \quad x \in \mathbb{R}\Big\}.$$

The solution of the problem with impulses (1), (2), (3) will be denoted by $x(t; t_0, \varphi)$ and the solution of the problem without impulses (1), (3) by $\chi(t; t_0, \varphi)$. It is easy to see that

$$x(t; t_0, \varphi) = \begin{cases} \varphi(t), & t_0 - h \leq t \leq t_0, \\ \chi(t; t_0, \varphi), & t_0 < t \leq \tau_1, \\ \chi(t; \tau_i, \varphi_i), & \tau_i < t \leq \tau_{i+1}, \quad i = 1, 2, \ldots, k \\ \chi(t; \tau_{k+1}, \varphi_{k+1}), & \tau_{k+1} < t \leq T, \end{cases}$$

where $\tau_i = \tau(x(\tau_i; t_0, \varphi))$, $i = 1, 2, \ldots, k+1$, $t_0 < \tau_1 < \tau_2 < \ldots < \tau_{k+1} < T$ and $t \neq \tau(x(t; t_0, \varphi))$ for $t \neq \tau_i$, $i = 1, 2, \ldots, k+1$.

In other words, $\tau_1, \tau_2, \ldots, \tau_{k+1}$, are the successive moments when the integral curve of the problem (1), (2), (3) meets the curve σ. The functions φ_i, $i = 1, 2, \ldots, k+1$ are defined as follows

$$\varphi_i(t) = \begin{cases} x(t; t_0, \varphi), & \tau_i - h \le t < \tau_i, \\ x(t; t_0, \varphi) + I(x(t; \tau_0, \varphi)), & t = \tau_i. \end{cases}$$

Let us note that there are three possible cases.

Case 1. $k = 0$. In this case, the integral curve of the problem (1), (2), (3) meets only once the curve σ. This is the basic case in our further investigations.

Case 2. $k > 0$. In this case, the integral curve $(t, x(t; t_0, \varphi))$ of the problem (1), (2), (3) meets the curve σ finitely many times. This phenomenon is called "beating".

Case 3. $k = \infty$. In this case, the integral curve meets the curve σ infinitely many times. This phenomenon is also called "beating". In the presence of this phenomenon, it is possible that the solution of the problem under consideration cannot be continued up to T. The following example illustrates the situation.

Example 1. *Let* $f(t, x(t), x(t-h)) = 0$, $\tau(x) = \operatorname{arctg} x$, $I(x) = 1$, $\varphi(x) = 1$, $t_0 = 0$, $h = 1$. *It is easy to see that under the assumptions of this example, the solution of the problem* (1), (2), (3) *has the form*

$$x(t; 0, 1) = \begin{cases} 1, & -1 \le t \le 0, \\ i, & \operatorname{arctg}(i-1) < t \le \operatorname{arctg} i, \quad i = 1, 2, \ldots. \end{cases}$$

Moreover, it is clear that the above solution cannot be continued to the right from

$$\lim_{i \to \infty} \operatorname{arctg} i = \pi/2.$$

The integral curve of the problem (1), (2), (3), *satisfying the assumptions of the example, meets infinitely many times the curve*

$$\sigma = \Big\{(t, x): \quad t = \operatorname{arctg} x\Big\}$$

at the moments $\tau_i = \operatorname{arctg} i$, $i = 1, 2, \ldots$.

We shall give sufficient conditions under which the phenomenon "beating" is absent. Denote by (H1) the following conditions:

H1.1. $f \in C[[t_0, T] \times \mathbb{R} \times \mathbb{R}]$;

H1.2. The function f satisfies the Lipschitz condition in the last two arguments, i.e., there exist constants $L_1 > 0$ and $L_2 > 0$ such that for each two points (x', y'), $(x'', y'') \in \mathbb{R} \times \mathbb{R}$ and for each t, $t_0 \le t \le T$, we have

$$\Big| f(t, x', y') - f(t, x'', y'') \Big| < L_1 |x' - x''| + L_2 |y' - y''|;$$

H1.3. There exists a constant $M > 0$ such that

$$\left|f(t,x,y)\right| \le M, \qquad (t,x,y) \in [t_0,T] \times \mathbb{R} \times \mathbb{R};$$

H1.4. The function $\tau : \mathbb{R} \to (t_0,T)$ satisfies the Lipschitz condition with a constant L, $0 \le L < 1/M$;

H1.5. The inequality

$$\tau(x + I(x)) < \tau(x), \quad x \in \mathbb{R}$$

is valid;

H1.6. $\varphi \in C[t_0 - h, t_0]$.

Lemma 1. *If the conditions* (H1) *are fulfilled, then:*

1. *The integral curve of the problem* (1), (2), (3) *meets only once the curve* σ *for* $t_0 < t < T$;

2. *The solution of the problem* (1), (2), (3) *exists in the interval* $[t_0 - h, T]$.

Proof: It follows from conditions H1.1, H1.2 and H1.6 that the solution of the problem without impulses (1), (3) exists and is unique in the interval $[t_0 - h, T]$. Let us suppose that the integral curve $(t, \chi(t;t_0,\varphi))$ of the problem (1), (3) does not meet the curve σ. Then

$$x(t;t_0,\varphi) = \chi(t;t_0,\varphi) \quad \text{for} \quad t_0 - h \le t \le T$$

and there are no impulses. Therefore, the function $\psi(t) = t - \tau(x(t;t_0,\varphi))$ is continuous in the interval $[t_0,T]$. On the other hand, the inequalities

$$\psi(t_0) = t_0 - \tau(x(t_0;t_0,\varphi)) < 0$$
$$\psi(T) = T - \tau(x(T;t_0,\varphi)) > 0$$

are valid. Thus, we get the conclusion that there exists a point τ_1, $t_0 < \tau_1 < T$, such that $\psi(\tau_1) = 0$, i.e.,

$$\tau_1 = \tau(x(\tau_1;t_0,\varphi)).$$

The last equality shows that the integral curve of the problem (1), (2), (3) meets the curve σ at the moment τ_1.

Let us suppose that the integral curve $(t, x(t;t_0,\varphi))$ meets the curve σ more than once and let τ_1 and τ_2 $(\tau_1 < \tau_2)$ be two successive moments of impulse effect. Using conditions H1.3, H1.4 and H1.5 we get the following contradiction

$$\begin{aligned}
\tau_2 - \tau_1 &= \tau(x(\tau_2;t_0,\varphi)) - \tau(x(\tau_1;t_0,\varphi)) \\
&< \tau(x(\tau_2;t_0,\varphi)) - \tau(x(\tau_1;t_0,\varphi) + I(x(\tau_1;t_0,\varphi))) \\
&= \tau(x(\tau_2;t_0,\varphi)) - \tau(x(\tau_1 + 0;t_0,\varphi)) \\
&\le L\left|x(\tau_2;t_0,\varphi) - x(\tau_1 + 0;t_0,\varphi)\right| \\
&= L\left|\int_{\tau_1}^{\tau_2} f(s, x(s;t_0,\varphi), x(s-h;t_0,\varphi))ds\right| \\
&\le LM(\tau_2 - \tau_1) < \tau_2 - \tau_1.
\end{aligned}$$

Thus, we have proved assertion 1 of Lemma 1. The assertion 2 follows from conditions H1.1, H1.2, H1.6 and assertion 1. □

The study of the phenomenon "beating" has been object of several investigations. We shall mention [2], [3] and [10].

Let us note that for the impulsive differential equations, it is possible that two different solutions can merge after an impulse. For this reason the uniqueness of solutions is not claimed for such equations. The paper [1] is devoted to this problem. In the case when impulses are at fixed moments, the uniqueness of the solution is guaranteed by the condition of the reversibility of the function I.

Under the conditions (H1) and having in mind Lemma 1, we conclude that the solution of the problem (1), (2), (3) has the form

$$x(t;t_0,\varphi) = \begin{cases} \varphi(t), & t_0 - h \le t \le t_0; \\ \chi(t;t_0,\varphi), & t_0 < t \le \tau_1; \\ \chi(t;t_0,\varphi_1), & \tau_1 < t \le T. \end{cases}$$

Definition 1. *Let the conditions* (H1) *be fulfilled. We shall say that the function* $v: [t_0 - h, T] \to \mathbb{R}$ *is a lower solution of the problem* (1), (2), (3) *if*

$$\frac{dv}{dt} \le f(t, v(t), v(t-h)), \quad t_0 \le t \le T, \quad t \ne \tau(v(t));$$
$$v(t+0) \le v(t) + I(v(t)), \quad t = \tau(v(t));$$
$$v(t) \le \varphi(t), \quad t_0 - h \le t \le t_0.$$

The notions of upper solution of the problem (1), (2), (3), as well as lower and upper solutions of the problem without impulses (1), (3) are defined analogously.

For the lower and upper solutions of the problem (1), (2), (3), it is possible to get results analogous to those of Lemma 1. The following lemma is valid.

Lemma 2. *Let the following conditions hold:*

1. *Conditions* (H1) *are fulfilled.*
2. *The function* τ *is monotone increasing (decreasing) in* $\mathbb{R}$.

Then:

1. *The integral curve of a lower (upper) solution of the problem* (1), (2), (3) *meets the curve* σ *only once for* $t_0 < t < T$;
2. *The lower (upper) solutions of the problem* (1), (2), (3) *exist in the interval* $[t_0 - h, T]$.

Proof: We shall consider the case of lower solutions of the problem (1), (2), (3) and monotone increasing function τ. Let $v = v(t;t_0,\varphi)$ be a lower solution of the problem without impulses (1), (3). If we suppose that the integral curve $(t, v(t;t_0,\varphi))$ does not meet the curve σ, then for the function $\psi(t) = t - \tau(v(t;t_0,\varphi))$ which is continuous in the interval $[t_0, T]$, we get $\psi(t_0) < 0$ and $\psi(T) > 0$. Therefore, there exists a point $\tau_1 \in (t_0, T)$ such that $\psi(\tau_1) = 0$, i.e.,

$$\tau_1 = \tau(v(\tau_1;t_0,\varphi)).$$

It follows from the last equality that the curves $(t, v(t; t_0, \varphi))$ and σ cross each other for $t_0 < t < T$, i.e., the integral curve of the lower solution of the problem with impulses (1), (2), (3) meets the curve σ at least once.

Let us suppose that the points of intersection are at least two and let τ_1 and τ_2 be successive points of intersection ($\tau_1 < \tau_2$). From the fact that τ is a monotone increasing function in $\mathbb{R}$ and from the inequality

$$0 < \tau_2 - \tau_1 = \tau(v(\tau_2; t_0, \varphi)) - \tau(v(\tau_1; t_0, \varphi))$$

it follows that $v(\tau_2; t_0, \varphi) > v(\tau_1; t_0, \varphi)$. Moreover, from condition H1.5 we have $\tau(x) > \tau(x + I(x))$, i.e., $x > x + I(x)$. The last inequality implies $I(x) < 0$ for $x \in \mathbb{R}$. Then the inequality

$$\begin{aligned} v(\tau_2; t_0, \varphi) &- v(\tau_1 + 0; t_0, \varphi) \\ &= v(\tau_2; t_0, \varphi) - v(\tau_1; t_0, \varphi) - I(v(\tau_1; t_0, \varphi)) > 0 \end{aligned}$$

is valid and hence we get the following contradiction

$$\begin{aligned} \tau_2 - \tau_1 &= \tau(v(\tau_2; t_0, \varphi)) - \tau(v(\tau_1; t_0, \varphi)) \\ &< \tau(v(\tau_2; t_0, \varphi)) - \tau(v(\tau_1; t_0, \varphi) + I(v(\tau_1; t_0, \varphi))) \\ &= \tau(v(\tau_2; t_0, \varphi)) - \tau(v(\tau_1 + 0; t_0, \varphi)) \\ &\leq L\Big|v(\tau_2; t_0, \varphi) - v(\tau_1 + 0; t_0, \varphi)\Big| \\ &= L\Big(v(\tau_2; t_0, \varphi) - v(\tau_1 + 0; t_0, \varphi)\Big) \\ &\leq L \int_{\tau_1}^{\tau_2} f(s, v(s; t_0, \varphi), v(s - h; t_0, \varphi))ds \\ &\leq LM(\tau_2 - \tau_1) < \tau_2 - \tau_1. \end{aligned}$$

Thus, we have proved assertion 1 of Lemma 2. The assertion 2 follows from conditions H1.1, H1.2, H1.6 and assertion 1.

The case of upper solutions (τ is monotone decreasing function) can be considered analogously. □

Lemma 3. *Let the following conditions hold:*

1. *Conditions* H1.1–H1.4 *and* H1.6 *are fulfilled.*
2. *The function τ is monotone increasing (decreasing) in $\mathbb{R}$.*

Then the integral curve of the lower (upper) solution of the problem without impulses (1), (3) *meets the curve σ only once for $t_0 < t < T$.*

The proof of the above lemma is already given in the proof of Lemma 2.

We shall denote by $P[J]$, $J \subset \mathbb{R}$ the set of all functions $u: J \to \mathbb{R}$, which have at most one point of discontinuity $\tau_1 \in J$, where the function u is left-continuous

($\lim_{t\to\tau_1-0} u(t) = u(\tau_1)$) and the right limit exists in this point and it is finite ($\lim_{t\to\tau_1+0} u(t) = \text{const} \neq \pm\infty$). We denote by $P^1[J]$ the set of all functions $u \in P[J]$ which are continuously differentiable at the points where u is continuous. Moreover, if the point τ_1 is a point of discontinuity of u, then there exists the limit

$$\frac{du^-}{dt} = \lim_{t\to\tau_1-0} \frac{u(t)-u(\tau_1)}{t-\tau_1}.$$

Let functions $v,\ w \in P^1[J]$ be given such that $v(t) \le w(t)$ for $t \in J$. Introduce the notation

$$S(J,v,w) = \Big\{u \in P^1[J]\colon \quad v(t) \le u(t) \le w(t), \quad t \in J\Big\}.$$

Consider the following differential inequality

$$\frac{dx(t)}{dt} \le -ax(t) - bx(t-h), \qquad t_0 \le t \le T \tag{4}$$

with initial condition

$$x(t) = 0, \qquad t_0 - h \le t \le t_0. \tag{5}$$

We need the following lemma.

Lemma 4. *Let the following conditions hold:*

1. *The constants a and b are positive.*
2. $(a+b)(T-t_0) \le 1$.

Then each solution of the problem (4), (5) *is nonpositive.*

Proof: Let $x = x(t)$ be an arbitrary solution of the problem (4), (5). For

$$t_1 = \inf\Big\{t\colon \quad t_0 \le t \le T, \quad x(t) \ne 0\Big\},$$

it is clear that $x(t_1) = 0$.

Suppose that there exists a point t_2, $t_1 < t_2 < T$ such that $x(t_2) > 0$. We shall consider the following three cases.

Case 1. There exists a constant h_1, $0 < h_1 < T - t_1$ such that $x(t) \le 0$ for $t_1 \le t \le t_1 + h_1$. Since $x(t_2) > 0$, we conclude that there exists at least one point in the interval $[t_1 + h_1, t_2)$, where the solution $x(t)$ is zero. For

$$t_3 = \inf\Big\{t\colon \quad t_1 + h_1 \le t < t_2, \quad x(t) = 0\Big\},$$

it is clear that $x(t_3) = 0$. For $m = \min\{x(t)\colon\ t_1 \le t \le t_3\}$, it is easy to see that $m < 0$. Let $m = x(t_4)$, $t_1 < t_4 < t_3$.

The inequalities

$$\begin{aligned}-m = & -x(t_4) = x(t_3) - x(t_4) \\ = & \int_{t_4}^{t_3} \frac{dx(t)}{dt} dt \leq \int_{t_4}^{t_3} (-ax(t) - bx(t-h))dt \\ \leq & \int_{t_4}^{t_3} (-am - bm)dt = -m(a+b)(t_3 - t_4) \\ < & -m(a+b)(T - t_0),\end{aligned}$$

are valid and hence we get

$$1 < (a+b)(T - t_0)$$

which contradicts condition 2 of the lemma.

Case 2. There exists a constant h_1, $0 < h_1 < \min\{h, T - t_1\}$ such that $x(t) \geq 0$ for $t_1 \leq t \leq t_1 + h_1$. Moreover, the solution $x(t)$ is decreasing in the interval since

$$\frac{dx(t)}{dt} \leq -ax(t) - bx(t-h) = -ax(t) \leq 0. \tag{6}$$

Moreover, having in mind that there exist points $t \in [t_1, t_1 + h_1]$, where $x(t) > 0$ (see the choice of the point t_1 and the conditions of this case) we conclude that at these points inequality (6) is strict. From this fact and from the equality $x(t_1) = 0$ we conclude that $x(t) < 0$ for $t_1 \leq t \leq t_1 + h_1$, which contradicts the assumption of this case.

Case 3. There exists a sequence $t_1^-, t_2^-, \ldots$, such that $t_1^- > t_2^- > \ldots$, $\lim_{n\to\infty} t_n^- = t_1$, $x(t_n^-) \leq 0$, $n = 1, 2 \ldots$. Moreover, there exists a sequence $t_1^+, t_2^+, \ldots$, such that $t_n^- < t_n^+ < t_{n+1}^-$, $x(t_n^+) > 0$, $n = 1, 2 \ldots$. It is easy to see that in this case, there exists a sequence $t_1^0, t_2^0, \ldots$, for which $t_n^- < t_n^0 < t_{n+1}^-$, $x(t_n^0) > 0$, $\frac{dx(t_n^0)}{dt} = 0$, $n = 1, 2, \ldots$. Since $\lim_{n\to\infty} t_n^0 = t_1$, there is a number n_0 such that $t_1 < t_n^0 < t_1 + h$ for $n > n_0$. Then, in view of inequality (4), we get for $n > n_0$ the following contradiction

$$0 = \frac{dx(t_n^0)}{dt} \leq -ax(t_n^0) - bx(t_n^0 - h) = -ax(t_n^0) < 0.$$

This completes the proof of Lemma 4. □

3. Main Results

We shall obtain sufficient conditions for the existence of a solution of the initial value problem with impulses (1), (2), (3). An algorithm for finding this solution will be given. To this end we shall use lower and upper solutions of the respective problem without impulses (1), (3) and monotone iterative technique.

Introduce the conditions (H2):

H2.1. The function $v_0 \in C^1[[t_0, T]]$ is a lower solution of the problem (1), (3);

H2.2. The function $w_0 \in C^1[[t_0, T]]$ is a upper solution of the problem (1), (3) $(v_0(t) \le w_0(t),\ t_0 \le t \le T)$;

H2.3. $(L_1 + L_2)(T - t_0) \le 1$;

H2.4. $I \in C[\mathbb{R}]$.

The following theorem is a main result in the present paper.

Theorem 1. *Let the following conditions hold:*

1. *Conditions* (H1) *and* (H2) *are fulfilled.*
2. *The function* τ *is monotone increasing in* $\mathbb{R}$.

Then there exists a sequence of functions $u_1,\ u_2, \ldots :\ [t_0 - h, T] \to \mathbb{R}$ *such that:*

1. *The functions* $u_1,\ u_2, \ldots$ *are lower solutions of the problem* (1), (2), (3);
2. $u_1(t) \le u_2(t) \le \ldots,\ t_0 - h \le t \le T$;
3. *There exists* $\lim_{n\to\infty} u_n(t) = u(t),\ t_0 - h \le t \le T$;
4. *The function* $u(t)$ *is a solution of the impulsive problem* (1), (2), (3).

Proof: Let $t_1 \in [t_0, T]$, $\varphi_1 \in P[[t_1 - h, t_1]]$ and $x_1 \in S([t_1 - h, T], v_0, w_0)$. Consider the initial value problem

$$\frac{dx(t)}{dt} = -L_1 x(t) - L_2 x(t-h) + F(t, x_1), \quad t_1 \le t \le T; \tag{7}$$

$$x(t) = \varphi_1(t), \quad t_1 - h \le t \le t_1, \tag{8}$$

where

$$F(t, x_1) = f(t, x_1(t), x_1(t-h)) + L_1 x_1(t) + L_2 x_1(t-h).$$

The solution of the problem (7), (8) is denoted by $x_2(t)$.

Define the operator A via the equality

$$x_2 = A(t_1, \varphi_1, x_1),$$

i.e., to each triple $(t_1, \varphi_1, x_1) \in [t_0, T] \times P[[t_1 - h, t_1]] \times S([t_1 - h, T], v_0, w_0)$ we assign the unique solution of the problem (7), (8).

Consider the initial value problem

$$\frac{dx(t)}{dt} = f(t, x(t), x(t-h)), \quad t_1 \le t \le T; \tag{9}$$

$$x(t) = \varphi_1(t), \quad t_1 - h \le t \le t_1. \tag{10}$$

Let $x_1(t)$ be a lower solution of the problem (9), (10). It is not difficult to show that the operator A possesses the following properties:

(i) If $x_2 = A(t_1, \varphi_1, x_1)$, then $x_2(t) \geq x_1(t)$, $t_1 - h \leq t \leq T$;

(ii) The function x_2 is a lower solution of the problem (9), (10);

(iii) If x_1', $x_1'' \in S([t_1 - h, t_1], v_0, w_0)$, $x_1'(t) \leq x_1''(t)$ for $t_1 - h \leq t \leq t_1$ and $x_2' = A(t_1, \varphi_1, x_1')$, $x_2'' = A(t_1, \varphi_1, x_1'')$, then $x_2'(t) \leq x_2''(t)$ for $t_1 - h \leq t \leq T$.

In order to prove the property (i) we set $x(t) = x_1(t) - x_2(t)$, $t_1 - h \leq t \leq T$. We have

$$\begin{aligned}\frac{dx(t)}{dt} &= \frac{dx_1(t)}{dt} - \frac{dx_2(t)}{dt}\\ &\leq f(t, x_1(t), x_1(t-h)) + L_1 x_2(t) + L_2 x_2(t-h) - F(t, x_1)\\ &= -L_1\Big(x_1(t) - x_2(t)\Big) - L_2\Big(x_1(t-h) - x_2(t-h)\Big)\\ &= -L_1 x(t) - L_2 x(t-h), \qquad t_1 < t \leq T.\end{aligned}$$

Moreover, it is clear that

$$x(t) = 0, \qquad t_1 - h \leq t \leq t_1.$$

Then, by Lemma 4 $x(t) \leq 0$ for $t_1 - h \leq t \leq T$ and thus property (i) is proved. The proofs of the other two properties are trivial.

According to Lemma 3, there exists a unique point $\tau_{v_0} \in (t_0, T)$ such that

$$\tau_{v_0} = \tau(v_0(\tau_{v_0})),$$

i.e., the integral curve $(t, v_0(t))$ meets the curve σ only at the moment τ_{v_0} in the interval (t_0, T).

Let $v_0^+(t)$ be the lower solution of the problem (9), (10), where $t_1 = \tau_{v_0}$ and the function φ_1 has the form

$$\varphi_1(t) = \begin{cases} v_0(t), & \tau_{v_0} - h \leq t < \tau_{v_0};\\ v_0(t) + I(v_0(t)), & t = \tau_{v_0}.\end{cases}$$

We have $v_0^+(t) = v_0(t)$ for $\tau_{v_0} - h \leq t < \tau_{v_0}$. As in the proof of Lemma 2, from condition H1.5 and the monotonicity of the function τ, it follows that $I(x) < 0$ for $x \in \mathbb{R}$. Therefore,

$$v_0^+(\tau_{v_0}) = \varphi_1(\tau_{v_0}) = v_0(\tau_{v_0}) + I(v_0(\tau_{v_0})) < v_0(\tau_{v_0}).$$

The above inequality enables us to suppose that the lower solution $v_0^+(t)$ of the problem (9), (10) is chosen such that

$$v_0^+(t) \leq v_0(t), \qquad \tau_{v_0} - h \leq t \leq T.$$

Set

$$u_0(t) = \begin{cases} v_0(t), & t_0 - h \le t < \tau_{v_0}; \\ v_0^+(t), & \tau_{v_0} \le t \le T. \end{cases}$$

The function $u_0(t)$ is a lower solution of the impulsive problem (1), (2), (3).

Since v_0 is a lower solution of the problem (9), (10), it is possible to define the function $v_1 = A(t_0, \varphi, v_0)$, i.e., v_1 is a solution of the problem (7), (8) for $t_1 = t_0$, $\varphi_1(t) = \varphi(t)$ and $x_1(t) = v_0(t)$. It follows from the properties of the operator A that v_1 is a lower solution of the problem (9), (10) for $t_1 = t_0$ and $\varphi_1(t) = \varphi(t)$. Moreover, the inequality

$$v_1(t) \ge v_0(t), \qquad t_0 - h \le t \le T \tag{11}$$

is valid.

According to Lemma 3, there exists a unique point $\tau_{v_1} \in (t_0, T)$ such that

$$\tau_{v_1} = \tau(v_1(\tau_{v_1})),$$

i.e., the integral curve $(t, v_1(t))$ meets the curve σ at the moment τ_{v_1}.

We will prove that $\tau_{v_1} > \tau_{v_0}$. If we suppose the contrary, i.e., $\tau_{v_1} \le \tau_{v_0}$, then from (11) and the fact that τ is a monotone increasing function, we get

$$\psi(\tau_{v_1}) = \tau_{v_1} - \tau(v_0(\tau_{v_1})) = \tau(v_1(\tau_{v_1})) - \tau(v_0(\tau_{v_1})) \ge 0.$$

Moreover,

$$\psi(t_0) = t_0 - \tau(v_0(t_0)) < 0.$$

Therefore, there exists a point τ^*, $t_0 < \tau^* \le \tau_{v_1}$ such that $\psi(\tau^*) = 0$, i.e.,

$$\tau^* = \tau(v_0(\tau^*)),$$

which means that the integral curve $(t, v_0(t))$ meets the curve σ at the point $\tau^* \le \tau_{v_1} < \tau_{v_0}$. We get that $(t, v_0(t))$ meets σ at two moments τ^* and τ_{v_0} which contradicts to the assertion of Lemma 3.

Consider the function $v_1^+ = A(\tau_{v_1}, \varphi_1, v_0^+)$, where

$$\varphi_1(t) = \begin{cases} v_1(t), & \tau_{v_1} - h \le t < \tau_{v_1}; \\ v_1(t) + I(v_1(t)), & t = \tau_{v_1}. \end{cases} \tag{12}$$

According to property (ii) of the operator A, the function v_1^+ is a lower solution of the initial value problem (9), (10) for $t_1 = \tau_{v_1}$ and φ_1 defined by (12). From property (i) of the operator A, it follows that $v_1^+(t) \ge v_0^+(t)$ for $t \in [\tau_{v_1}, T]$.

Define the function

$$u_1(t) = \begin{cases} v_1(t), & t_0 - h \le t < \tau_{v_1}; \\ v_1^+(t), & \tau_{v_1} \le t \le T. \end{cases}$$

Let us note that the function u_1 is a lower solution of the impulsive problem (1), (2), (3) and it satisfies the inequality

$$u_1(t) \geq u_0(t), \qquad t_0 - h \leq t \leq T.$$

Suppose that for $n \in \mathbb{N}$, we have found the functions $v_{n-1} : \ [t_0 - h, T] \to \mathbb{R}$, $v_{n-1}^+ : \ [\tau_{v_{n-1}}, T] \to \mathbb{R}$,

$$u_{n-1}(t) = \begin{cases} v_{n-1}(t), & t_0 - h \leq t < \tau_{v_{n-1}}; \\ v_{n-1}^+(t), & \tau_{v_{n-1}} \leq t \leq T. \end{cases}$$

Then we set $v_n = A(t_0, \varphi, v_{n-1})$. Let $\tau_{v_n} = \tau(v_n(\tau_{v_n}))$. According to Lemma 3, there is only one point in the interval (t_0, T), where the function $\psi(t) = t - \tau(v_n(t))$ is zero. It is not difficult to show that the inequality

$$\tau_{v_n} > \tau_{v_{n-1}} \tag{13}$$

is valid. We obtain the function $v_n^+ = A(\tau_{v_n}, \varphi_1, v_{n-1}^+)$, where

$$\varphi_1(t) = \begin{cases} v_n(t), & \tau_{v_n} - h \leq t < \tau_{v_n}; \\ v_n(t) + I(v_n(t)), & t = \tau_{v_n}. \end{cases} \tag{14}$$

Finally, we set

$$u_n(t) = \begin{cases} v_n(t), & t_0 - h \leq t \leq \tau_{v_n}; \\ v_n^+(t), & \tau_{v_n} < t \leq T. \end{cases}$$

The function $u_n(t)$ is a lower solution of the impulsive initial value problem (1), (2), (3). The inequality

$$u_n(t) \geq u_{n-1}(t), \qquad t_0 - h \leq t \leq T \tag{15}$$

is valid. Thus, we obtain a sequence of impulsive moments $\tau_{v_0}, \ \tau_{v_1}, \ldots,$ and sequence of lower solutions of the problem (1), (2), (3) $u_0, \ u_1, \ldots,$ which satisfy inequalities (13) and (15), respectively.

Since the sequence $\tau_{v_0}, \ \tau_{v_1}, \ldots$ is monotone increasing and bounded from above by the constant T, it is convergent. Let $\tau_v \in (t_0, T]$ be the limit.

Consider the sequence of functions $v_0, \ v_1, \ldots,$ which are continuous in the interval $[t_0 - h, T]$. Having in mind their definitions and property (i) of the operator A, we conclude that this sequence is monotone increasing, i.e., $v_n(t) \geq v_{n-1}(t)$, $t_0 - h \leq t \leq T$, $n = 1, 2, \ldots$. On the other hand, the sequence $v_0, \ v_1, \ldots$ is bounded from above. We have $v_n(t) = \varphi(t)$, $n = 1, 2, \ldots,$ for $t_0 - h \leq t \leq t_0$. Therefore,

$$v_n(t) \leq M_\varphi = \max\Big\{\varphi(t) \colon \quad t_0 - h \leq t \leq t_0\Big\}.$$

Moreover, for $t \in (t_0, T]$, we have

$$\begin{aligned} v_n(t) = \ & v_n(t_0) + \int_{t_0}^{t} \Big(L_1(v_{n-1}(s) - v_n(s)) + L_2(v_{n-1}(s-h) - v_n(s-h))\Big) ds \\ & + \int_{t_0}^{t} f(s, v_{n-1}(s), v_{n-1}(s-h)) ds \end{aligned}$$

$$\leq \varphi(t_0) + \int_{t_0}^{t} f(s, v_{n-1}(s), v_{n-1}(s-h))ds$$

$$\leq \varphi(t_0) + M(t - t_0) \leq M_\varphi + M(T - t_0) = M_v.$$

Thus, we get

$$v_n(t) \leq M_v, \quad t_0 - h \leq t \leq T.$$

Therefore, the sequence $v_0, v_1, \ldots$ is uniformly convergent on the interval $[t_0 - h, T]$. Let $v(t) = \lim_{n\to\infty} v_n(t)$. It is clear that the function v is continuous and that it is a solution of the problem (1), (3) in the interval $[t_0 - h, T]$. Actually, passing to the limit $n \to \infty$ in the equality $v_n = A(t_0, \varphi, v_{n-1})$, we get $v = A(t_0, \varphi, v)$, which means that v is a solution of the initial value problem (1), (3). Moreover,

$$v(t) \geq v_n(t), \qquad t_0 - h \leq t \leq T, \quad n = 1, 2, \ldots.$$

Taking $n \to \infty$ in the equality $\tau_{v_n} = \tau(v_n(\tau_{v_n}))$, we obtain $\tau_v = \tau(v(\tau_v))$. Therefore, the integral curve of the solution $v(t)$ of the problem (1), (3) meets the curve σ at the moment τ_v. According to Lemma 3 this moment is unique.

It is clear that the function v_n^+ is defined for $\tau_v \leq t \leq T$, $n = 1, 2, \ldots$, since $\tau_{v_n} \leq \tau_v$. Moreover, in view of (14), the sequence of functions

$$v_1^+,\ v_2^+,\ \ldots, \qquad \tau_v \leq t \leq T \tag{16}$$

is monotone increasing in their common definition domain. We shall prove that the sequence of functions (16) is bounded from above. Indeed, for $t \in [t_{v_n} - h, t_{v_n})$, we have

$$v_n^+(t) = v_n(t) \leq M_v, \quad n = 1, 2, \ldots.$$

For $t \in [t_{v_n}, T]$, the estimate

$$\begin{aligned} v_n^+(t) &= v_n(t_{v_n}) + I(v_n(t_{v_n})) \\ &\quad + \int_{t_{v_n}}^{t} \Big(L_1(v_{n-1}^+(s) - v_n^+(s)) + L_2(v_{n-1}^+(s-h) - v_n^+(s-h)) \\ &\qquad + f(s, v_{n-1}^+(s), v_{n-1}^+(s-h))\Big)ds \\ &\leq M_v + \int_{t_{v_n}}^{t} f(s, v_{n-1}^+(s), v_{n-1}^+(s-h))ds \\ &\leq M_v + M(t - t_{v_n}) \leq M_v + M(T - t_0) = M_v^+ \end{aligned}$$

is valid.

Therefore, the sequence (16) is uniformly convergent as $\tau_v \leq t \leq T$. Let $v^+(t) = \lim_{n\to\infty} v_n^+(t)$. We have

$$v^+(t) \geq v_n^+(t), \qquad \tau_v \leq t \leq T, \quad n = 1, 2, \ldots.$$

Since

$$v_n^+(\tau_{v_n}) = v_n(\tau_{v_n}) + I(v_n(\tau_{v_n})), \quad n = 1, 2, \ldots,$$

passing to the limit $n \to \infty$, we obtain

$$v^+(\tau_v) = v(\tau_v) + I(v(\tau_v)).$$

Taking $n \to \infty$ in the equality $v_n^+ = A(t_{v_n}, \varphi_1, v_{n-1}^+)$, we conclude that the function v^+ satisfies the equation

$$\frac{dv^+(t)}{dt} = A(t_v, \varphi_1, v^+) = f(t, v^+(t), v^+(t-h))$$

with initial condition

$$\varphi_1(t) = \begin{cases} v(t), & \tau_v - h \leq t < \tau_v; \\ v(t) + I(v(t)), & t = \tau_v. \end{cases}$$

Set

$$u(t) = \begin{cases} v(t), & t_0 - h \leq t < \tau_v; \\ v^+(t), & \tau_v \leq t \leq T. \end{cases}$$

The function u is a solution of the impulsive initial value problem (1), (2), (3).

This completes the proof of Theorem 1. □

Theorem 2. *Let the following conditions hold:*

1. *Conditions* (H1), H2.2, H2.3 *and* H2.4 *are fulfilled.*
2. *The function τ is monotone decreasing in $\mathbb{R}$.*

Then there exists a sequence of functions $u_1, u_2, \ldots : [t_0 - h, T] \to \mathbb{R}$ such that:

1. *The functions $u_1, u_2, \ldots$ are upper solutions of the problem* (1), (2), (3)*;*
2. $u_1(t) \geq u_2(t) \geq \ldots$, $t_0 - h \leq t \leq T$*;*
3. *There exists* $\lim_{n\to\infty} u_n(t) = u(t)$, $t_0 - h \leq t \leq T$*;*
4. *The function $u(t)$ is a solution of the impulsive problem* (1), (2), (3).

The proof of the above theorem is analogous to the proof of Theorem 1 and therefore is omitted.

Acknowledgements. The present investigation was supported by the Bulgarian Ministry of Education, Science and Technologies under Grant MM–511.

References

1. Bainov D. and Dishliev A., Quasiuniqueness, uniqueness and continuability of the solutions of impulsive functional differential equations, Rendiconti di Matematica, Serie VII, **15** (1995) Roma, 391–404.
2. Dishliev A. and Bainov D., Sufficient conditions for absence of "beating" in systems of differential equations with impulses, Applicable Analysis, **18** (1984), 67–73.
3. Dishliev A. and Bainov D., Conditions for the absence of the phenomenon "beating" for systems of impulsive differential equations, Bulletin of the Institute of Mathematics Academia Sinica, **13** (1985), no. 3, 237–256.
4. Hristova S. and Bainov D., A projection-iterative method for finding periodic solutions of nonlinear systems of difference-differential equations with impulses, J. Approx. Theory, **49** (1987), no. 4, 311–320.
5. Hristova S. and Bainov D., Numeric-analytic method for finding the periodic solutions of nonlinear differential-difference equations with impulses, Computing, **38** (1987), 363–368.
6. Hristova S. and Bainov D., Application of the monotone-iterative techniques of V. Lakshmikantham for solving the initial value problem for impulsive differential-difference equations, Rocky Mount. J. of Mathematics, **23** (1993), no. 2, 1–10.
7. Ladde G., Lakshmikantham V., and Vatsala A., *Monotone Iterative Techniques for Nonlinear Differential Equations*, Pitman, Belmonth, 1985.
8. Lakshmikantham V., Leela S., and Oguztoreli M., Quasi-solutions, vector Lyapunov functions and monotone method, *IEEE Trans. Automat. Control*, Vol. 26, 1986.
9. Lakshmikantham V. and Vatsala A., Quasi-solutions and monotone method for systems of nonlinear boundary value problems, J. Math. Anal. Appl., **79** (1981), 38–47.
10. Lakshmikantham V., Bainov D., and Simeonov P., *Theory of Impulsive Differential Equations*, World Scientific Publishers, Singapore, 1989.

Drumi Bainov
Medical University of Sofia
P. O. Box 45, Sofia 1504
Bulgaria

Angel Dishliev
University of Chemical
Technology and Metallurgy
Sofia
Bulgaria

Snezhana Hristova
Plovdiv University "Paissii Hilendarski"
Plovdiv
Bulgaria

Multivariate Cosine Wavelets

Kai Bittner, Charles K. Chui, and Jürgen Prestin

Abstract. We construct bivariate biorthogonal cosine wavelets on a two-overlapping rectangular grid with bell functions not necessary of tensor product type. The biorthogonal system as well as frame and Riesz basis conditions are given explicitly. Our methods are based on the properties of bivariate total folding and unfolding operators.

1. Introduction

Recently, local trigonometric bases have been investigated by many authors. Let us mention here only Malvar [8], Coifman and Meyer [5], Daubechies, Jaffard and Journee [6], Auscher, Weiss and Wickerhauser [1], Wickerhauser [10], Jawerth and Sweldens [7], Matviyenko [9], Xia and Suter [11], Chui and Shi [3,4], and the literature cited there. In the meantime the idea to use windowed sinusoids as an orthonormal basis for $L^2(\mathbb{R})$ is widely applied, e.g. in image processing to eliminate blocking effects in transform coding. Therefore, not only univariate bases but also bivariate constructions were investigated in more detail.

In this paper we are interested in bivariate biorthogonal cosine wavelet bases generated on arbitrary 'two-overlapping' rectangular grids. In particular, the corresponding bell functions need not have tensor product structure. Our approach is based on univariate results for a nonuniform grid obtained in [3,4]. As a main tool for our purposes we introduce a bivariate folding operator. Such an operator maps a function on $\mathbb{R}^2$ into a function which satisfies certain parity conditions on the gridlines of the underlying rectangular grid. Then, frame and Riesz basis properties of the cosine wavelets could be studied in terms of boundedness of the corresponding total folding and unfolding operators. In particular, it is pointed out that in this context frame and Riesz basis are equivalent conditions. The best possible constants in the inequalities could be given explicitly as minimal and maximal eigenvalue of certain matrices. In a final example we compare a linear bell function of pyramid type with a bilinear bell of tensor product type.

Multivariate Approximation and Splines
G. Nürnberger, J. W. Schmidt, and G. Walz (eds.), pp. 29–43.

ISBN 3-7643-5654-5.

2. The bivariate folding operator

Let $(a_r)_{r\in\mathbb{Z}}$, $(a_r^+)_{r\in\mathbb{Z}}$, $(a_r^-)_{r\in\mathbb{Z}}$ and $(b_s)_{s\in\mathbb{Z}}$, $(b_s^+)_{s\in\mathbb{Z}}$, $(b_s^-)_{s\in\mathbb{Z}}$ be given sequences with

$$a_r < a_r^+ \le a_{r+1}^- < a_{r+1} \quad \text{and} \quad a_r - a_r^- = a_r^+ - a_r \quad \text{for } r \in \mathbb{Z}$$

and analogous conditions for b_s, so these numbers generate a rectangular grid for $\mathbb{R}^2$ (see **Figure 1**).

We consider now bivariate bell functions $w_{r,s} : \mathbb{R}^2 \to \mathbb{C}$ with support

$$[a_r, a_{r+1}] \times [b_s, b_{s+1}] \subset \operatorname{supp} w_{r,s} \subset [a_r^-, a_{r+1}^+] \times [b_s^-, a_{s+1}^+]. \tag{1}$$

Then

$$\operatorname{supp} w_{r,s} \cap \operatorname{supp} w_{p,q} = \emptyset \quad \text{if } |r-p| > 1 \text{ or } |s-q| > 1,$$

which means that at most 4 bell functions overlap. Further we define the functions

$$C^k_{[\alpha,\beta]}(x) := \sqrt{\frac{2}{\beta-\alpha}} \cos\left(\left(k+\tfrac{1}{2}\right)\frac{x-\alpha}{\beta-\alpha}\pi\right), \quad k, l \in \mathbb{N}_0$$

which form an orthonormal basis of $L^2([\alpha,\beta])$. The main properties of these particular trigonometric polynomials used in the sequel are that $C^k_{[\alpha,\beta]}$ is even with respect to α, i.e.

$$C^k_{[\alpha,\beta]}(\alpha - x) = C^k_{[\alpha,\beta]}(\alpha + x)$$

and odd with respect to β, i.e.

$$C^k_{[\alpha,\beta]}(\beta - x) = -C^k_{[\alpha,\beta]}(\beta + x).$$

Now we introduce the bivariate cosine wavelets

$$\Psi^{k,l}_{r,s}(x,y) := w_{r,s}(x,y)\, C^k_{[a_r,a_{r+1}]}(x)\, C^l_{[b_s,b_{s+1}]}(y), \quad r,s \in \mathbb{Z},\ k,l \in \mathbb{N}_0.$$

To investigate the basis properties of $\Psi^{k,l}_{r,s}$ we define the matrices

$$\boldsymbol{K}_{r,s}(x,y) := \boldsymbol{K}^w_{r,s}(x,y) := \begin{pmatrix} w_{r,s}(x,y) & w_{r,s}(u_r,y) \\ -w_{r-1,s}(x,y) & w_{r-1,s}(u_r,y) \end{pmatrix},$$

$$\boldsymbol{L}_{r,s}(x,y) := \boldsymbol{L}^w_{r,s}(x,y) := \begin{pmatrix} w_{r,s}(x,y) & w_{r,s}(x,v_s) \\ -w_{r,s-1}(x,y) & w_{r,s-1}(x,v_s) \end{pmatrix}$$

and

$$\boldsymbol{M}_{r,s}(x,y) := \boldsymbol{M}^w_{r,s}(x,y)$$
$$:= \begin{pmatrix} w_{r,s}(x,y) & w_{r,s}(u_r,y) & w_{r,s}(x,v_s) & w_{r,s}(u_r,v_s) \\ -w_{r-1,s}(x,y) & w_{r-1,s}(u_r,y) & -w_{r-1,s}(x,v_s) & w_{r-1,s}(u_r,v_s) \\ -w_{r,s-1}(x,y) & -w_{r,s-1}(u_r,y) & w_{r,s-1}(x,v_s) & w_{r,s-1}(u_r,v_s) \\ w_{r-1,s-1}(x,y) & -w_{r-1,s-1}(u_r,y) & -w_{r-1,s-1}(x,v_s) & w_{r-1,s-1}(u_r,v_s) \end{pmatrix}$$

where $u_r := 2a_r - x$ and $v_s := 2b_s - y$. For further discussion we consider the partition of $\mathbb{R}^2$ into the disjoint rectangles (see **Figure 1**)

$$\begin{aligned} R_1 &:= R_1^{r,s} := [a_r^+, a_{r+1}^-) \times [b_s^+, b_{s+1}^-) , \\ R_2 &:= R_2^{r,s} := [a_r^-, a_r^+) \times [b_s^+, b_{s+1}^-) , \\ R_3 &:= R_3^{r,s} := [a_r^+, a_{r+1}^-) \times [b_s^-, b_s^+) , \\ R_4 &:= R_4^{r,s} := [a_r^-, a_r^+) \times [b_s^-, b_s^+) , \end{aligned}$$

i.e.

$$\mathbb{R}^2 = \bigcup_{r,s\in\mathbb{Z}} R_1^{r,s} \cup R_2^{r,s} \cup R_3^{r,s} \cup R_4^{r,s} .$$

In this way we obtain that on R_1 only the function $w_{r,s}$ is not zero, on R_2 resp. R_3 we have two non-vanishing functions $w_{r,s}$ and $w_{r-1,s}$ resp. $w_{r,s}$ and $w_{r,s-1}$ and finally on R_4 only the four functions $w_{r,s}$, $w_{r-1,s}$, $w_{r,s-1}$ and $w_{r-1,s-1}$ are not zero.

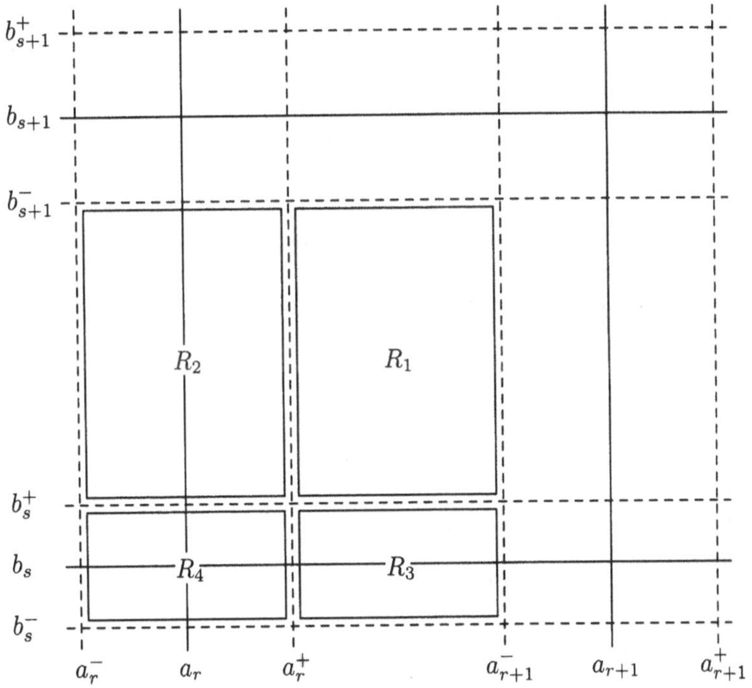

Fig. 1. Grid for the support of the bell function $w_{r,s}$.

With the help of these matrices we introduce the total folding operator $\mathcal{T}_w$. For a given measurable function f we define $\mathcal{T}_w f$ separately in every of the rectangles

$R_i^{r,s}$, $i = 1,2,3,4$; $r, s \in \mathbb{Z}$ in the following way

$$\begin{aligned}
\mathcal{T}_w f(x,y) &:= w_{r,s}(x,y)\, f(x,y) && \text{for } (x,y) \in R_1^{r,s}, \\
\begin{pmatrix} \mathcal{T}_w f(x,y) \\ \mathcal{T}_w f(u_r,y) \end{pmatrix} &:= \boldsymbol{K}_{r,s}(x,y) \begin{pmatrix} f(x,y) \\ f(u_r,y) \end{pmatrix} && \text{for } \begin{matrix} a_r < x < a_r^+, \\ b_s^+ < y < b_{s+1}^-, \end{matrix} \\
\begin{pmatrix} \mathcal{T}_w f(x,y) \\ \mathcal{T}_w f(x,v_s) \end{pmatrix} &:= \boldsymbol{L}_{r,s}(x,y) \begin{pmatrix} f(x,y) \\ f(x,v_s) \end{pmatrix} && \text{for } \begin{matrix} a_r^+ < x < a_{r+1}^-, \\ b_s < y < b_s^+, \end{matrix} \\
\begin{pmatrix} \mathcal{T}_w f(x,y) \\ \mathcal{T}_w f(u_r,y) \\ \mathcal{T}_w f(x,v_s) \\ \mathcal{T}_w f(u_r,v_s) \end{pmatrix} &:= \boldsymbol{M}_{r,s}(x,y) \begin{pmatrix} f(x,y) \\ f(u_r,y) \\ f(x,v_s) \\ f(u_r,v_s) \end{pmatrix} && \text{for } \begin{matrix} a_r < x < a_r^+, \\ b_s < y < b_s^+. \end{matrix}
\end{aligned} \tag{2}$$

Note that this definition determines $\mathcal{T}_w f$ uniquely a.e. on $\mathbb{R}^2$.

We want to emphasize that the introduction of $\mathcal{T}_w$ is motivated by the following observation. Since $C^k_{[\alpha,\beta]}$ is even with respect to α and odd with respect to β one can 'fold' $\Psi^{k,l}_{r,s}$ into $[a_r, a_{r+1}] \times [b_s, b_{s+1}]$ to obtain

$$\int_{\mathbb{R}^2} \Psi^{k,l}_{r,s}(x,y)\, f(x,y)\, dx\, dy = \int_{a_r}^{a_{r+1}} \int_{b_s}^{b_{s+1}} C^k_{[a_r,a_{r+1}]}(x)\, C^l_{[b_s,b_{s+1}]}(y)\, \mathcal{T}_w f(x,y)\, dx\, dy \tag{3}$$

if the integrals are well-defined. Only for simplicity we have restricted ourselves to the particular cosine functions $C^k_{[\alpha,\beta]}$ which then determine the structure of the folding operator $\mathcal{T}_w f$. Other trigonometric bases would work as well but lead to modified folding operators.

Applying formula (3) we will deduce basis properties of the $\Psi^{k,l}_{r,s}$ from norm estimates of $\mathcal{T}_w$. In particular, we are interested in the L^2-boundedness of $\mathcal{T}_w$ and in the existence and boundedness of $\mathcal{T}_w^{-1}$. For this we consider the following functions

$$\begin{aligned}
\Delta_K(x,y) &:= |w_{r,s}(x,y)|^2 + |w_{r,s}(u_r,y)|^2 + |w_{r-1,s}(x,y)|^2 + |w_{r-1,s}(u_r,y)|^2, \\
\Delta_L(x,y) &:= |w_{r,s}(x,y)|^2 + |w_{r,s}(x,v_s)|^2 + |w_{r,s-1}(x,y)|^2 + |w_{r,s-1}(x,v_s)|^2
\end{aligned}$$

and constants

$$\begin{aligned}
B_1^{r,s} &:= \operatorname*{ess\,sup}_{(x,y)\in R_1^{r,s}} |w_{r,s}(x,y)|^2, \\
B_2^{r,s} &:= \operatorname*{ess\,sup}_{\substack{a_r<x<a_r^+ \\ b_s^+<y<b_{s+1}^-}} \|\boldsymbol{K}_{r,s}(x,y)\|_2^2 \\
&= \operatorname*{ess\,sup}_{\substack{a_r<x<a_r^+ \\ b_s^+<y<b_{s+1}^-}} \frac{\Delta_K(x,y)}{2} + \sqrt{\frac{\Delta_K^2(x,y)}{4} - |\det \boldsymbol{K}_{r,s}(x,y)|^2},
\end{aligned}$$

$$\begin{aligned} B_3^{r,s} &:= \operatorname*{ess\,sup}_{\substack{a_r^+ < x < a_{r+1}^- \\ b_s < y < b_s^+}} \|\boldsymbol{L}_{r,s}(x,y)\|_2^2 \\ &= \operatorname*{ess\,sup}_{\substack{a_r^+ < x < a_{r+1}^- \\ b_s < y < b_s^+}} \frac{\Delta_L(x,y)}{2} + \sqrt{\frac{\Delta_L^2(x,y)}{4} - |\det \boldsymbol{L}_{r,s}(x,y)|^2}\,, \\ B_4^{r,s} &:= \operatorname*{ess\,sup}_{\substack{a_r < x < a_r^+ \\ b_s < y < b_s^+}} \|\boldsymbol{M}_{r,s}(x,y)\|_2^2\,. \end{aligned}$$

Here

$$\|\boldsymbol{A}\|_2 := \sup_{|\boldsymbol{x}|=1} |\boldsymbol{A}\boldsymbol{x}| = \sqrt{\rho(\boldsymbol{A}^H \boldsymbol{A})}$$

is the spectral norm of a matrix $\boldsymbol{A}$.

Now we can establish the following assertion.

Lemma 1. *Let $\{w_{r,s}\}$ be measurable functions such that*

$$B_0 := \sup_{r,s\in\mathbb{Z}} \max\{B_1^{r,s}, B_2^{r,s}, B_3^{r,s}, B_4^{r,s}\} < \infty\,. \tag{4}$$

Then $\mathcal{T}_w$ is a bounded operator from $L^2(\mathbb{R}^2) \to L^2(\mathbb{R}^2)$ with

$$\|\mathcal{T}_w\|_{L^2(\mathbb{R}^2)\to L^2(\mathbb{R}^2)}^2 = B_0\,.$$

Proof: For arbitrary $f \in L^2(\mathbb{R}^2)$, we have

$$\|\mathcal{T}_w f\|_{L^2(\mathbb{R}^2)}^2 = \sum_{r,s\in\mathbb{Z}} \|\mathcal{T}_w f\|_{L^2(R_1^{r,s})}^2 + \|\mathcal{T}_w f\|_{L^2(R_2^{r,s})}^2 + \|\mathcal{T}_w f\|_{L^2(R_3^{r,s})}^2 + \|\mathcal{T}_w f\|_{L^2(R_4^{r,s})}^2\,.$$

We consider now every term separately. For R_1 we obtain immediately, by Hölder inequality

$$\|\mathcal{T}_w f\|_{L^2(R_1^{r,s})}^2 = \iint\limits_{R_1^{r,s}} |w_{r,s}(x,y)|^2\, |f(x,y)|^2\, dx\, dy \le B_1^{r,s}\, \|f\|_{L^2(R_1^{r,s})}^2\,,$$

where $B_1^{r,s}$ is the best possible constant. For R_2 we have

$$\begin{aligned} \|\mathcal{T}_w f\|_{L^2(R_2^{r,s})}^2 &= \int_{b_s^+}^{b_{s+1}^-} \int_{a_r}^{a_r^+} \left| \Big(\mathcal{T}_w f(x,y), \mathcal{T}_w f(u_r,y)\Big)^T \right|^2 dx\, dy \\ &\le \int_{b_s^+}^{b_{s+1}^-} \int_{a_r}^{a_r^+} \|\boldsymbol{K}_{r,s}(x,y)\|_2^2 \left| \Big(f(x,y), f(u_r,y)\Big)^T \right|^2 dx\, dy\,. \end{aligned}$$

In particular, equality is attained if $f \in L^2(R_2^{r,s})$ is given by

$$\begin{pmatrix} f(x,y) \\ f(u_r,y) \end{pmatrix} = a(x,y) \begin{pmatrix} g_1(x,y) \\ g_2(u_r,y) \end{pmatrix}, \quad (x,y) \in (a_r, a_r^+) \times (b_s^+, b_{s+1}^-)$$

with an arbitrary function $a \in L^2\left((a_r, a_r^+) \times (b_s^+, b_{s+1}^-)\right)$ and a normalized eigenvector $((g_1(x,y), g_2(x,y))^T$ corresponding to the maximal eigenvalue of $\boldsymbol{K}_{r,s}(x,y)^T \boldsymbol{K}_{r,s}(x,y)$. Again $B_2^{r,s}$ is the best possible constant in the Hölder inequality

$$\|\mathcal{T}_w f\|^2_{L^2(R_2^{r,s})} \leq B_2^{r,s} \, \|a\|^2_{L^2\left((a_r,a_r^+)\times(b_s^+,b_{s+1}^-)\right)} = B_2^{r,s} \, \|f\|^2_{L^2(R_2^{r,s})} \,.$$

Analogously we obtain

$$\|\mathcal{T}_w f\|^2_{L^2(R_3^{r,s})} \leq B_3^{r,s} \, \|f\|^2_{L^2(R_3^{r,s})}$$

and

$$\|\mathcal{T}_w f\|^2_{L^2(R_4^{r,s})} \leq B_4^{r,s} \, \|f\|^2_{L^2(R_4^{r,s})}$$

with best possible constants $B_3^{r,s}$ and $B_4^{r,s}$. Therefore,

$$\begin{aligned} &\|\mathcal{T}_w f\|^2_{L^2(\mathbb{R}^2)} \\ &\quad \leq \sum_{r,s\in\mathbb{Z}} B_1^{r,s} \, \|f\|^2_{L^2(R_1^{r,s})} + B_2^{r,s} \, \|f\|^2_{L^2(R_2^{r,s})} + B_3^{r,s} \, \|f\|^2_{L^2(R_3^{r,s})} + B_4^{r,s} \, \|f\|^2_{L^2(R_4^{r,s})} \\ &\quad \leq \sup_{r,s\in\mathbb{Z}} \max\left\{B_1^{r,s}, B_2^{r,s}, B_3^{r,s}, B_4^{r,s}\right\} \, \|f\|^2_{L^2(\mathbb{R}^2)} \,. \end{aligned}$$

Hence, $\|\mathcal{T}_w\|^2_{L^2(\mathbb{R}^2)\to L^2(\mathbb{R}^2)} = B_0$. □

Remark 2. *Note that $B_0 < \infty$ iff $\sup_{r,s\in\mathbb{Z}} \|w_{r,s}\|_{L^\infty(\mathbb{R}^2)} < \infty$, which is obviously equivalent to*

$$\Big\| \sum_{r,s\in\mathbb{Z}} |w_{r,s}|^2 \Big\|_{L^\infty(\mathbb{R}^2)} < \infty.$$

Let us now assume that

$$w_{r,s} \neq 0 \text{ a.e. on } R_1^{r,s}\,, \tag{5a}$$

$$\det \boldsymbol{K}_{r,s} \neq 0 \text{ a.e. on } R_2^{r,s}\,, \tag{5b}$$

$$\det \boldsymbol{L}_{r,s} \neq 0 \text{ a.e. on } R_3^{r,s}\,, \tag{5c}$$

$$\det \boldsymbol{M}_{r,s} \neq 0 \text{ a.e. on } R_4^{r,s}\,. \tag{5d}$$

In the same way as in (2) we define an operator $\mathcal{V}_w$ by replacing $w_{r,s}$, $\boldsymbol{K}^w_{r,s}$, $\boldsymbol{L}^w_{r,s}$ and $\boldsymbol{M}^w_{r,s}$ by $w_{r,s}^{-1}$, $(\boldsymbol{K}^w_{r,s})^{-1}$, $(\boldsymbol{L}^w_{r,s})^{-1}$ and $(\boldsymbol{M}^w_{r,s})^{-1}$, respectively. If the conditions

(5a)-(5d) are satisfied the operator $\mathcal{V}_w$ is well–defined a.e. and we obtain by simple inspection that

$$\mathcal{V}_w \mathcal{T}_w f = \mathcal{T}_w \mathcal{V}_w f = f \quad \text{a.e.}.$$

Analogously to **Lemma 1** we introduce the following constants

$$A_1^{r,s} := \operatorname*{ess\,inf}_{(x,y)\in R_1^{r,s}} |w_{r,s}(x,y)|^2 \tag{6}$$

$$A_2^{r,s} := \operatorname*{ess\,inf}_{\substack{a_r<x<a_r^+ \\ b_s^+<y<b_{s+1}^-}} \|\boldsymbol{K}_{r,s}^{-1}(x,y)\|_2^{-2}$$

$$= \operatorname*{ess\,inf}_{\substack{a_r<x<a_r^+ \\ b_s^+<y<b_{s+1}^-}} \frac{\Delta_K(x,y)}{2} - \sqrt{\frac{\Delta_K^2(x,y)}{4} - |\det \boldsymbol{K}_{r,s}(x,y)|^2} \tag{7}$$

$$A_3^{r,s} := \operatorname*{ess\,inf}_{\substack{a_r^+<x<a_{r+1}^- \\ b_s<y<b_s^+}} \|\boldsymbol{L}_{r,s}^{-1}(x,y)\|_2^{-2}$$

$$= \operatorname*{ess\,inf}_{\substack{a_r^+<x<a_{r+1}^- \\ b_s<y<b_s^+}} \frac{\Delta_L(x,y)}{2} - \sqrt{\frac{\Delta_L^2(x,y)}{4} - |\det \boldsymbol{L}_{r,s}(x,y)|^2} \tag{8}$$

$$A_4^{r,s} := \operatorname*{ess\,inf}_{\substack{a_r<x<a_r^+ \\ b_s<y<b_s^+}} \|\boldsymbol{M}_{r,s}^{-1}(x,y)\|_2^{-2}.$$

Note that for an invertible matrix it holds $\|\boldsymbol{A}^{-1}\|_2^{-2} = \min \sigma(\boldsymbol{A}^H \boldsymbol{A})$.

Lemma 3. *Let* $\{w_{r,s}\}$ *be measurable functions and the conditions* (5a)-(5d) *be satisfied. If*

$$A_0 := \inf_{r,s\in\mathbb{Z}} \min\{A_1^{r,s}, A_2^{r,s}, A_3^{r,s}, A_4^{r,s}\} > 0, \tag{9}$$

then $\mathcal{V}_w$ *is a bounded operator from* $L^2(\mathbb{R}^2) \to L^2(\mathbb{R}^2)$ *with*

$$\|\mathcal{V}_w\|^2_{L^2(\mathbb{R}^2)\to L^2(\mathbb{R}^2)} = A_0^{-1}$$

and $\mathcal{T}_w^{-1} = \mathcal{V}_w$.

The proof is analogous to the proof of **Lemma 1**.

Let us mention here that in the bivariate setting the condition $A_0 > 0$ could not be rewritten as easy as $B_0 < \infty$ in **Remark 2**. To explain this let us restrict ourselves to nonnegative bell functions $w_{r,s}$. As in the univariate setting and analogous to **Remark 2** it follows from

$$\operatorname*{ess\,inf}_{x,y\in\mathbb{R}} \sum_{r,s\in\mathbb{Z}} w_{r,s}(x,y) > 0 \tag{10}$$

that

$$\inf_{r,s\in\mathbb{Z}} \min\{A_1^{r,s}, A_2^{r,s}, A_3^{r,s}\} > 0.$$

This can be easily derived from (6)–(8). However, if the bell functions are not tensor products of univariate functions, then $A_0 > 0$ is a stronger assumption than (10). This can be seen by the following example.

Let the bell functions in $R_4^{r,s}$ be uniquely defined by setting them piecewise constant ξ or 1 for all $(x,y) \in (a_r, a_r^+) \times (b_s, b_s^+)$ and such that

$$h(\xi) := \det \boldsymbol{M}^w_{r,s}(x,y) = \det \begin{pmatrix} 1 & \xi & \xi & \xi \\ -\xi & \xi & -1 & \xi \\ -\xi & -1 & \xi & \xi \\ \xi & -\xi & -\xi & 1 \end{pmatrix}.$$

Then we obtain $h(\xi) = (\xi+1)^3(3\xi-1)$, i.e., $h(1/3) = 0$. In other words, this yields an example that all four bell functions supported in R_4 are greater or equal 1/3 but $A_4^{r,s}$ is unbounded.

3. Biorthogonal bases

We start this section with the definition of a bell function $\tilde{w}_{r,s}$ with

$$\operatorname{supp} \tilde{w}_{r,s} \subset [a_r^-, a_{r+1}^+] \times [b_s^-, b_{s+1}^+].$$

We define $\tilde{w}_{r,s}$ by giving the complex conjugate of $\tilde{w}_{r,s}$ in the following way

$$\overline{\tilde{w}_{r,s}(x,y)} = \begin{cases} 1/w_{r,s}(x,y), & \text{if } (x,y) \in R_1^{r,s}, \\ w_{r-1,s}(u_r,y)/\det \boldsymbol{K}_{r,s}(x,y), & \text{if } (x,y) \in R_2^{r,s}, \\ w_{r+1,s}(u_{r+1},y)/\det \boldsymbol{K}_{r+1,s}(x,y), & \text{if } (x,y) \in R_2^{r+1,s}, \\ w_{r,s-1}(x,v_s)/\det \boldsymbol{L}_{r,s}(x,y), & \text{if } (x,y) \in R_3^{r,s}, \\ w_{r+1,s}(x,v_{s+1})/\det \boldsymbol{L}_{r,s+1}(x,y), & \text{if } (x,y) \in R_3^{r,s+1}, \\ \det \boldsymbol{M}_{r,s}^{1,1}(x,y)/\det \boldsymbol{M}_{r,s}(x,y), & \text{if } (x,y) \in R_4^{r,s}, \\ \det \boldsymbol{M}_{r+1,s}^{2,1}(x,y)/\det \boldsymbol{M}_{r+1,s}(x,y), & \text{if } (x,y) \in R_4^{r+1,s}, \\ \det \boldsymbol{M}_{r,s+1}^{3,1}(x,y)/\det \boldsymbol{M}_{r,s+1}(x,y), & \text{if } (x,y) \in R_4^{r,s+1}, \\ \det \boldsymbol{M}_{r+1,s+1}^{4,1}(x,y)/\det \boldsymbol{M}_{r+1,s+1}(x,y), & \text{if } (x,y) \in R_4^{r+1,s+1}, \\ 0 & \text{otherwise} \end{cases}$$

where $\boldsymbol{M}_{r,s}^{i,j}(x,y)$ is the minor of the matrix $\boldsymbol{M}_{r,s}(x,y)$ canceling the i-th row and the j-th column.

Let us mention here that the continuity of $w_{r,s}$, for all $r, s \in \mathbb{Z}$ implies the continuity of $\tilde{w}_{r,s}$.

Remark 4. *If the bell functions $w_{r,s}$ are tensor products of univariate bells, i.e.*

$$w_{r,s} = w_r^1 \otimes w_s^2$$

we obtain that

$$\begin{aligned} \boldsymbol{M}_{r,s}(x,y) &= \boldsymbol{M}_r^1(x) \otimes \boldsymbol{M}_s^2(y), \\ \boldsymbol{K}_{r,s}(x,y) &= \boldsymbol{M}_r^1(x), \\ \boldsymbol{L}_{r,s}(x,y) &= \boldsymbol{M}_s^2(y). \end{aligned}$$

In this case the bell functions $\tilde{w}_{r,s}$ are also tensor products of univariate bell functions.

With $\tilde{w}_{r,s}$ defined above we can describe biorthogonal functions

$$\tilde{\Psi}_{r,s} := \tilde{w}_{r,s}\, C^k_{[a_r,a_{r+1}]} \otimes C^l_{[b_s,b_{s+1}]}\,.$$

Lemma 5. *Let $w_{r,s} \in L^2(\mathbb{R}^2)$ for all $r,s \in \mathbb{Z}$. Let the functions $\{\tilde{w}_{r,s}\}$ be well defined, i.e. (5a)-(5d) are satisfied. If $\tilde{\Psi}^{k,l}_{r,s} \in L^2(\mathbb{R}^2)$ for all $r,s \in \mathbb{Z},\ k,l \in \mathbb{N}_0$, then the system $\{\tilde{\Psi}^{k,l}_{r,s}\}$ is biorthogonal to $\{\Psi^{k,l}_{r,s}\}$, i.e.*

$$\langle \Psi^{k',l'}_{r',s'}\,,\ \tilde{\Psi}^{k,l}_{r,s}\rangle_{L^2(\mathbb{R}^2)} = \delta_{r,r'}\delta_{s,s'}\delta_{k,k'}\delta_{l,l'}\,. \tag{11}$$

Proof: If $\min(|r-r'|,|s-s'|) \geq 2$, then the condition (11) follows immediately from

$$\operatorname{supp} \tilde{w}_{r,s} \subset [a_r^-, a_{r+1}^+] \times [b_s^-, b_{s+1}^+]\,.$$

Now we show (11) for $r' = r-1$ and $s' = s-1$. Because $C^{k'}_{[a,b]}\, C^k_{[b,c]}$ is an odd function with respect to b we conclude

$$\begin{aligned}
&\langle \Psi^{k',l'}_{r-1,s-1}\,,\ \tilde{\Psi}^{k,l}_{r,s}\rangle_{L^2(\mathbb{R}^2)}\\
&= \iint\limits_{R_4^{r,s}} w_{r-1,s-1}(x,y)\,\overline{\tilde{w}_{r,s}(x,y)}\\
&\qquad\times C^{k'}_{[a_{r-1},a_r]}(x)\, C^k_{[a_r,a_{r+1}]}(x)\, C^{l'}_{[b_{s-1},b_s]}(y)\, C^l_{[b_s,b_{s+1}]}(y)\,dx\,dy\\
&= \int_{a_r}^{a_r^+}\int_{b_s}^{b_s^+} \Big(w_{r-1,s-1}(x,y)\overline{\tilde{w}_{r,s}(x,y)} - w_{r-1,s-1}(u_r,y)\overline{\tilde{w}_{r,s}(u_r,y)}\\
&\qquad - w_{r-1,s-1}(x,v_s)\overline{\tilde{w}_{r,s}(x,v_s)} + w_{r-1,s-1}(u_r,v_s)\overline{\tilde{w}_{r,s}(u_r,v_s)}\Big)\\
&\qquad\times C^{k'}_{[a_{r-1},a_r]}(x)\, C^k_{[a_r,a_{r+1}]}(x)\, C^{l'}_{[b_{s-1},b_s]}(y)\, C^l_{[b_s,b_{s+1}]}(y)\,dx\,dy\,.
\end{aligned}$$

From the definition of $\tilde{w}_{r,s}(x,y)$ on $R_4^{r,s}$ it follows that

$$\begin{aligned}
&\det \boldsymbol{M}_{r,s}(x,y)\Big(w_{r-1,s-1}(x,y)\overline{\tilde{w}_{r,s}(x,y)} - w_{r-1,s-1}(u_r,y)\overline{\tilde{w}_{r,s}(u_r,y)}\\
&\qquad - w_{r-1,s-1}(x,v_s)\overline{\tilde{w}_{r,s}(x,v_s)} + w_{r-1,s-1}(u_r,v_s)\overline{\tilde{w}_{r,s}(u_r,v_s)}\Big)\\
&= w_{r-1,s-1}(x,y)\det \boldsymbol{M}^{1,1}_{r,s}(x,y) - w_{r-1,s-1}(u_r,y)\det \boldsymbol{M}^{1,1}_{r,s}(u_r,y)\\
&\qquad - w_{r-1,s-1}(x,v_s)\det \boldsymbol{M}^{1,1}_{r,s}(x,v_s) + w_{r-1,s-1}(u_r,v_s)\det \boldsymbol{M}^{1,1}_{r,s}(u_r,v_s)\\
&= w_{r-1,s-1}(x,y)\det \boldsymbol{M}^{1,1}_{r,s}(x,y) + w_{r-1,s-1}(u_r,y)\det \boldsymbol{M}^{1,2}_{r,s}(x,y)\\
&\qquad - w_{r-1,s-1}(x,v_s)\det \boldsymbol{M}^{1,3}_{r,s}(x,y) - w_{r-1,s-1}(u_r,v_s)\det \boldsymbol{M}^{1,4}_{r,s}(x,y)\\
&= \det M^{4\to 1}_{r,s}(x,y) = 0\,,
\end{aligned}$$

where $M_{r,s}^{j\to i}$ is the matrix $M_{r,s}$ with its i-th row replaced by the j-th row. Therefore we obtain

$$\langle \Psi_{r-1,s-1}^{k',l'}, \tilde{\Psi}_{r,s}^{k,l}\rangle_{L^2(\mathbb{R}^2)} = 0 .$$

Analogously we can show (11) for all other r', s' with $\max(|r-r'|, |s-s'|) = 1$.

Finally, since $C_{[a,b]}^{k'} C_{[a,b]}^{k}$ is even with respect to a and b we obtain

$$\begin{aligned}
&\langle \Psi_{r',s'}^{k',l'}, \tilde{\Psi}_{r,s}^{k,l}\rangle_{L^2(\mathbb{R}^2)} \\
&= \int_{a_r}^{a_{r+1}} C_{[a_r,a_{r+1}]}^{k'}(x)\, C_{[a_r,a_{r+1}]}^{k}(x)\, dx \int_{b_s}^{b_{s+1}} C_{[b_s,b_{s+1}]}^{l'}(y)\, C_{[b_s,b_{s+1}]}^{l}(y)\, dy \\
&= \delta_{k,k'}\delta_{l,l'} .
\end{aligned}$$

□

We define now the total unfolding operator $\mathcal{U}_w$ by replacing in (2) $w_{r,s}$, $\boldsymbol{K}_{r,s}^{w}$, $\boldsymbol{L}_{r,s}^{w}$ and $\boldsymbol{M}_{r,s}^{w}$ by $\overline{w_{r,s}}$, $(\boldsymbol{K}_{r,s}^{w})^H$, $(\boldsymbol{L}_{r,s}^{w})^H$ and $(\boldsymbol{M}_{r,s}^{w})^H$, respectively. From the definition of $\tilde{w}_{r,s}$ we deduce

$$\left(\boldsymbol{M}_{r,s}^{w}\right)^{-1} = \left(\boldsymbol{M}_{r,s}^{\tilde{w}}\right)^{H} .$$

Therefore we obtain $\mathcal{T}_w^{-1}=\mathcal{U}_{\tilde{w}}$.

Furthermore, the norm of $\mathcal{U}_w$ can be determined in the following way. For a non-singular matrix $\boldsymbol{A}$ it holds that $\boldsymbol{A}\boldsymbol{A}^H = \boldsymbol{A}\boldsymbol{A}^H\boldsymbol{A}\boldsymbol{A}^{-1}$. Hence $\boldsymbol{A}\boldsymbol{A}^H$ and $\boldsymbol{A}^H\boldsymbol{A}$ have the same eigenvalues and thus $\|\boldsymbol{A}^{\pm 1}\|_2 = \|\boldsymbol{A}^{\pm H}\|_2$. Therefore, it follows that $\|\mathcal{U}_w\| = \|\mathcal{T}_w\|$ and $\|\mathcal{U}_w^{-1}\| = \|\mathcal{T}_w^{-1}\|$ if $\mathcal{T}_w$ and $\mathcal{T}_w^{-1}$ are bounded operators.

For many applications one needs that $\{\Psi_{r,s}^{k,l}\}$ is a frame, i.e. there exist numbers $0 < A \le B < \infty$ such that

$$A\|f\|_{L^2(\mathbb{R}^2)}^2 \le \sum_{r,s,k,l} \left|\langle f, \Psi_{r,s}^{k,l}\rangle\right|^2 \le B\|f\|_{L^2(\mathbb{R}^2)}^2$$

or that $\{\Psi_{r,s}^{k,l}\}$ is a Riesz basis, i.e. it is complete in $L^2(\mathbb{R}^2)$ and

$$A \sum_{r,s,k,l} \left|\left\langle f, \tilde{\psi}_{r,s}^{k,l}\right\rangle\right|^2 \le \|f\|_{L^2(\mathbb{R}^2)}^2 \le B \sum_{r,s,k,l} \left|\left\langle f, \tilde{\psi}_{r,s}^{k,l}\right\rangle\right|^2$$

with $0 < A \le B < \infty$. For cosine wavelets, it turns out that there is no difference between Riesz bases and frames, as stated in the following.

Theorem 6. *Let $\{w_{r,s}\}$ be measurable functions such that $\Psi_{r,s}^{k,l} \in L^2(\mathbb{R}^2)$. Then the following statements are equivalent.*

(i) *The system $\{\Psi_{r,s}^{k,l}\}$ is a frame of $L^2(\mathbb{R}^2)$ with frame bounds A and B.*

(ii) *The system $\{\Psi_{r,s}^{k,l}\}$ is a Riesz basis of $L^2(\mathbb{R}^2)$ with Riesz bounds A and B.*

(iii) *It holds $0 < A \le A_0 \le B_0 \le B < \infty$ with A_0 and B_0 from (9) and (4).*

Proof: First we show (i) $\iff$ (iii). It is known that the functions $C^k_{[\alpha,\beta]}$, $k \in \mathbb{N}_0$ form an orthonormal basis of $L^2([\alpha,\beta])$. Therefore with (3) we obtain

$$\sum_{r,s\in\mathbb{Z}} \sum_{k,l\in\mathbb{N}_0} \left|\langle f, \Psi^{k,l}_{r,s}\rangle\right|^2 = \|\mathcal{T}_w f\|^2_{L^2(\mathbb{R}^2)} .$$

Hence, $\{\Psi^{k,l}_{r,s}\}$ satisfies the frame inequality with the best possible frame bounds $A_0 = \|\mathcal{T}_w^{-1}\|^{-2}$ and $B_0 = \|\mathcal{T}_w\|^2$.

Now we show that (i, iii) $\iff$ (ii). In particular, (i) follows from (ii) immediately. Furthermore, if the implication

$$\sum_{r,s,k,l} a^{k,l}_{r,s} \Psi^{k,l}_{r,s} = 0 \;\Rightarrow\; a^{k,l}_{r,s} = 0, \;\forall r,s,k,l \tag{12}$$

is proved, the statement (ii) follows from (i) (see e.g. [2] Theorem 2.1). To show (12) let us notice that

$$\Psi^{k,l}_{r,s} = \mathcal{U}_w\left(\chi_{[a_r,a_{r+1}]\times[b_s,b_{s+1}]}\, C^k_{[a_r,a_{r+1}]} \otimes C^l_{[b_s,b_{s+1}]}\right) ,$$

where $\chi_{[a_r,a_{r+1}]\times[b_s,b_{s+1}]}$ is the characteristic function of $[a_r,a_{r+1}]\times[b_s,b_{s+1}]$. From (iii) it follows, that $\mathcal{U}_w^{-1}$ is a bounded operator. Therefore we obtain that

$$\sum_{r,s\in\mathbb{Z}} \sum_{k,l\in\mathbb{N}_0} a^{k,l}_{r,s} \Psi^{k,l}_{r,s} = 0$$

implies

$$\begin{aligned} 0 &= \mathcal{U}_w^{-1} \sum_{r,s\in\mathbb{Z}} \sum_{k,l\in\mathbb{N}_0} a^{k,l}_{r,s}\, \mathcal{U}_w\left(\chi_{[a_r,a_{r+1}]\times[b_s,b_{s+1}]}\, C^k_{[a_r,a_{r+1}]} \otimes C^l_{[b_s,b_{s+1}]}\right) \\ &= \sum_{r,s\in\mathbb{Z}} \chi_{[a_r,a_{r+1}]\times[b_s,b_{s+1}]} \sum_{k,l\in\mathbb{N}_0} a^{k,l}_{r,s}\, C^k_{[a_r,a_{r+1}]} \otimes C^l_{[b_s,b_{s+1}]} . \end{aligned}$$

This means that for all $r,s \in \mathbb{Z}$ it holds that

$$\sum_{k,l\in\mathbb{N}_0} a^{k,l}_{r,s}\, C^k_{[a_r,a_{r+1}]} \otimes C^l_{[b_s,b_{s+1}]} = 0 \quad \text{a.e on } [a_r,a_{r+1}]\times[b_s,b_{s+1}] .$$

Since $\{C^k_{[a_r,a_{r+1}]} \otimes C^l_{[b_s,b_{s+1}]} \;:\; k,l \in \mathbb{N}_0\}$ is a basis of $L^2([a_r,a_{r+1}]\times[b_s,b_{s+1}])$ it follows that $a^{k,l}_{r,s} = 0$ for all $r,s \in \mathbb{Z}$ and $k,l \in \mathbb{N}_0$. Hence, (12) is proved. □

Finally we present conditions on the bells to obtain orthonormal bases. Obviously the $\Psi^{k,l}_{r,s}$ constitute an orthonormal basis iff

$$w_{r,s} = \tilde{w}_{r,s} \quad \text{for all } r,s \in \mathbb{Z} .$$

Because $\boldsymbol{M}^{\tilde{w}}_{r,s} = (\boldsymbol{M}^{w}_{r,s})^{-H}$ this is equivalent to

$$\begin{aligned} |w_{r,s}| &=1 \quad \text{a.e on } R_1^{r,s}, \\ \boldsymbol{K}^H_{r,s}\boldsymbol{K}_{r,s} &=\boldsymbol{I} \quad \text{a.e on } R_2^{r,s}, \\ \boldsymbol{L}^H_{r,s}\boldsymbol{L}_{r,s} &=\boldsymbol{I} \quad \text{a.e on } R_3^{r,s}, \\ \boldsymbol{M}^H_{r,s}\boldsymbol{M}_{r,s} &=\boldsymbol{I} \quad \text{a.e on } R_4^{r,s}. \end{aligned}$$

It is easy to see, that the last three equations can be written as

$$\boldsymbol{K}^H_{r,s}\begin{pmatrix} w_{r,s} \\ w_{r-1,s}\end{pmatrix} = \begin{pmatrix} 1 \\ 0 \end{pmatrix} \quad \text{a.e on } R_2^{r,s},$$

$$\boldsymbol{L}^H_{r,s}\begin{pmatrix} w_{r,s} \\ w_{r,s-1}\end{pmatrix} = \begin{pmatrix} 1 \\ 0 \end{pmatrix} \quad \text{a.e on } R_3^{r,s},$$

$$\boldsymbol{M}^H_{r,s}\begin{pmatrix} w_{r,s} \\ w_{r-1,s} \\ w_{r,s-1} \\ w_{r-1,s-1}\end{pmatrix} = \begin{pmatrix} 1 \\ 0 \\ 0 \\ 0 \end{pmatrix} \quad \text{a.e on } R_4^{r,s}$$

for all $r, s \in \mathbb{Z}$. Similar conditions can be found also in [11].

4. Examples

First we want to consider a continuous, piecewise linear non-tensor product bell function in more detail. We restrict ourselves to the simple rectangular pyramid $w^1_{r,s} = w^1_{0,0}(\cdot - r, \cdot - s)$ with

$$w^1_{0,0}(x,y) = \begin{cases} y, & \text{if } 0 \le x \le 2 \text{ and } 0 \le y \le \min(x, 2-x), \\ 1-y, & \text{if } 0 \le x \le 2 \text{ and } \max(x, 2-x) \le y \le 2, \\ x, & \text{if } 0 \le y \le 2 \text{ and } 0 \le x \le \min(y, 2-y), \\ 1-x, & \text{if } 0 \le y \le 2 \text{ and } \max(y, 2-y) \le y \le 2, \\ 0 & \text{otherwise} \end{cases}$$

for the grid $a_r = r + \frac{1}{2}$, $b_s = s + \frac{1}{2}$, $a^+_{r-1} = a^-_r = r$ and $b^+_{s-1} = b^-_s = s$ (see **Fig. 2**). As dual bell $\tilde{w}_{r,s}$ we obtain a piecewise rational, continuous function (see **Fig. 2**) with

$$\tilde{w}^1_{0,0}(x,y) = \frac{y\left(2y^2 - 2x^2 + 4x - 1\right)}{4x^4 - 8x^3 + 8x^2 - 4x + 8xy^2 - 8x^2y^2 + 4y^4 + 1}$$

for $0 \le x \le 1$ and $0 \le y \le \min(x, 1-x)$ and

$$\tilde{w}^1_{0,0}(x,y) =$$

$$\frac{2x^2 - 4x - 2x^2y + 4xy - y - 2y^2 + 2y^3 + 2}{4y^4 - 8y^3 + 4y + 16xy^2 - 8x^2y^2 + 8x^2y - 16xy - 16x + 24x^2 - 16x^3 + 4x^4 + 5}$$

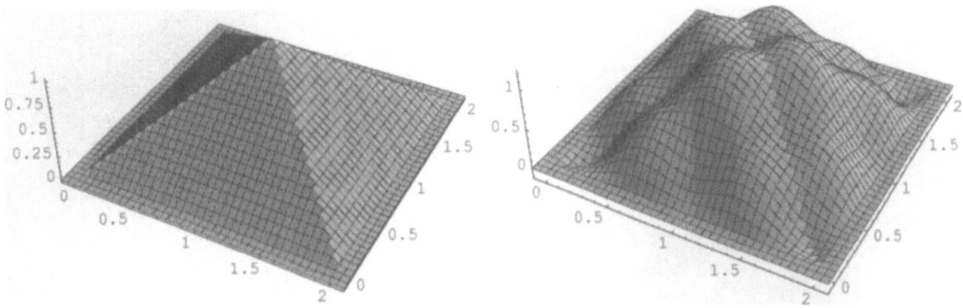

Fig. 2. The piecewise linear bell $w^1_{0,0}$ and the dual bell $\tilde{w}^1_{0,0}$.

for $\frac{1}{2} \le x \le 1$ and $1 - x \le y \le x$. The remaining values of the function can be determined by symmetry arguments as follows

$$\begin{aligned} \tilde{w}^1_{0,0}(x,y) &= \tilde{w}^1_{0,0}(y,x) \,, && \text{if } 0 \le x \le 1 \text{ and } x \le y \le 1, \\ \tilde{w}^1_{0,0}(x,y) &= \tilde{w}^1_{0,0}(-x,y), && \text{if } 1 \le x \le 2 \text{ and } 0 \le y \le 1, \\ \tilde{w}^1_{0,0}(x,y) &= \tilde{w}^1_{0,0}(x,-y), && \text{if } 0 \le x \le 2 \text{ and } 1 \le y \le 2. \end{aligned}$$

To compute the best possible Riesz bounds for this bell function following **Theorem 6** we have to consider the eigenvalues of the corresponding matrix $(\boldsymbol{M}^{w^1}_{0,0})^T \boldsymbol{M}^{w^1}_{0,0}$. By the above mentioned symmetry we only have to deal with x, y from the triangle $0 \le x \le 1$, $0 \le y \le \min(x, 1-x)$. In this case the eigenvalues are

$$\begin{aligned} \lambda_1(x,y) = \lambda_2(x,y) &= 2(x - y - \tfrac{1}{2})^2 + \tfrac{1}{2}\,, \\ \lambda_3(x,y) = \lambda_4(x,y) &= 2(x + y - \tfrac{1}{2})^2 + \tfrac{1}{2}\,. \end{aligned}$$

By easy calculations we obtain the Riesz bounds $A_0 = \frac{1}{2}$ and $B_0 = 1$.

The second example is the tensor product

$$w^2_{0,0}(x,y) := h(x)h(y)$$

of the linear B-spline

$$h(x) := \begin{cases} x, & \text{if } 0 \le x < 1, \\ 2 - x, & \text{if } 1 \le x < 2, \\ 0 & \text{otherwise} \end{cases}$$

(see **Fig. 3**).

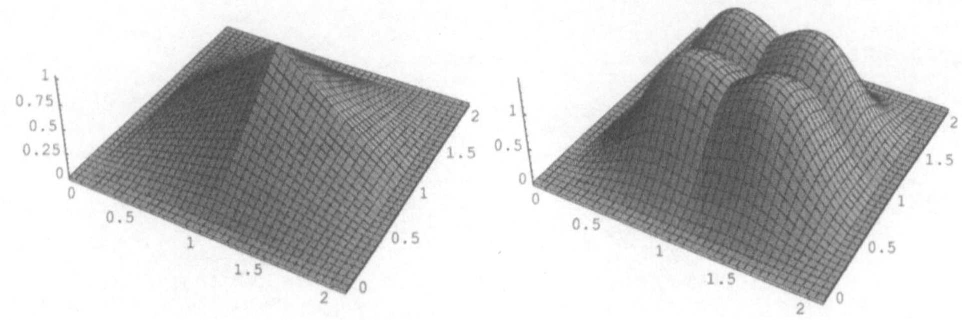

Fig. 3. The piecewise bilinear bell $w^2_{0,0}$ and the dual bell $\tilde{w}^2_{0,0}$.

The corresponding dual bell function is given by

$$\tilde{w}^2_{0,0}(x, y) = \frac{x\,y}{(2x^2 - 2x + 1)\,(2y^2 - 2y + 1)}$$

for $0 \leq x, y \leq 1$. The remaining values can be determined by the same symmetry conditions as in the first example.

In the tensor product case the four eigenvalues of $\boldsymbol{M}^T_{0,0}\boldsymbol{M}_{0,0}$ are products of the eigenvalues of $(\boldsymbol{M}^1_0)^T\boldsymbol{M}^1_0$ and $(\boldsymbol{M}^2_0)^T\boldsymbol{M}^2_0$ (see **Remark 4**). Easily one computes

$$\lambda_1(x, y) = \lambda_2(x, y) = \lambda_3(x, y) = \lambda_4(x, y) = \left(2(x - \tfrac{1}{2})^2 + \tfrac{1}{2}\right)\left(2(y - \tfrac{1}{2})^2 + \tfrac{1}{2}\right).$$

Here we obtain the best possible Riesz bounds $A_0 = \frac{1}{4}$ and $B_0 = 1$.

Acknowledgements. The research of C. K. Chui and J. Prestin was supported by NATO CRG Programme, Grant No. 950681. In addition, the research of C. K. Chui was sponsored by NSF Grant #DMS-95-05460 and ARO Grant #DAAH 04-95-10193. The research of K. Bittner was supported by the Deutsche Forschungsgemeinschaft.

References

1. Auscher P., Weiss G., and Wickerhauser M. V., Local sine and cosine bases of Coifman and Meyer and the construction of smooth wavelets, in *Wavelets - A Tutorial in Theory and Applications*, Chui C. K. (ed.), Academic Press, Boston, 1992, 237–256.
2. Christensen O., Frames containing a Riesz basis and approximation of the frame coefficients using finite-dimensional methods, J. Math. Anal. Appl. **199** (1996), 256–270.

3. Chui C. K. and Shi X., A study of biorthogonal sinusoidal wavelets, in *Curves and Surfaces* III, A. LeMehaute, C. Rabut, and L. L. Schumaker (eds.), to appear.
4. Chui C. K. and Shi X., Characterization and construction of biorthogonal wavelets. CAT Report #375, Texas A&M University, 1996.
5. Coifman R. R. and Meyer Y., Remarques sur l'analyse de Fourier á fenêtre, C. R. Acad. Sci. Paris **312** (1991), 259–261.
6. Daubechies I., Jaffard S., and Journé J. L., A simple Wilson orthonormal basis with exponential decay, SIAM J. Math. Anal. **22** (1991), 554–572.
7. Jawerth B. and Sweldens W., Biorthogonal smooth local trigonometric bases, J. Fourier Anal. Appl. **2** (1995), 109–133.
8. Malvar H., Lapped transforms for efficient transform/subband coding, IEEE Trans. Acoustic, Speech, and Signal Processing **38** (1990), 969–978.
9. Matviyenko G., Optimized local trigonometric bases, Appl. Comput. Harmonic Anal. **3** (1996), 301-323.
10. Wickerhauser M. V., *Adapted Wavelet Analysis from Theory to Software*, A. K. Peters, Wellesley, MA, 1994.
11. Xia X.-G. and Suter B. W., A family of two-dimensional nonseparable Malvar wavelets, Appl. Comput. Harmonic Anal. **2** (1995), 243–256.

Kai Bittner
FB Mathematik
Universität Rostock
Universitätsplatz 1
D-18051 Rostock
Germany
kai.bittner@cks1.uni-rostock.de

Charles K. Chui
Center for Approximation Theory
Texas A&M University
College Station, TX 77843-3368
cchui@tamu.edu

Jürgen Prestin
FB Mathematik
Universität Rostock
Universitätsplatz 1
D-18051 Rostock
Germany
prestin@mathematik.uni-rostock.d400.de

On Almost Interpolation by Multivariate Splines

Oleg Davydov, Manfred Sommer, and Hans Strauss

Abstract. A survey on some recent developments in multivariate interpolation, including characterizations of almost interpolation sets with respect to finite-dimensional spaces by conditions of Schoenberg-Whitney type, is given.

1. Introduction

Let U denote a finite-dimensional subspace of real valued functions defined on some set K. The problem of describing those configurations $T = \{t_1, \ldots, t_n\} \subset K, n = \dim U$, such that for any given data $\{y_1, \ldots, y_n\}$ there exists a unique function $u \in U$ satisfying

$$u(t_i) = y_i, \ i = 1, \ldots, n,$$

has attracted considerable interest in recent years, especially for the case when $K \subset \mathbb{R}^k, k \geq 2$. In contrast to the univariate case $K \subset \mathbb{R}$, where all interpolation sets T with respect to a spline space can be characterized by the well-known Schoenberg-Whitney condition [17] (see Section 2), it seems to be no reasonably simple way to characterize interpolation sets in the multivariate case (see [6, p. 136]). Therefore, several sufficient conditions and methods to construct such configurations for multivariate interpolation have been developed (see [3, 5, 6, 15] and references therein).

A new approach to multivariate interpolation has been found by Sommer and Strauss [23] introducing the concept of almost interpolation. A set $T = \{t_1, \ldots, t_s\} \subset K, s \leq \dim U$ is called an *almost interpolation set* (*AI-set*) *with respect to* U if for any system of neighborhoods B_i of t_i, $i = 1, \ldots, s$ there exist points $t_i' \in B_i$ such that $T' = \{t_1', \ldots, t_s'\}$ is an *interpolation set* (*I-set*) *with respect to* U; i.e.,

$$\dim U|_{T'} = s.$$

Otherwise, T' is called an *NI-set w.r.t.* U.

Multivariate Approximation and Splines
G. Nürnberger, J. W. Schmidt, and G. Walz (eds.), pp. 45–58.

ISBN 3-7643-5654-5.

It is shown in [23] that for a wide class of generalized spline spaces defined on polyhedral partitions AI-sets can be characterized by conditions of Schoenberg-Whitney type (Section 3).

Davydov [8] has considered AI-sets in the case of any finite-dimensional space U of real valued functions defined on an arbitrary topological space K. Using the notion of *local dimension* (see Section 4.1) he has shown that under some minor additional hypotheses on K any U has a piecewise almost Chebyshev structure (Sections 4.2 and 4.3), and AI-sets w.r.t. U can be characterized by a Schoenberg-Whitney type condition (Section 4.4) which extends the results in [23].

In Section 5 we present some results on how to transform a given AI-set into an I-set for the case of multivariate polynomial splines.

In the sequel we shall use the notations I-set and AI-set w.r.t. a space U, respectively, as we have defined them above. We denote by $F(K)$ the linear space of all real valued functions on a topological space K and by $C(K)$ its subspace consisting of all continuous functions. Moreover, we define, for a function $u \in F(K)$

$$\operatorname{supp} u := \overline{\{t \in K : u(t) \neq 0\}},$$

and denote by card M the number of elements of a finite set M.

2. Schoenberg-Whitney Type Conditions for Univariate Spline Interpolation

In this section we shall present some well-known results on univariate spline interpolation.

Assume that $K = [a, b] \subset \mathbb{R}$ and $\Delta : a = x_0 < \ldots < x_{r+1} = b$ denote any partition on K. Let $m \in \mathbb{N}$. The *linear space of polynomial spline functions of degree m with r fixed knots* is defined by

$$U := S_m(\Delta) := \{u \in C^{m-1}[a, b] : u|_{[x_i, x_{i+1}]} \in \pi_m, 0 \leq i \leq r\}$$

where π_m denotes the linear space of polynomials of degree at most m. Then $n := \dim U = m + r + 1$ and interpolation sets w.r.t. U can be characterized by an interlacing property due to Schoenberg and Whitney [17] as follows.

Interlacing property. *If $T = \{t_1, \ldots, t_n\} \subset [a, b]$, then T is an I-set w.r.t. U if and only if*

$$t_i < x_i < t_{i+m+1}, \quad i = 1, \ldots, r. \tag{2.1}$$

An equivalent statement to (2.1) is given in terms of a basis of functions in U with minimal support, the so-called B-spline functions (see e.g. [19]).

Support property. *Let $\{B_1, \ldots, B_n\}$ denote the B-spline basis for U. If $T = \{t_1, \ldots, t_n\} \subset [a, b]$, then T is an I-set w.r.t. U if and only if*

$$t_i \in \{t \in K : B_i(t) \neq 0\}, \quad i = 1, \ldots, n. \tag{2.2}$$

A generalization of this support property to the multivariate case plays an important role in the problem of determining AI-sets, especially for locally linearly independent systems of functions (see Theorems 3.7, 4.12 and [10, Theorem 2.3]).

It is easily seen that (2.1) can be reformulated in terms of the restriction of U to certain knot intervals.

Dimension property. *If* $T = \{t_1, \ldots, t_n\} \subset [a, b]$, *then* T *is an* I*-set w.r.t.* U *if and only if*

$$\operatorname{card}(T \cap [x_i, x_j]) \leq \dim U|_{[x_i,x_j]}, \quad i, j = 0, \ldots, r,\ i < j. \tag{2.3}$$

Fig. 2.1.

A property like (2.3) on the dimension behavior of U on certain "subcells" of the partition Δ will enable us to derive Schoenberg-Whitney type conditions for multivariate interpolation. In fact, for that case a more general dimension property as (2.3) will be better suitable.

Strong dimension property. *If* $T = \{t_1, \ldots, t_n\} \subset [a, b]$, *then* T *is an* I*-set w.r.t.* U *if and only if*

$$\operatorname{card}(T \cap M_P) \leq \dim U|_{M_P} \tag{2.4}$$

for any $P \subset \{0, \ldots, r\}$ *where* $M_P := \bigcup_{i \in P} [x_i, x_{i+1}]$.

Fig. 2.2.

Remark. Schoenberg-Whitney type conditions can be used for the characterization of I-sets with respect to some other spaces of univariate functions. Interlacing property (2.1) and support property (2.2), respectively, have been extended in [16, 21] to spaces of generalized splines. An extension of the support property to locally linearly independent weak Descartes systems of functions has been found in [4]. Extensions of the dimension properties (2.3) and (2.4) to weak Chebyshev spaces have been given in [7, 9, 22].

3. Schoenberg-Whitney Type Conditions for Almost Interpolation on Polyhedral Partitions

In this section we shall present some recent results on almost interpolation of multivariate functions defined on polyhedral partitions in $\mathbb{R}^k$. The conditions which even characterize *AI*-sets are extensions of (2.2) and (2.4), respectively and therefore, can be considered as conditions of Schoenberg-Whitney type.

Let us begin by introducing the spaces of interest. Assume that $\mathcal{K}$ denotes a finite family of l-dimensional simplices in $\mathbb{R}^k$ where $k \in \mathbb{N}, l \in \mathbb{N} \cup \{0\}$ and $l \leq k$ satisfying the following properties:

1) If the simplex s belongs to $\mathcal{K}$, then every face of s belongs also to $\mathcal{K}$.
2) If $s, \tilde{s} \in \mathcal{K}$, then the intersection of s and $\tilde{s}$ is empty or a common face.

The point-set union of all simplices of the family $\mathcal{K}$ is called a *polyhedron* in $\mathbb{R}^k$ (see [18]).

Let

$$K := \bigcup_{i \in I} K_i$$

where every K_i is a polyhedron in $\mathbb{R}^k$ and I denotes a finite set. Assume that $K_i \not\subset \bigcup_{j \in I \setminus \{i\}} K_j$. Moreover, assume that K is *regular*; i.e., the set $\{K_i\}_{i \in I}$ of polyhedrons in K satisfies also property 2) above.

Example 3.1. *If $k = 1$, then $K = [a,b]$ and $K_i = [x_i, x_{i+1}]$, $i = 0, \dots, r$ where $a = x_0 < \dots < x_{r+1} = b$.*

Example 3.2. (Regular triangulation) *Let $K = \bigcup_{i \in I} K_i \subset \mathbb{R}^2$ where $\{K_i\}_{i \in I}$ is a set of triangles with the property that no vertex of K_i lies on the interior of K_j or on the interior of a side of $K_j, i, j \in I$.*

Example 3.3. (Rectangular partition) *Let $K = [a,b] \times [c,d] \subset \mathbb{R}^2$ and $a = x_0 < \dots < x_{r+1} = b$, $c = y_0 < \dots < y_{s+1} = d$. Then $K = \bigcup_{(i,j) \in I} K_{ij}$ where $K_{ij} = [x_i, x_{i+1}] \times [y_j, y_{j+1}]$, $I = \{(i,j) : i \in \{0, \dots, r\}, j \in \{0, \dots, s\}\}$.*

If the rectangular partition is refined by drawing in all diagonals with positive slope or both diagonals in every K_{ij}, the resulting partition is called *type-1* or *type-2 triangulation*, respectively (see [6, p. 27]).

Let $p \in \mathbb{N} \cup \{0\}$. For every $i \in I$, assume that U_i denotes a finite-dimensional subspace of $C^p(K_i)$ satisfying the *L-property*: Let $u \in U_i$ and $\tilde{t} \in K_i$ be a zero of u. If there exists $\epsilon > 0$ such that $u(t) = 0$ for every $t \in K_i$ satisfying $||\tilde{t} - t|| < \epsilon$, then $u \equiv 0$ on K_i. (In the special case when $K_i \subset \mathbb{R}$, the most important examples of U_i are Haar subspaces.)

We define the *linear space S of generalized spline functions of smoothness p* by $S := \{s \in C^p(K) :$ for every $i \in I$ there exists $s_i \in U_i$ such that $s|_{K_i} = s_i\}$. Suppose that $\{u_1, \dots, u_n\}$ denotes a system of linearly independent functions in S. Set

$$U := \text{span } \{u_1, \dots, u_n\}.$$

Recently, Sommer and Strauss [23] were concerned with the question of when a subset $T = \{t_1, \ldots, t_s\}$ of $K, s \leq n$ is an AI-set w.r.t. U. For that they gave an extension of condition (2.4) as follows.

Definition 3.4. *Let* $T = \{t_1, \ldots, t_s\} \subset K, s \leq n$. *Then* T *is said to satisfy a* condition of Schoenberg-Whitney type *or* T *is called an* SWT-set w.r.t. U *if*

$$\operatorname{card}(T \cap \operatorname{int}_K M_P) \leq \dim U|_{M_P} \tag{3.1}$$

for any $P \subset I$ *where* $M_P := \bigcup_{i \in P} K_i$ *and* $\operatorname{int}_K M_P := K \setminus \bigcup_{i \in I \setminus P} K_i$.

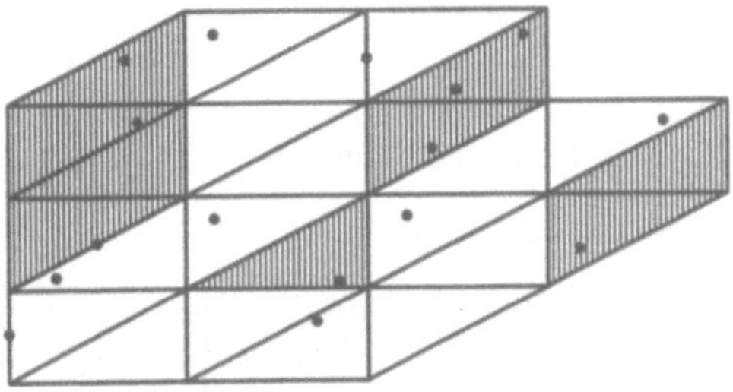

Fig. 3.1.

Using this condition a characterization of all AI–sets w.r.t. U was given in [23].

Theorem 3.5. *Let* $T = \{t_1, \ldots, t_s\} \subset K, s \leq n$. *Then* T *is an* AI–set w.r.t. U *if and only if* T *is an* SWT-set w.r.t. U.

It is a nice consequence of this result that in practice it should suffice to use AI-sets for interpolation problems. In fact, in [23] the following result was shown.

Corollary 3.6. *If* $\mathcal{T} := \{T = \{t_1, \ldots, t_s\} \subset K : T$ *is an* AI-set w.r.t. $U\}$, *and* $\tilde{\mathcal{T}} := \{T \in \mathcal{T} : T$ *is an* NI-set w.r.t. $U\}$, *then* $\tilde{\mathcal{T}}$ *is a set of first category in* $\mathcal{T}$.

The following result which gives an extension of (2.2) is also due to [23].

Theorem 3.7. *Let* $T = \{t_1, \ldots, t_s\} \subset K, s \leq n$. *The following conditions are equivalent.*

1) T *is an* AI-set w.r.t. U.
2) *For each basis* $\{u_1, \ldots, u_n\}$ *of* U *there is some permutation* σ *of* $\{1, \ldots, n\}$ *such that*

$$t_i \in \operatorname{supp} u_{\sigma(i)}, \quad i = 1, \ldots, s.$$

4. Almost Interpolation by Functions Defined on Topological Spaces

Here we survey some results by Davydov [8], who has shown that a Schoenberg-Whitney type characterization of AI-sets holds in fact for any finite-dimensional linear space of continuous functions on a topological space satisfying some minor restrictions. Particularly, this is true for the spaces of multivariate splines with respect to non-polyhedral partitions.

4.1. Local Dimension

Assume that K is a topological space and U denotes a finite-dimensional subspace of $F(K), \dim U = n$.

Definition 4.1. *[8] Let K' be any subset of K. By the* local dimension of U on K' *we mean*

$$\text{l-dim}_{K'} U := \inf \{\dim U|_B : K' \subset B, B \text{ open} \}.$$

With the help of local dimension it is possible to give a „local" characterization of almost interpolation sets with respect to any finite-dimensional space U.

Theorem 4.2. *[8] Let $T = \{t_1, \ldots, t_s\} \subset K, s \leq n$. Then T is an AI–set* w.r.t. *$U \subset F(K)$ if and only if*

$$\text{l-dim}_{T'} U \geq \text{card}\, T'$$

for any choice of a nonempty subset $T' \subset T$.

We write $\text{l-dim}_t U$ instead of $\text{l-dim}_{\{t\}} U$. The function $\varphi : K \to \mathbb{Z}_+$ defined by $\varphi(t) := \text{l-dim}_t U$ is evidently upper semicontinuous. Moreover, it is continuous on an open everywhere dense subset $G_U \subset K$. Figure 4.1 presents the graph of the local dimension of the space of univariate splines $S_m(\Delta)$.

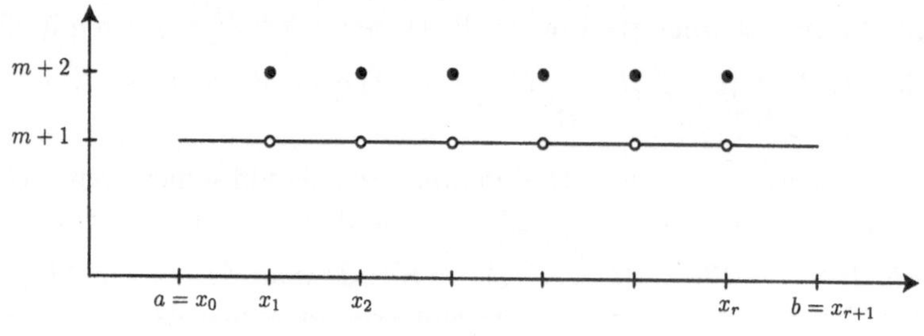

Fig. 4.1.

As another example consider the space of linear bivariate splines on the triangulation in Figure 4.2. We have

$$\text{l-dim}_{t_1} U = 3, \ \text{l-dim}_{t_2} U = 4, \ \text{l-dim}_{t_3} U = 5.$$

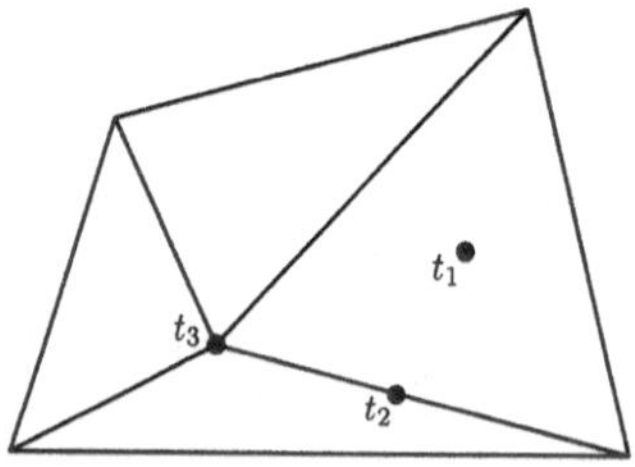

Fig. 4.2.

4.2. Almost Chebyshev Systems

It is well-known that Chebyshev systems (T-systems) play an important role in the approximation theory. Recall that a system of functions $u_1, \ldots, u_n \in F(K)$ is said to be a *Chebyshev system* if every nonzero function $u \in U = \text{ span } \{u_1, \ldots, u_n\}$ has at most $n-1$ zeros. The linear span of a Chebyshev system is called a *Haar space.* (Some authors prefer the notation "Chebyshev space".) It is an important feature of Haar spaces that they are as good for interpolation as possible: any set $T = \{t_1, \ldots, t_n\} \subset K$ is an interpolation set w.r.t. such a space. In fact, this property can be taken as a definition of a Haar space or a Chebyshev system.

Mairhuber's theorem [14] shows that the existence of a Haar space $U \subset F(K)$ of dimension $n \geq 2$ implies some severe restrictions on K. Particularly, K cannot be homeomorphic to a subset of $\mathbb{R}^k, k \geq 2$, with nonempty interior. Hence, Chebyshev systems cannot be used for approximation of multivariate functions. Because of this we consider an "almost interpolation" analogue of Chebyshev systems.

Definition 4.3. *A system of functions $u_1, \ldots, u_n \in F(K)$ is said to be an* almost Chebyshev system *if any set $T = \{t_1, \ldots, t_n\} \subset K$ is an AI–set w.r.t. $U =$* span $\{u_1, \ldots, u_n\}$. *The linear span U of an almost Chebyshev system is called an* almost Haar space.

In the next theorem we give some characteristic properties of almost Haar spaces.

Theorem 4.4. *Let K be a topological space and let $U \subset F(K)$ denote a finite-dimensional linear space, $\dim U = n$.*

1) *U is an almost Haar space if and only if for any nonempty open set $B \subset K$,*

$$\dim U|_B = \min \{n, \operatorname{card} B\}.$$

2) *Suppose that every nonempty open set $B \subset K$ is infinite. Then U is an almost Haar space if and only if no nonzero function $u \in U$ can vanish identically on an open subset B of K.*
3) *Suppose that K is a compact metric space and $U \subset C(K)$. Then U is an almost Haar space if and only if it is an* almost Chebyshev subspace *of the normed space $C(K)$ in the sense that the set of elements $f \in C(K)$ for which there exists a unique best approximation to f from U, is of the second category in $C(K)$.*
4) *Suppose that K is connected and satisfies T_1-axiom of separation. Then U is an almost Haar space if and only if* l-$\dim_t U =$ *constant* , $t \in K$.

The notion of almost Chebyshev subspaces mentioned in 3) was introduced by Stechkin [24]. Garkavi [12, 13] showed that there exist almost Chebyshev subspaces of arbitrary finite dimensions in any separable Banach space. Parts 1) (in the case of K being a compact metric space) and 3) of the above theorem are due to Garkavi [13]. 2) is an immediate consequence of 1). 4) is proved in [8] with the help of the following result.

Proposition 4.5. *[8] Under the hypotheses of Theorem 4.4, let K' be a connected subset of K. If* $\text{l-dim}_t\, U = m$, $t \in K'$, *then* $\text{l-dim}_{K'}\, U = m$.

It is easily seen from Theorem 4.4 that the class of almost Haar spaces is rather wide. For example, any finite-dimensional space of analytic functions on a domain $K \subset \mathbb{R}^k$ is a space of this type. In the case $K \subset \mathbb{R}$ the same is true for any subspace of a Haar space.

4.3. Piecewise Almost Chebyshev Structure

Consider again the function $\varphi(t) = \text{l-dim}_t\, U$, where $U \subset F(K)$ is a finite-dimensional linear space and K denotes a topological linear space. Denote by G_U the set of all points of continuity of $\varphi(t)$ and decompose G_U into the union of its connected components,

$$G_U = \bigcup_{i \in I} K_i.$$

Then G_U is open and everywhere dense in K, so that

$$\overline{\bigcup_{i \in I} K_i} = K.$$

Because of this we consider the set $\{K_i : i \in I\}$ as a *partition* of K. The *cells* K_i of this partition are disjoint and connected.

Since $\varphi(t)$ takes only integer values, it remains constant on each cell $K_i, i \in I$. Theorem 4.4 then shows that $U|_{K_i}$ is an almost Haar space if K_i is not a singleton, satisfies T_1-axiom of separation and, additionally,

$$\text{l-dim}_t\, U = \text{l-dim}_t\, U|_{K_i}, \ t \in K_i.$$

This last condition can be guaranteed by imposing some restrictions on K.

Theorem 4.6. *[8] Let K be a locally connected T_1-space and let $U \subset F(K)$ be a finite-dimensional linear space. Define the partition $K = \overline{\cup_{i \in I} K_i}$ as above. Then the following conditions hold.*

1) *The cells K_i are open and connected subsets of K.*
2) *$U|_{K_i}$ is an almost Haar space for any $i \in I$.*

Thus, under the hypotheses of Theorem 4.6, U is generated on the cells by some almost Chebyshev systems and, hence, may be thought of as a "piecewise almost Chebyshev" space.

In the case $K \subset \mathbb{R}$ we obtain a similar result without requiring that K is locally connected.

Theorem 4.7. *[8] Let K be any subset of $\mathbb{R}$ and let $U \subset F(K)$ denote a finite-dimensional linear space. Define the partition $K = \overline{\cup_{i \in I} K_i}$ as above. Then the following is true.*

1) *Each cell K_i is either a singleton or an (finite or infinite) open, closed or half-open interval. In particular, I is countable.*
2) *$U|_{K_i}$ is an almost Haar space for any $i \in I$.*

We can say more about $U|_{K_i}$ in the case when $K = [a, b]$ and $U \subset C[a, b]$ is a *weak Chebyshev space*; i.e., every nonzero function $u \in U$ has at most $n - 1$ sign changes ($n = \dim U$). By Theorem 4.6, K_i are open connected subsets of $[a, b]$, so that

$$G_U = [a, a') \cup \bigcup_{j \in J} (\alpha_j, \beta_j) \cup (b', b],$$

where $a \leq a'$, $b' \leq b$ (we mean $[x, x) = (x, x] = \emptyset$), (α_j, β_j), $j \in J$, are disjoint open subintervals of (a', b'), $\cup_{j \in J} (\alpha_j, \beta_j)$ is everywhere dense in (a', b').

We say that a point $t \in K$ is *essential* w.r.t. $U \subset F(K)$ if there exists $u \in U$ such that $u(t) \neq 0$.

Theorem 4.8. *Let $U \subset C[a, b]$ be a weak Chebyshev space. Suppose that any point $t \in [a, b]$ is essential w.r.t. U. Then*

$$U|_{(a,a')},\ U|_{(b',b)},\ U|_{(\alpha_j,\beta_j)},\ j \in J,$$

are Haar spaces.

Proof: Indeed, by [20, Theorem 1.4] these spaces are weak Chebyshev because U is weak Chebyshev. Theorem 4.6 states that they are also almost Haar spaces, so that no nonzero function vanishes identically on a nondegenerate proper subinterval of $[a, a']$, $[b', b]$ or $[\alpha_j, \beta_j]$ respectively. Since every point $t \in [a, b]$ is essential w.r.t. U, it follows that each of $U|_{[a,a']}$, $U|_{[b',b]}$ and $U|_{[\alpha_j,\beta_j]}$, $j \in J$, has Chebyshev rank at most $n - 1$. Then Remark i in [20, p. 59] implies that the restrictions of U to corresponding open intervals are in fact Haar spaces. □

The statement of Theorem 4.8 can be strengthened for the important case when a weak Chebyshev subspace U of $C[a, b]$ does not contain functions with "arbitrarily small" zero intervals. Following Bartelt [1] we say that U satisfies *condition* (I) if there exists $\delta > 0$ such that if $u \in U$ and $u \equiv 0$ on $[c, d] \subset [a, b]$, $c, d \in \operatorname{supp} u \cup \{a, b\}$, then $d - c \geq \delta$. This implies a "spline-like" behavior as the following result shows.

Theorem 4.9. *[20] Let U be a weak Chebyshev subspace of $C[a, b]$ and suppose that U satisfies condition (I). The following statements hold.*

1) *There exists a finite set of points $a = x_0 < \ldots < x_{r+1} = b$ such that for each $i = 0, \ldots, r$,*

$$U|_{[x_i,x_{i+1}]}$$

is an almost Haar subspace of $C[x_i, x_{i+1}]$.

2) If in addition every $t \in [a, b]$ is essential w.r.t. U, then there exists a finite set of points $a = x_0 < \ldots < x_{r+1} = b$ such that for each $i = 0, \ldots, r$,

$$U|_{[x_i, x_{i+1}]}$$

is even a Haar subspace of $C[x_i, x_{i+1}]$.

4.4. Schoenberg-Whitney Type Conditions

Suppose that K is a locally connected T_1-space and $U \subset F(K)$ is a finite-dimensional linear space, $\dim U = n$. Define the partition $K = \overline{\bigcup_{i \in I} K_i}$ as in the previous subsection. Then Theorem 4.6 can be applied so that $U|_{K_i}$ is an almost Haar space when K_i is not a singleton. Assuming additionally that $U \subset C(K)$, we give a Schoenberg-Whitney type characterization of almost interpolation sets through an extension of conditions (2.4) and (3.1). The next two theorems are immediate consequences of Theorem 3.10 and Corollary 4.18 in [8].

Theorem 4.10. *Suppose that $U \subset C(K)$ and let $T = \{t_1, \ldots, t_s\} \subset K, s \leq n$. Then T is an AI-set w.r.t. U if and only if*

$$\operatorname{card}(T \cap \operatorname{int} M_P) \leq \dim U|_{M_P} \tag{4.1}$$

for any $P \subset I$ where $M_P := \overline{\cup_{i \in P} K_i}$ and $\operatorname{int} M_P$ denotes the set of all interior points of M_P w.r.t. topology on K.

When I is infinite, we have in (4.1) an infinite set of inequalities. However, we are able to show that for each fixed $T = \{t_1, \ldots, t_s\}$ it is enough to check (4.1) for a finite number of M_P's.

Theorem 4.11. *Under the hypotheses of Theorem 4.10, let $B_1, \ldots, B_s$ be open L–neighborhoods of $t_1, \ldots, t_s$, respectively; i.e.,*

$$\dim U|_{B_j} = \text{l-dim}_{t_j} U, \quad j = 1, \ldots, s.$$

In order for $T = \{t_1, \ldots, t_s\}$ to be an AI–set w.r.t. U it is sufficient that (4.1) holds for any $P \subset I$ of the form

$$P = \bigcup_{j \in Q} P_j, \quad Q \subset \{1, \ldots, s\},$$

where $P_j := \{i \in I : K_i \cap B_j \neq \emptyset\}, \quad j = 1, \ldots, s.$

It is easily seen that the Theorems 4.10 and 4.11 can be applied to the spaces of generalized splines considered in Section 3 as well as to any space of continuous piecewise polynomial functions with respect to an arbitrary partition of a domain $K \subset \mathbb{R}^k$.

A general version of Theorem 3.7 is also true.

Theorem 4.12. *[11] Suppose that K is a topological space and $U \subset F(K)$ is a finite-dimensional linear space, $\dim U = n$. Let $T = \{t_1, \ldots, t_s\} \subset K, s \leq n$. Then the following conditions are equivalent.*

1) *T is an AI-set w.r.t. U.*
2) *For each basis $\{u_1, \ldots, u_n\}$ of U there exists some permutation σ of $\{1, \ldots, n\}$ such that $t_i \in \operatorname{supp} u_{\sigma(i)}$, for all $i = 1, \ldots, s$.*

5. Transforming AI-sets into I-set

In the preceding sections we have considered the problem of characterizing AI-set w.r.t. finite-dimensional subspaces U of $F(K)$.

By Corollary 3.6 we know, at least for spaces of generalized splines U, that if $\mathcal{T}$ denotes the set of all AI-sets w.r.t. U and $\tilde{\mathcal{T}}$ its subset of NI-sets w.r.t. U, then $\tilde{\mathcal{T}}$ is a set of first category in $\mathcal{T}$.

Hence the question arises whether it is possible to find simple methods for transforming AI-sets T into I-sets in some neighborhood of T.

Let $T = \{t_1, \ldots, t_s\}$, $s \leq n$, $n = \dim U$, be an AI-set w.r.t. U and let some neighborhoods $B_1, \ldots, B_s$ of the points $t_1, \ldots, t_s$, respectively, be given. Set $n_i := \dim U|_{B_i}$, $i = 1, \ldots, s$. It is always possible to choose some points $t_{i,j} \in B_i$, $j = 1, \ldots, n_i$, $i = 1, \ldots, s$, in such a way that $T_i := \{t_{i,1}, \ldots, t_{i,n_i}\}$ is an I-set w.r.t. $U|_{B_i}$, $i = 1, \ldots, s$.

Theorem 5.1. *[8] For any $i \in \{1, \ldots, s\}$ there exists $\mu(i) \in \{1, \ldots, n_i\}$ such that $\{t_{i,\mu(i)} : i = 1, \ldots, s\}$ is an I-set w.r.t. U.*

Assume now that S denotes the linear space of polynomial splines of smoothness p and degree m defined as in Section 3 on the set $K = \cup_{i \in I} K_i \subset \mathbb{R}^k$ where K_i is a convex polyhedron for all $i \in I$, such that $S|_{K_i}$ coincides with the set of polynomials of total degree at most m, $i \in I$. We consider the following situation.

Let a set $T = \{t_1, \ldots, t_n\} \subset \mathbb{R}^k$ where $n = \dim S$ be given. Assume that T is an AI-set w.r.t. S satisfying $t_i \in K_{j_i}$, $i = 1, \ldots, n$. Moreover, let $V = \{v_1, \ldots, v_n\}$ be an I-set w.r.t. S such that $v_i \in K_{j_i}$, $i = 1, \ldots, n$. Notice that by the definition of AI-sets every neighborhood of T contains such an I-set. We now define the straight lines through t_i and v_i,

$$l_i := \{t \in K_{j_i} : \text{ there exists } \lambda \in \mathbb{R} \text{ such that } t = t_i(\lambda) = (1 - \lambda)t_i + \lambda v_i\}.$$

Since K_{j_i} is convex, we have $t_i(\lambda) \in K_{j_i}$ for all $0 \leq \lambda \leq 1$.

Under these assumptions we obtain the following result.

Theorem 5.2. *[22] Let $T(\lambda) := \{t_1(\lambda), \ldots, t_n(\lambda)\}$. Then $T(\lambda)$ is an I-set w.r.t. S for all $0 \leq \lambda \leq 1$ with the exception of a finite number of points $0 \leq \lambda_1 < \ldots \lambda_q \leq 1$ where $0 \leq q \leq mn$.*

Corollary 5.3. *[22] Let the assumptions of Theorem 5.2 be given. Then there exists a real number $\lambda_0 > 0$ such that $T(\lambda)$ are I-sets w.r.t. S for all $0 < \lambda \leq \lambda_0$.*

Remark 5.4. 1) Let $V = \{v_1, \dots, v_n\}$ be an I-set w.r.t. S such that $v_i \in K_{j_i}$. Then it follows from Theorem 3.5 that every set $T = \{t_1, \dots, t_n\}$ satisfying $t_i \in K_{j_i}$, $i = 1, \dots, n$, is an AI-set w.r.t. S. Hence we choose an arbitrary set $\tilde{T} = \{\tilde{t_1}, \dots, \tilde{t_n}\}$ satisfying $\tilde{t_i} \in K_{j_i}$, $i = 1, \dots, n$. It follows from Corollary 5.3 that there is a real number $\lambda_0 > 0$ such that $\{(1-\lambda)\tilde{t_i} + \lambda v_i\}_{i=1}^n$ is an I–set w.r.t. S for all $0 < \lambda < \lambda_0$. This means the following: If we have an I–set V such that $v_i \in K_{j_i}$, $i = 1, \dots, n$, then we can move the points v_i to arbitrary points $\tilde{t_i}$ in the same polyhedron and we always have I–sets on the lines connecting v_i and $\tilde{t_i}$ in a neighborhood of $\tilde{t_i}$, $i = 1, \dots, n$. But this is not true if both T and V are AI–sets which fail to be I–sets. It can be shown by simple examples that $\{(1-\lambda)t_i + \lambda v_i\}_{i=1}^n$ can be NI–sets for all $0 \le \lambda \le 1$. Therefore, starting with some special interpolation configuration (a variety of methods of constructing them can be found in [3, 5, 6, 15]), we can apply the above method in order to obtain interpolation configurations with desirable location. For example, for the space of continuous bivariate spline functions on regular triangulations an initial I–set can be easily constructed by well-known finite-element methods (see e.g. [2, p. 155]).
2) Let us consider the case of Theorem 5.2 such that V is an I-set and T is an AI-set. We shall give an example where $T(\lambda)$ is an NI-set for some $0 \le \lambda < 1$: Define a set of vertices in $\mathbb{R}^2$ by $e_1 = (1,0), e_2 = (0,1), e_3 = (-1,0)$ and $e_4 = (0,-1)$ and let $K = K_1 \cup K_2$ be a triangulation such that K_1 is the convex hull of $\{e_1, e_2, e_4\}$ and K_2 is the convex hull of $\{e_2, e_3, e_4\}$. Assume that S is the space of linear continuous splines defined on K. Then the set $V = \{e_1, \dots, e_4\}$ is an I-set w.r.t. S. The set $T = \{t_1, \dots, t_4\}$ given by $t_1 = (1/2, 0), t_2 = (1/2, -1/2), t_3 = (-1/2, 0)$ and $t_4 = (-1/2, 1/2)$ is an AI-set, but fails to be an I-set. We now define the lines

$$t_i(\lambda) = (1-\lambda)t_i + \lambda e_i, \ i = 1, \dots, 4.$$

For $\lambda = 1/3$ all points $t_i(1/3)$, $i = 1, \dots, 4$ are contained in the x-axis. Hence $\{t_i(\lambda)\}_{i=1}^4$ is an NI-set for $\lambda = 0$ and $\lambda = 1/3$. It is easily seen that for any other $\lambda \in (0, 1/3) \cup (1/3, 1]$, $\{t_i(\lambda)\}_{i=1}^4$ is an I-set.

Finally we give a description of a wide class of I-sets for linear bivariate splines.

Theorem 5.5. *[22] Let $K = \cup_{i \in I} K_i \subset \mathbb{R}^2$ be a regular triangulation and S denote the space of linear continuous splines on K. Assume that $\{e_1, \dots, e_n\}$ denotes the set of vertices of K. Then $S = \operatorname{span}\{u_1, \dots, u_n\}$ where $u_i \in S$ is defined by $u_i(e_j) = \delta_{ij}$, $i, j = 1, \dots, n$. Let us define a set M_i by*

$$M_i := \{t \in K : u_i(t) > \frac{1}{2}\}, \ i = 1, \dots, n.$$

Then every set $\{t_1, \dots, t_n\}$ satisfying $t_i \in M_i$ for $i = 1, \dots, n$ is an I-set w.r.t. S.

Note that the result is no longer true if we replace each set M_i by its closure.

Acknowledgements. O. Davydov was supported by the Alexander von Humboldt Foundation, under Research Fellowship.

References

1. Bartelt M. W., Weak Chebyshev sets and splines, J. Approx. Theory **14** (1975), 30–37.
2. Becker E. B., Carey G. F., and Oden J. T., *Finite Elements: An Introduction*, Vol. I, Prentice-Hall, New Jersey, 1981.
3. Bojanov B. D., Hakopian H. A., and Sahakian A. A., *Spline Functions and Multivariate Interpolations*, Kluwer Academic Publishers, Dordrecht, 1993.
4. Carnicer J. M. and Peña J. M., Spaces with almost strictly totally positive bases, Math. Nachrichten **169** (1994), 69–79.
5. Cheney E. W., *Multivariate Approximation Theory: Selected Topics*, CBMS–SIAM, Philadelphia, 1986.
6. Chui C. K., *Multivariate Splines*, CBMS–SIAM, Philadelphia, 1988.
7. Davydov O., A class of weak Chebyshev spaces and characterization of best approximations, J. Approx. Theory **81** (1995), 250–259.
8. Davydov O., On almost interpolation, J. Approx. Theory, to appear.
9. Davydov O. and Sommer M., Interpolation by weak Chebyshev spaces, preprint.
10. Davydov O., Sommer M., and Strauss H., Locally linearly independent systems and almost interpolation, this volume.
11. Davydov O., Sommer M., and Strauss H., On almost interpolation and locally linearly independent bases, preprint.
12. Garkavi A. L., On Chebyshev and almost Chebyshev subspaces, Izv. Akad. Nauk SSSR Ser. Mat. **28** (1964), 799–818 [in Russian]; English translation in Amer. Math. Soc. Transl. (2) **96** (1970), 153–175.
13. Garkavi A. L., Almost Chebyshev systems of continuous function, Izv. Vyssh. Uchebn. Zaved. Mat. (2) **45** (1965), 36–44 [in Russian]; English translation in Amer. Math. Soc. Transl. (2) **96** (1970), 177–187.
14. Mairhuber J. C., On Haar's theorem concerning Chebyshev approximation problems having unique solutions, Proc. Amer. Math. Soc. **7** (1956), 609–615.
15. Nürnberger G., Approximation by univariate and bivariate splines, in *Second International Colloquium on Numerical Analysis* (Bainov D. and Covachev V., Eds.), VSP, 1994, 143–153.
16. Nürnberger G., Schumaker L. L., Sommer M., and Strauss H., Interpolation by generalized splines, Numer. Math. **42** (1983), 195–212.
17. Schoenberg I. J. and Whitney A., On Polya frequency functions III. The positivity of translation determinants with application on the interpolation problem by spline curves, Trans. Amer. Math. Soc. **74** (1953), 246–259.
18. Schubert H., *Topologie*, Teubner, Stuttgart, 1975.

19. Schumaker L. L., *Spline Functions: Basic Theory* Wiley-Interscience, New York, 1981.

20. Sommer M., Weak Chebyshev spaces and best L_1-approximation, J. Approx. Theory **39** (1983), 54–71.

21. Sommer M. and Strauss H., Weak Descartes systems in generalized spline spaces, Constr. Approx. **4** (1988), 133–145.

22. Sommer M. and Strauss H., Interpolation by uni- and multivariate generalized splines, J. Approx. Theory **83** (1995), 423–447.

23. Sommer M. and Strauss H., A condition of Schoenberg-Whitney type for multivariate spline interpolation, Advances in Comp. Math. **5** (1996), 381–397.

24. Stechkin S. B., Approximation properties of sets in normed linear spaces, Rev. Math. Pures Appl. **8** (1963), 5–18 [in Russian].

Oleg Davydov
Department of Mechanics and Mathematics
Dnepropetrovsk State University
pr.Gagarina 72
Dnepropetrovsk GSP 320625, Ukraine
davydov@euklid.math.uni-mannheim.de

Manfred Sommer
Mathematisch–Geographische Fakultät
Katholische Universität Eichstätt
85071 Eichstätt, Germany
manfred.sommer@ku-eichstaett.de

Hans Strauß
Institut für Angewandte Mathematik
Universität Erlangen–Nürnberg
91058 Erlangen, Germany
strauss@am.uni-erlangen.de

Locally Linearly Independent Systems and Almost Interpolation

Oleg Davydov, Manfred Sommer, Hans Strauss

Abstract. A simple method for constructing almost interpolation sets in the case of existence of locally linearly independent systems of basis functions is presented. Various examples of such systems, including translates of box splines and finite-element splines, are considered.

1. Introduction

In [16] we have shown how the well-known Schoenberg-Whitney condition for interpolation by univariate polynomial splines can be extended to multivariate splines or even to the general setting of real functions defined on a topological space. For this case it characterizes *almost interpolation sets* (*AI–sets*); *i.e.*, those configurations T such that in every neighborhood of T there exists a configuration $\tilde{T}$ (*I–set*) which admits Lagrange interpolation.

In practice it is clearly quite important to have algorithms of constructing I-sets. For instance, for a system $\{B_1, \dots, B_n\}$ of univariate polynomial B-splines it is well-known that any set $T = \{t_1, \dots, t_n\}$ such that $t_i \in \{t \in \mathbb{R} : B_i(t) \neq 0\}$, $i = 1, \dots, n$ is an I-set w.r.t. span $\{B_1, \dots, B_n\}$ (*support property*).

Since general methods of transforming AI-sets into an I-sets are available (see [16, Section 5]), and, on the other hand, no simple characterization of I-sets seems possible in the multivariate case, it would be desirable to find simple construction methods for AI-sets. In the above example of univariate polynomial B-splines, it can be easily seen that AI-sets are characterized by the condition

$$t_i \in \operatorname{supp} B_i := \overline{\{t \in \mathbb{R} : B_i(t) \neq 0\}}, \quad i = 1, \dots, n.$$

A certain extension of this *weak support property* to multivariate splines has been found in [25] (see also [16, Theorem 3.7 and Theorem 4.12]). However, it substantially differs from its univariate source. The disadvantage is that *each* basis

Multivariate Approximation and Splines
G. Nürnberger, J. W. Schmidt, and G. Walz (eds.), pp. 59–72.

ISBN 3-7643-5654-5.

of a multivariate spline space U has to be examined in order to check whether a configuration T is an AI–set.

Fortunately, this drawback can be overcome if U admits a *locally linearly independent basis* (*LI–basis*). Local linear independence was considered by de Boor and Höllig [6], Dahmen and Micchelli [13] and Jia [19] as a property of the integer translates of a box spline, and further investigated in [2,8,9,14,15,17,20,26]. (Particularly, Carnicer and Peña [9] have shown that I-sets with respect to a finite-dimensional space spanned by a locally linearly independent weak Descartes system of univariate continuous functions can be characterized by the support property.) The importance of this notion for the study of AI-sets follows from the fact that for a space U spanned by an LI–basis $\{u_1, \dots, u_n\}$, a set $T = \{t_1, \dots, t_s\}$, $s \leq n$ is an AI-set w.r.t. U if and only if there exists some permutation σ of $\{1, \dots, n\}$ such that $t_i \in \operatorname{supp} u_{\sigma(i)}$, $i = 1, \dots, s$ (Theorem 2.3).

For constructing AI-sets we are therefore interested in spaces which admit LI-bases. In Section 3 we present various examples of such spaces, including univariate generalized splines, translates of box splines, tensor product splines, continuous multivariate splines on simplex partitions and finite-element bivariate splines.

In this paper we shall follow the notations of [16].

2. Locally Linearly Independent Systems

In this section we describe some properties of LI–bases and their relationship to the problem of constructing almost interpolation sets. Although we are mostly interested in finite systems of functions and finite-dimensional linear spaces spanned by them, the theory of locally linear independence can be developed for certain infinite systems.

Let K be a topological space. We say that a system of nonzero functions $\{u_i\}_{i \in I} \subset F(K)$, is *locally finite* if for any $t \in K$ there exists a neighborhood $B(t)$ such that the set

$$\{i \in I : B(t) \cap \text{ supp } u_i \neq \emptyset\}$$

is finite. Particularly, we can consider the infinite series

$$\sum_{i \in I} a_i u_i(x), \quad x \in K,$$

taking into account the fact that for each fixed $x \in K$ only a finite number of terms is nonzero.

It is also quite clear that the local dimension

$$\text{l-dim}_{K'} U := \inf \{\dim U|_B : K' \subset B, \ B \text{ open}\}$$

is finite when $U \subset F(K)$ denotes a linear space spanned by a locally finite system of functions and K' is a finite set. Particularly, $\varphi(t) := \text{l-dim}_t U := \text{l-dim}_{\{t\}} U$ is well-defined for such spaces.

Definition 2.1. *A locally finite system $\{u_i\}_{i\in I} \subset F(K)$ is said to be* locally linearly independent *(LI-system) if for any $t \in K$ and any neighborhood $B(t)$ of t there exists an open set B' such that $t \in B' \subset B(t)$ and the subsystem*

$$\{u_i : B' \cap \operatorname{supp} u_i \neq \emptyset\}$$

is linearly independent. The linear span of an LI-system is called LI-space.

The next theorem gives some equivalent definitions of LI-systems.

Theorem 2.2. *[17] Let $\{u_i\}_{i\in I} \subset F(K)$ be a locally finite system of functions and let $U = \operatorname{span}\{u_i : i \in I\}$. The following conditions are equivalent.*

1) $\{u_i\}_{i\in I}$ *is a locally linearly independent system.*
2) $\text{l-dim}_t\, U = \operatorname{card}\{i \in I : t \in \operatorname{supp} u_i\}$, $t \in K$.
3) $\text{l-dim}_{K'}\, U = \operatorname{card} I_{K'}$, $K' \subset K$ *finite, where*

$$I_{K'} := \{i \in I : K' \cap \operatorname{supp} u_i \neq \emptyset\}.$$

4) $\dim U|_B = \operatorname{card} I_B$, B *open,* *if I_B is finite.*
5) *Given any open $B \subset K$,*

$$\sum_{i\in I} a_i u_i(x) = 0,\ x \in B, \quad \textit{implies} \quad a_i = 0 \quad \textit{for all} \ \ i \in I_B\,.$$

6) $\operatorname{supp}\big(\sum_{i\in I} a_i u_i\big) = \bigcup_{\substack{i\in I\\ a_i\neq 0}} \operatorname{supp} u_i.$

We note that the equivalence of 5) and 6) has been shown by Carnicer and Peña [8, Proposition 3.2] for the case when $\{u_i\}_{i\in I}$ is a finite system of functions.

It turns out that the statement of [16, Theorem 4.12] can be substantially simplified in the case that U is a finite-dimensional LI-space. Thus, we obtain a characterization of almost interpolation sets w.r.t. such spaces through a support property similar to [16, Support property (2.2)].

Theorem 2.3. *[17] Let $\{u_1, \ldots, u_n\} \subset F(K)$ be a locally linearly independent system and $U = \operatorname{span}\{u_1, \ldots, u_n\}$. A finite set $T = \{t_1, \ldots, t_s\} \subset K, s \leq n$, is an AI-set w.r.t. U if and only if there exists some permutation σ of $\{1, \ldots, n\}$ such that*

$$t_i \in \operatorname{supp} u_{\sigma(i)},\ i = 1, \ldots, s.$$

When a particular system of functions is to be checked whether it is an LI-system or not, it is often helpful to make use of the following theorem.

Theorem 2.4. *[17] Let $\{u_i : i \in I\} \subset F(K)$ be a locally finite system of functions, $U = \operatorname{span}\{u_i : i \in I\}$. Assume that*

$$\overline{\operatorname{int}\operatorname{supp} u_i} = \operatorname{supp} u_i, \ i \in I. \tag{2.1}$$

If $\{u_{i|G_U} : i \in I\}$ is an LI-system, where G_U denotes the set of all points of continuity of $\varphi(t) = \text{l-dim}_t U$, then $\{u_i : i \in I\}$ is also an LI-system.

Note that (2.1) holds for any system of *continuous* functions u_i, since in that case $\{t \in K : u_i(t) \neq 0\}$ is open and everywhere dense in $\operatorname{supp} u_i$.

Some further properties of LI-spaces can be found in [17].

3. Examples of LI-systems

In this section we shall present various examples of LI-systems of uni- and multivariate functions. In every particular case when the construction of the LI-system is given, it is quite easy to characterize almost interpolation sets with the help of Theorem 2.3.

3.1. Univariate Polynomial B-Splines and Generalized Splines

It is well known that for the n-dimensional space $S_m(\Delta)$ of polynomial spline functions of degree m with r fixed knots there exists a basis $\{B_1, \ldots, B_n\}$ of functions with minimal support, the so-called B-splines (see e.g. [22]). In view of the properties of these functions, it is obvious that $\mathcal{U} := \{B_1, \ldots, B_n\}$ forms an LI-system for $S_m(\Delta)$.

This system can even be extended by an infinite knot sequence $\{x_i\}_{i=-\infty}^{\infty}$ to a system $\tilde{\mathcal{U}} := \{B_i\}_{i=-\infty}^{\infty}$ of B-splines of degree m, which is also locally linearly independent. Moreover, $\tilde{\mathcal{U}}$ forms a *totally positive system*; i.e., if $j_1 < \ldots < j_m$ are any integers, then $\det(B_{j_i}(t_k)) \geq 0$ for all points $t_1 < \ldots < t_m$ in $\mathbb{R}$, and strict positivity holds if and only if $t_i \in$ int supp $B_{j_i}, i = 1, \ldots, m$. That case clearly implies that $\{t_1, \ldots, t_m\}$ is an I-set w.r.t. span $\{B_{j_1}, \ldots, B_{j_m}\}$.

Sommer and Strauss introduced in [24] a class of generalized spline spaces which retains most of the features of the polynomial splines and includes important subclasses of spline spaces such as polynomial splines, generalized Chebyshevian splines and subspaces of splines in tension. The main result of [24] consists in constructing a basis for the generalized spline space which forms a weak Descartes system, admits a recursion relation and, as it can be easily seen from [24, Theorem 2.1], is locally linearly independent.

Carnicer and Peña [9] showed that a finite system of continuous functions on a closed interval of the real line is a locally linearly independent weak Descartes system if and only if its collocation matrices are almost strictly totally positive.

3.2. Translates of Box Splines

Let X be an arbitrary set of vectors (not necessarily distinct) containing a basis for $\mathbb{R}^k$,

$$X = \{x^i\}_{i=1}^n \subset \mathbb{R}^k \setminus \{0\}, \text{ span } X = \mathbb{R}^k.$$

(X will also be used to denote a $k \times n$ matrix.) Moreover, let

$$X(t) := \sum_{i=1}^{n} t_i x^i, \text{ if } t = \{t_1, \dots, t_n\} \in \mathbb{R}^n,$$

$$\mathbb{B}(X) := \{V \subset X : \text{ card } V = \dim \text{ span } V = k\}.$$

Definition 3.1. *The* box spline $B(\cdot|X)$ *is a function defined by the rule*

$$\int_{\mathbb{R}^k} f(x)B(x|X)dx = \int_{[0,1)^n} f(X(t))dt \quad (f \in C_0(\mathbb{R}^k)),$$

where $[0,1)^n$ *denotes the halfopen unit n-cube.*

The box splines have some important properties, including local support and piecewise polynomial structure (see e.g. [3,11,7]).

Theorem 3.2. *The following statements are true.*

1) $\operatorname{supp} B(\cdot|X) = X([0,1]^n)$.
2) $\int_{\mathbb{R}^k} B(x|X)dx = 1$.
3) $B(\cdot|X) \in C^{d(X)-1}(\mathbb{R}^k) \setminus C^{d(X)}(\mathbb{R}^k)$ *where* $d(X) := \min\{\operatorname{card} Y : Y \subset X, \operatorname{span} X \setminus Y \neq \mathbb{R}^k\} - 1$.
4) *Set*

$$B_X := \{\sum_{j=1}^{k-1} c_j x^{i_j} + \sum_{j \in I'} b_j x^{i'_j} : 0 \le c_j \le 1,\ b_j = \pm 1,$$
$$1 \le i_1 < \dots < i_{k-1} \le n\}$$

where $I' = \{i'_j\}$ *denotes the complementary set of* $\{i_j\}_{j=1}^{k-1}$ *w.r.t.* $\{1, \dots, n\}$. *Then the restriction of* $B(\cdot|X)$ *to each component of the complement of* B_X *is a polynomial of total degree* $n-k$ *(*B_X *is called the* grid partition *of* $B(\cdot|X)$*).*

Now we consider the space $S(X)$ spanned by the integer translates of the box spline $B(\cdot|X)$,

$$S(X) := \text{ span } \{B(\cdot - \alpha|X) : \alpha \in \mathbb{Z}^k\}.$$

To determine AI-sets w.r.t. finite-dimensional subspaces of $S(X)$ we are interested in the question of whether the system of translates

$$\{B(\cdot - \alpha|X) : \alpha \in \mathbb{Z}^k\}$$

represents an LI-system. In fact, the following characterization due to de Boor and Höllig [5,6], Dahmen and Micchelli [12,13] and Jia [18,19] is true.

Theorem 3.3. *Let* $X = \{x^i\}_{i=1}^n \subset \mathbb{Z}^k \setminus \{0\}$ *with* $\operatorname{span} X = \mathbb{R}^k$. *The following conditions are equivalent.*

1) X *is unimodular; i.e.,*

$$|\det V| = 1 \text{ for all } V \subset \mathbb{B}(X).$$

2) $\{B(\cdot - \alpha|X) : \alpha \in \mathbb{Z}^k\}$ *is an LI-system.*
3) $\{B(\cdot - \alpha|X) : \alpha \in \mathbb{Z}^k\}$ *is a globally linearly independent system of functions.*

We note that several characterizations of unimodular matrices are available. (See, for example, [21, Chapters 19–21] and, in the case $k = 2$, [7, (II.29), p. 41].)

Let us now consider the case $k = 2$ more detailed. Suppose that the matrix $X \subset \mathbb{Z}^2$ contains the unit vectors e^1 and e^2. Following the notation of [11] we set

$$B_{tuvw}(\cdot) = B(\cdot|X_{tuvw})$$

with

$$X_{tuvw} = \{\underbrace{e^1, \ldots, e^1}_{t}, \underbrace{e^2, \ldots, e^2}_{u}, \underbrace{e^1 + e^2, \ldots, e^1 + e^2}_{v}, \underbrace{e^2 - e^1, \ldots, e^2 - e^1}_{w}\},$$

t, u, v, w nonnegative integers, $t, u \geq 1$. It follows easily from Theorem 3.2 that the grid partition w.r.t. B_{tuvw} is

1) a rectangular partition if $v = w = 0$ and $t, u \in \mathbb{N}$;
2) a type-1 triangulation if $w = 0$ and $t, u, v \in \mathbb{N}$ or $v = 0$ and $t, u, w \in \mathbb{N}$;
3) a type-2 triangulation if $t, u, v, w \in \mathbb{N}$.

Moreover, the following result is an immediate consequence of Theorem 3.3 and [7, (II.29)].

Proposition 3.4. *Assume that* $t, u \geq 1$. *Then the following statements hold.*

1) X_{tuvw} *is unimodular if and only if* $v = 0$ *or* $w = 0$.
2) *The system of translates*

$$\{B_{tuvw}(\cdot - \alpha_1, \cdot - \alpha_2) : \alpha_1, \alpha_2 \in \mathbb{Z}\}$$

is locally linearly independent if and only if $v = 0$ *or* $w = 0$.

3) *Each function* $B_{tuvw}(\cdot - \alpha_1, \cdot - \alpha_2)$ *is a bivariate spline function of total degree* $t + u + v + w - 2$ *and smoothness* $t + u + v + w - \max\{t, u, v, w\} - 2$.

Example 3.5. We consider two special cases.

1) Let $t = 2, u = 1, v = 2, w = 0$. Then the resulting system of translates of the box spline B_{2120} is an LI-system. The support of B_{2120} is given by the first figure in Fig. 3.1.
2) Let $t = u = v = w = 1$. Then the resulting system of translates fails to be an LI-system. The support of the corresponding box spline B_{1111} is given by the second figure in Fig. 3.1.

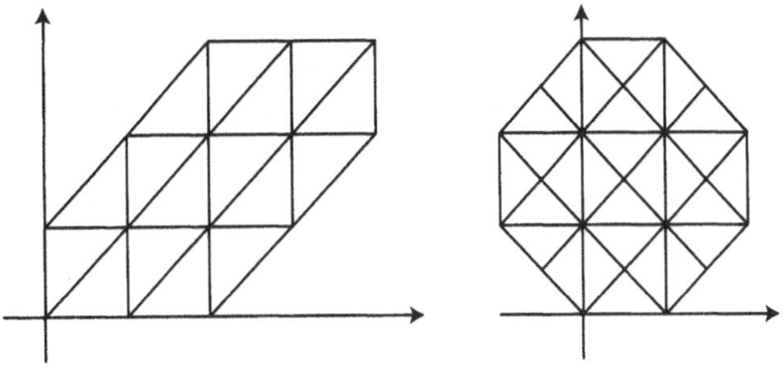

Fig. 3.1.

3.3. Tensor-Product Spaces

Let K_1 and K_2, respectively be topological spaces. Assume that $\mathcal{B}_1 = \{B_i\}_{i\in I_1}$ and $\mathcal{B}_2 = \{\tilde{B}_i\}_{i\in I_2}$ denote LI-systems in $F(K_1)$ and $F(K_2)$, respectively where I_1 and I_2 are index sets. Set

$$K := K_1 \times K_2,$$

the topological product of K_1 and K_2, and

$$\mathcal{B} := \{B_i\tilde{B}_j\}_{i\in I_1\, j\in I_2}.$$

Theorem 3.6. *The system $\mathcal{B}$ is an LI-system in $F(K)$.*

Proof: Assume that $\mathcal{B}$ fails to be an LI-system. Then there exist $t \in K$ and an open neighborhood M of t such that for some $(\tilde{i},\tilde{j}) \in I_1 \times I_2$,

$$B_{\tilde{i}}\tilde{B}_{\tilde{j}} = \sum_{i\in\tilde{I}_1}\sum_{j\in\tilde{I}_2} \alpha_{ij} B_i \tilde{B}_j \text{ on } M,$$

where $\tilde{I}_1$ and $\tilde{I}_2$ are finite index sets such that $(\tilde{I}_1 \times \tilde{I}_2) \cup \{\tilde{i},\tilde{j}\} = \{(i,j) \in I_1 \times I_2 : M \cap \text{ supp } B_i\tilde{B}_j \neq \emptyset\}$.

Without loss of generality assume that $M = M_1 \times M_2$ for some open $M_i \subset K_i$, $i = 1,2$. Let $\tilde{t} = (\tilde{t}_1, \tilde{t}_2) \in M$ such that $\tilde{t}_i \in K_i$, $i = 1,2$ and $\tilde{B}_{\tilde{j}}(\tilde{t}_2) \neq 0$. Then

$$B_{\tilde{i}}(\tilde{t}_1) = \sum_{i\in\tilde{I}_1}\left(\sum_{j\in\tilde{I}_2} \alpha_{ij} B_j(\tilde{t}_2)/\tilde{B}_{\tilde{j}}(\tilde{t}_2)\right) B_i(\tilde{t}_1).$$

This is even true for any $\tilde{t}_1 \in M_1$, contradicting the properties of the LI-system $\mathcal{B}_1$. □

Remark 3.7. As an important application of the preceding theorem let us consider the case when $K_1 = [a,b] \subset \mathbb{R}$, $K_2 = [c,d] \subset \mathbb{R}$, and Δ and $\tilde{\Delta}$ are some knot partitions of K_1 and K_2, respectively. Assume that $S_m(\Delta)$ and $S_l(\tilde{\Delta})$ denote the linear spaces of polynomial splines on K_1 and K_2, respectively as has been defined in [16, Section 2]. By Section 3.1 the B-spline bases of $S_m(\Delta)$ and $S_l(\tilde{\Delta})$, denoted by

$$\mathcal{B}_1 = \{B_1, \dots, B_n\} \text{ and } \mathcal{B}_2 = \{\tilde{B}_1, \dots, \tilde{B}_{\tilde{n}}\},$$

respectively are LI-systems in $F(K_1)$ and $F(K_2)$, respectively. Hence by Theorem 3.6 the tensor-product space of bivariate polynomial spline functions of degree m in the first variable and degree l in the second one,

$$\mathcal{U} := \text{span}\,\{B_i\tilde{B}_j\}_{i=1}^{n}{}_{j=1}^{\tilde{n}}$$

has a locally linearly independent basis, $\{B_i\tilde{B}_j\}_{i=1}^{n}{}_{j=1}^{\tilde{n}}$.

3.4. Multivariate Splines on Simplex Partitions

Let $x^0, \dots, x^k \in \mathbb{R}^k$, $k \geq 1$. The simplex

$$[x^0, \dots, x^k] := \{\sum_{i=0}^{k} \lambda_i x^i : \sum_{i=0}^{k} \lambda_i = 1,\ \lambda_i \geq 0\}$$

with vertices $x^0, \dots, x^k$ is called a *k-simplex*, if its volume in $\mathbb{R}^k$ is nonzero. Let

$$\Delta = \{S_i\}_{i \in I},$$

a family of finitely many k-simplices in $\mathbb{R}^k$ which satisfy the following property: If $S_i, S_j \in \Delta$, then $S_i \cap S_j$ is empty or a common face. Set

$$K = \bigcup_{i \in I} S_i.$$

For given integers r and d $(0 \leq r < d)$ we consider

$$S_d^r(\Delta) := \{s \in C^r(K) : s \text{ restricted to each } k\text{-simplex is a polynomial of total degree } d\},$$

the space of *polynomial splines of degree d and smoothness r* on Δ.

Suppose now that $\tilde{S} := [x^0, \dots, x^k]$ is a k-simplex in Δ. Then any $x = (x_1, \dots, x_k) \in \mathbb{R}^k$ can be identified by the *barycentric coordinates* $(\lambda_0, \dots, \lambda_k)$ w.r.t. $\tilde{S}$, where

$$\lambda_i = \lambda_i(x) = \frac{\text{vol}_k[x^0, \dots, x^{i-1}, x, x^{i+1}, \dots, x^k]}{\text{vol}_k[x^0, \dots, x^k]}, \quad i = 0, \dots, k.$$

Hence, λ_i is a linear polynomial in x. For any $\beta = (\beta_0, \ldots, \beta_k) \in \mathbb{Z}_+^{k+1}$, as usual we set

$$\lambda^\beta = \lambda_0^{\beta_0} \ldots \lambda_k^{\beta_k}, \; \beta! = \beta_0! \ldots \beta_k!, \; |\beta| = \beta_0 + \ldots + \beta_k.$$

Then

$$\varphi_\beta^d(\lambda) = \frac{d!}{\beta!}\lambda^\beta, \quad |\beta| = d$$

is a polynomial of degree d, and the set of all such functions $\{\varphi_\beta^d(\lambda) : |\beta| = d\}$ is a basis of π_d^k, the space of all polynomials of total degree at most d with k variables. Let $s \in S_d^r$ and $x \in \tilde{S}$. Then $s|_{\tilde{S}} \in \pi_d^k$ and

$$s(x) = s|_{\tilde{S}}(x) = \sum_{|\beta|=d} a_\beta^d(\tilde{S})\varphi_\beta^d(\lambda)$$

which is called the *Bézier-Bernstein* form of s w.r.t. $\tilde{S}$ (see [4]).

In addition, the set $\{(P_\beta(\tilde{S}), a_\beta^d(\tilde{S})) : |\beta| = d\}$ is called the *Bézier-net* of s w.r.t. $\tilde{S}$ where

$$P_\beta(\tilde{S}) = \sum_{i=0}^{d} \frac{\beta_i}{d} x^i,$$

the *domain point* and each $a_\beta^d(\tilde{S})$ is called *Bézier-ordinate*, associated with $P_\beta(\tilde{S})$.

The Case $r = 0$

Let $T_l = [x^0, \ldots, x^l]$ be an l-simplex in $\mathbb{R}^k$ where $0 \leq l < k$ and let

$$S_1 = [x^0, \ldots, x^l, x^{l+1}, \ldots, x^k] \in \Delta,$$

$$S_2 = [x^0, \ldots, x^l, y^{l+1}, \ldots, y^k] \in \Delta$$

be two adjacent k-simplices with $S_1 \cap S_2 = T_l$. Suppose that $s \in S_d^0(\Delta)$. Then

$$s|_{S_1}(x) = \sum_{|\beta|=d} a_\beta^d(S_1)\varphi_\beta^d(\lambda), \quad x \in S_1,$$

$$s|_{S_2}(x) = \sum_{|\beta|=d} a_\beta^d(S_2)\varphi_\beta^d(\eta), \quad x \in S_2,$$

where $\lambda = (\lambda_0, \ldots, \lambda_k)$ and $\eta = (\eta_0, \ldots, \eta_k)$ are the barycentric coordinates w.r.t. S_1 and S_2, respectively. Clearly, $x \in T_l$ if and only if

$$x = \sum_{i=0}^{k} \lambda_i x^i = \sum_{i=0}^{k} \eta_i x^i$$

which implies that $\lambda_i = \eta_i, i = 0, \ldots, k$. Moreover, it follows that if $s|_{S_1 \cup S_2} \in C(S_1 \cup S_2)$,

$$a^d_\beta(S_1) = a^d_\beta(S_2)$$

for all $\beta = (\beta_0, \ldots, \beta_l, 0, \ldots, 0)$ with $\beta_0 + \ldots + \beta_l = d$.

Let $\mathcal{B}_d(\Delta)$ denote the set of all domain points of all k-simplices $S_i, i \in I$. Given a point $P = P_\beta(\tilde{S}) \in \mathcal{B}_d(\Delta)$, let λ_P be the linear functional defined on $S^0_d(\Delta)$ with the property that

$$\lambda_P s = a^d_\beta(\tilde{S}),$$

the Bézier-ordinate of s w.r.t. $\tilde{S}$ (associated with P).

The following dimension formula and explicit construction of a basis for $S^0_d(\Delta)$ are due to Alfeld, Schumaker and Sirvent [1].

Theorem 3.8. *The following statements hold.*

1) $\dim S^0_d(\Delta) = \operatorname{card} \mathcal{B}_d(\Delta)$.
2) *There exists a basis of* $S^0_d(\Delta)$ *given by*

$$\mathcal{B} := \{B_P \in S^0_d(\Delta) : \lambda_Q B_P = \delta_{QP} \quad \text{for all} \quad P, Q \in \mathcal{B}_d(\Delta)\}.$$

We are able to show that $\mathcal{B}$ is locally linearly independent.

Theorem 3.9. *The basis $\mathcal{B}$ of $S^0_d(\Delta)$ defined above is an LI-system.*

Proof: Assume that $\mathcal{B}$ fails to be locally linearly independent. Then we can find a k-simplex $\tilde{S}$ such that for some points $P_\mu, P_\rho \in \mathcal{B}_d(\Delta)$,

$$\mathcal{B}_{P_\mu} = \sum_\rho d_\rho \mathcal{B}_{P_\rho}$$

on $\tilde{S}$ where we clearly assume that $\tilde{S} \subset \operatorname{supp} \mathcal{B}_{P_\mu}, \tilde{S} \subset \operatorname{supp} \mathcal{B}_{P_\rho}, d_\rho \neq 0$ and $P_\mu \neq P_\rho$, all P_ρ. It is obvious that $P_\mu \in \tilde{S}$, because otherwise for the Bézier-ordinates $a^d_\beta(\tilde{S}; \mathcal{B}_{P_\mu})$ of $\mathcal{B}_{P_\mu}$,

$$0 = \lambda_Q \mathcal{B}_{P_\mu} = a^d_\beta(\tilde{S}; \mathcal{B}_{P_\mu}),$$

for all $Q = Q_\beta(\tilde{S}) \in \mathcal{B}_d(\Delta) \cap \tilde{S}$. This would imply that

$$\mathcal{B}_{P_\mu} \equiv 0 \quad \text{on} \quad \tilde{S},$$

a contradiction. Analogously we have that $P_\rho \in \tilde{S}$ for all ρ. Thus $P_\mu = P_{\tilde{\beta}}(\tilde{S})$ and

$$a^d_{\tilde{\beta}}(\tilde{S}; \mathcal{B}_{P_\mu}) = 1, \quad a^d_{\tilde{\beta}}(\tilde{S}; \mathcal{B}_{P_\rho}) = 0$$

where $a^d_{\tilde{\beta}}(\tilde{S};\mathcal{B}_{P_\rho})$ denotes the Bézier-ordinate of $\mathcal{B}_{P_\rho}$, associated with P_μ. Comparing now the coefficients of $\mathcal{B}_{P_\mu}$ and $\sum_\rho d_\rho \mathcal{B}_{P_\rho}$ on $\tilde{S}$, we obtain

$$1 = a^d_{\tilde{\beta}}(\tilde{S};\mathcal{B}_{P_\mu}) = \sum_\rho d_\rho a^d_{\tilde{\beta}}(\tilde{S};\mathcal{B}_{P_\rho}) = 0,$$

a contradiction. □

Example 3.10. Let $k = 2$ and assume that K is a simply connected polygonal domain in $\mathbb{R}^2$. Moreover, let Δ be a regular triangulation of K (see [16, Example 3.2]). Then by the above arguments the space $S^0_d(\Delta)$ of continuous spline functions of degree d admits an LI-basis $\mathcal{B} = \{B_P\}_{P\in\mathcal{B}_d(\Delta)}$ where B_P is defined as above using the Bézier-Bernstein form.

3.5 Finite-Element Bivariate Splines

Let $\Delta = \{\Delta_i : i = 1,\dots,N\}$ be a triangulation of a domain $K \subset \mathbb{R}^2$. Suppose that the vertices and the edges of Δ are denoted by $v_1,\dots,v_V$ and $\epsilon_1,\dots,\epsilon_E$, respectively. Following the notation of [23] we define the space of *super splines of degree d and smoothness r,ρ*, with $r \le \rho < d$, by

$$S^{r,\rho}_d(\Delta) = \{s \in C^r(K) : s|_{\Delta_i} \in \pi_d,\ i = 1,\dots,N,$$
$$s \in C^\rho(v_j),\ j = 1,\dots,V\},$$

where π_d is the space of bivariate polynomials of total degree d, and $C^\rho(v_j)$ denotes the set of functions which are ρ times continuously differentiable at the point v_j.

Suppose that $0 \le 2r \le \rho$ and $d \ge 2\rho + 1$. In this case it has been shown by Schumaker [23, Section 4] that a basis for $S^{r,\rho}_d(\Delta)$ can be constructed by using the classical finite-element method. We describe this basis as follows. For every vertex v_i, $i = 1,\dots,V$ consider Hermite interpolation conditions,

$$\frac{\partial^{\mu+\nu} s}{\partial x^\mu \partial y^\nu}(v_i) = a^i_{\mu,\nu}, \quad 0 \le \mu+\nu \le \rho. \tag{3.4}$$

On each edge ϵ_i, $i = 1,\dots,E$ choose points $\xi^{i,\nu}_1,\dots,\xi^{i,\nu}_{d-1-2\rho+\nu} \in \epsilon_i$, $\nu = 0,\dots,r$ and consider the following interpolation conditions,

$$\frac{\partial^\nu s}{\partial n^\nu_i}(\xi^{i,\nu}_l) = a^i_{\nu,l},\ \nu = 0,\dots,r,\ l = 1,\dots,d-1-2\rho+\nu, \tag{3.5}$$

where $\frac{\partial^\nu}{\partial n^\nu_i}$ denotes the ν-th order normal derivative w.r.t. the edge ϵ_i. Finally, take $N_{d,r,\rho} := \binom{d-3r-1}{2} - 3\binom{p-2r}{2}$ points $\zeta^i_1,\dots,\zeta^i_{N_{d,r,\rho}}$ inside each triangle Δ_i in such a way that Lagrange interpolation conditions

$$s(\zeta^i_l) = a^i_l, \quad l = 1,\dots,N_{d,r,\rho} \tag{3.6}$$

together with the conditions at the vertices and on the edges of the triangle Δ_i uniquely determine a polynomial of total degree d on that triangle. See [23, Theorem 4.1 and Remark 4] for the details concerning existence and construction of the points ζ_l^i (so-called "type-2 data").

In view of a theorem by Ženíšek [27], the piecewise polynomial function defined an each triangle Δ_i by conditions (3.4)–(3.6) necessarily belongs to $S_d^{r,\rho}(\Delta)$. Since the total number of conditions equals $n := \dim S_d^{r,\rho}(\Delta)$, we conclude that (3.4)-(3.6) is a well-posed Hermite interpolation problem for $S_d^{r,\rho}(\Delta)$. Therefore, the corresponding fundamental functions $s_1, \ldots, s_n$ form a basis for $S_d^{r,\rho}(\Delta)$. (Recall that the *fundamental functions* $u_1, \ldots, u_n$ of an interpolation problem $F_i(u) = a_i$, $i = 1, \ldots, n$, $u \in U$, $\dim U = n$, are determined by the conditions $F_i(u_j) = \delta_{ij}$ where F_i, $i = 1, \ldots, n$ are certain functionals and $\delta_{ii} = 1$, $\delta_{ij} = 0$ if $i \neq j$.)

Theorem 3.11. *The fundamental functions $s_1, \ldots, s_n$ of the interpolation problem (3.4)–(3.6) form a locally linearly independent basis for the space of super splines $S_d^{r,\rho}(\Delta)$, $0 \le 2r \le \rho$, $d \ge 2\rho + 1$.*

Proof: Consider first the supports of s_j, $j = 1, \ldots, n$. If s_j corresponds to a point v_i, $\xi_l^{i,\nu}$ or ζ_l^i, then supp s_j is evidently the union of all triangles with common vertex at v_i, the union of two triangles with common edge ϵ_i, or the triangle Δ_i, respectively. By Theorem 2.4 it suffice to check the locally linear independence of $\{s_j : j = 1 \ldots, n\}$ on the continuity set G_U of local dimension; *i.e.*, on the interiors of the triangles Δ_i. If now $x \in$ int Δ_{i_0}, then $x \in$ supp s_j exactly for those s_j which correspond to the points $v_i, \xi_l^{i,\nu}, \zeta_l^i$ lying on Δ_{i_0}. According to the construction of the conditions (3.4)–(3.6), the number of such s_j's equals the dimension of π_d and their restrictions to the triangle Δ_{i_0} are fundamental polynomials of a well-posed polynomial interpolation problem. Therefore, they are linearly independent in any neighborhood of x. This completes the proof. □

Acknowledgements. O. Davydov was supported by the Alexander von Humboldt Foundation, under Research Fellowship.

References

1. Alfeld P., Schumaker L. L., and Sirvent M., On dimension and existence of local bases for multivariate spline spaces, J. Approx. Theory **70** (1992), 243–264.

2. Ben-Artzi A. and Ron A., On the integer translates of a compactly supported functions: dual bases and linear projectors, SIAM J. Math. Anal. **21** (1990), 1550–1562.

3. Bojanov B. D., Hakopian H. A., and Sahakian A. A., *Spline Functions and Multivariate Interpolations*, Kluwer Academic Publishers, Dordrecht, 1993.

4. de Boor C., *B*-form basics, in *Geometric Modeling: Algorithms and New Trends* (Farin G. E., Ed.), SIAM, Philadelphia, 1987, 131–148.
5. de Boor C. and Höllig K., *B*-splines from parallelepipeds, J. Analyse Math. **42** (1982), 99–115.
6. de Boor C. and Höllig K., Bivariate box splines and smooth pp functions on a three direction mesh, J. Comput. Appl. Math. **9** (1983), 13–28.
7. de Boor C., Höllig K., and Riemenschneider S., *Box Splines*, Springer, New York, 1993.
8. Carnicer J. M. and Peña J. M., Least supported bases and local linear independence, Numer. Math. **67** (1994), 289–301.
9. Carnicer J. M. and Peña J. M., Spaces with almost strictly totally positive bases, Math. Nachrichten **169** (1994), 69–79.
10. Cheney E. W., *Multivariate Approximation Theory: Selected Topics*, CBMS–SIAM, Philadelphia, 1986.
11. Chui C. K., *Multivariate Splines*, CBMS–SIAM, Philadelphia, 1988.
12. Dahmen W. and Micchelli C. A., Translates of multivariate splines, Linear Algebra Appl. **52/53** (1983), 217–234.
13. Dahmen W. and Micchelli C. A., On the local linear independence of translates of a box spline, Studia Math. **82** (1985), 243–262.
14. Dahmen W. and Micchelli C. A., On multivariate *E*-splines, Advances in Math. **76** (1989), 33–93.
15. Davydov O. and Sommer M., Interpolation by weak Chebyshev spaces, preprint.
16. Davydov O., Sommer M., and Strauss H., On almost interpolation by multivariate splines, this volume.
17. Davydov O., Sommer M., and Strauss H., On almost interpolation and locally linearly independent bases, preprint.
18. Jia R.-Q., Linear independence of translates of a box spline, J. Approx. Theory **40** (1984), 158–160.
19. Jia R.-Q., Local linear independence of the translates of a box spline, Constr. Approx. **1** (1985), 175–182.
20. Ron A., Linear independence of the translates of an exponential box spline, Rocky Mountain J. Math. **22** (1992), 331–351.
21. Schrijver A., *Theory of linear and integer programming*, Wiley-Interscience, New York, 1986.
22. Schumaker L. L., *Spline Functions: Basic Theory*, Wiley-Interscience, New York, 1981.
23. Schumaker L. L., On super splines and finite elements, SIAM J. Numer. Anal. **26** (1989), 997–1005.

24. Sommer M. and Strauss H., Weak Descartes systems in generalized spline spaces, Constr. Approx. **4** (1988), 133–145.

25. Sommer M. and Strauss H., A condition of Schoenberg-Whitney type for multivariate spline interpolation, Advances in Comp. Math. **5** (1996), 381–397.

26. Sun Q., A note on the integer translates of a compactly supported distribution on $\mathbb{R}$, Arch. Math. **60** (1993), 359–363.

27. Ženíšek A., Interpolation polynomials on the triangle, Numer. Math. **15** (1970), 283–296.

Oleg Davydov
Department of Mechanics and Mathematics
Dnepropetrovsk State University
pr.Gagarina 72
Dnepropetrovsk GSP 320625, Ukraine
davydov@euklid.math.uni-mannheim.de

Manfred Sommer
Mathematisch–Geographische Fakultät
Katholische Universität Eichstätt
85071 Eichstätt, Germany
manfred.sommer@ku-eichstaett.de

Hans Strauß
Institut für Angewandte Mathematik
Universität Erlangen–Nürnberg
91058 Erlangen, Germany
strauss@am.uni-erlangen.de

Exponential-type Approximation in Multivariate Harmonic Hilbert Spaces

Franz-Jürgen Delvos

Abstract. Babuška introduced the concept of periodic Hilbert spaces for studying universally optimal quadrature formulas. Prager continued these investigations and discovered the relationship between optimal approximation of linear functionals on periodic Hilbert spaces and minimum norm interpolation (optimal periodic interpolation). In the case of a uniform mesh methods of periodic interpolation by translation are applicable and relations to periodic spline interpolation and approximation have been studied. It is the objective of this paper to introduce the concept of harmonic Hilbert space in a multivariate setting as an extension of periodic Hilbert space and to study approximation via Fourier partial integrals and exponential-type interpolation in these spaces.

1. Harmonic Hilbert Spaces

The concept of univariate harmonic Hilbert spaces was introduced in Delvos [1996] as an extension of periodic Hilbert space introduced by Babuška [1968] (see also Prager [1979] and Delvos [1990, 1995]). We will extend this construction to the multivariate case and discuss two special methods of generating multivariate harmonic Hilbert spaces from univariate ones. A harmonic Hilbert space $H_D(\mathbf{R}^n)$ is related to the nonnegative *defining function* $D \in L_1(\mathbf{R}^n) \cap L_\infty(\mathbf{R}^n)$. We consider measurable functions F such that $F/\sqrt{D} \in L_2(\mathbf{R}^n)$. Since $\sqrt{D} \in L_2(\mathbf{R}^n)$ we have $F \in L_1(\mathbf{R}^n)$ and

$$\|F\|_1 \le \left\|F/\sqrt{D}\right\|_2 \sqrt{\|D\|_1} \,. \tag{1.1}$$

The *Fourier integral* of F defines a function

$$f(x) = \int_{\mathbf{R}^n} F(t) \exp\left(i(t,x)\right) dt, \quad (t,x) = \sum_{j=1}^{n} t_j x_j \,,$$

Multivariate Approximation and Splines
G. Nürnberger, J. W. Schmidt, and G. Walz (eds.), pp. 73–82.

ISBN 3-7643-5654-5.

from $C_0(\mathbf{R}^n)$, the algebra of uniformly continuous functions on $\mathbf{R}^n$ which vanish at infinity. The norm of $C_0(\mathbf{R}^n)$ is given by

$$\|f\|_\infty = \max\{|f(x)| : x \in \mathbf{R}^n\}.$$

The Wiener algebra $A(\mathbf{R}^n)$ is defined as the subspace of functions f from $C_0(\mathbf{R}^n)$ possessing the Fourier integral representation

$$f(x) = \int_{\mathbf{R}^n} F(t) \exp(i(t,x))\, dt\ , F \in L_1(\mathbf{R}^n)\ .$$

The corresponding norm of $f \in A(\mathbf{R}^n)$ is given by

$$\|f\|_a = \int_{\mathbf{R}^n} |F(t)|\, dt.$$

Moreover we have

$$A(\mathbf{R}^n) \subseteq C_0(\mathbf{R}^n), \quad \|f\|_\infty \leq \|f\|_a .$$

The harmonic Hilbert space $H_D(\mathbf{R}^n)$ is now defined as the subspace of functions $f \in A(\mathbf{R}^n)$ such that $F/\sqrt{D} \in L_2(\mathbf{R}^n)$. The associated inner product is defined by

$$(f,g)_D = \int_{\mathbf{R}^n} F(t)\overline{G(t)}/D(t)dt\ .$$

The harmonic Hilbert space $H_D(\mathbf{R}^n)$ is *translation invariant:*

$$(f(\cdot - a), g(\cdot - a))_D = (f,g)_D\ , \quad a \in \mathbf{R}^n.$$

Moreover the following relations hold:

$$H_D(\mathbf{R}^n) \subseteq A(\mathbf{R}^n) \subseteq C_0(\mathbf{R}^n), \quad \|f\|_\infty \leq \|f\|_a \leq \|f\|_D \sqrt{\|D\|_1}. \tag{1.2}$$

The simplest function of $H_D(\mathbf{R}^n)$ is its *generating function*

$$d(x) = \int_{\mathbf{R}^n} D(t) \exp(i(t,x))dt.$$

We have

$$\|d(\cdot - a)\|_D^2 = \int_{\mathbf{R}^n} D(t)dt,\ a \in \mathbf{R}^n.$$

$H_D(\mathbf{R}^n)$ is a *reproducing kernel Hilbert space* since

$$f(x) = (f, d(\cdot - x))_D\ ,\ x \in \mathbf{R}^n. \tag{1.3}$$

In addition to the relation (1.2) we have

Proposition 1.1. *Let $f \in H_D(\mathbf{R}^n)$. Then $f \in L_2(\mathbf{R}^n)$ and*

$$H_D(\mathbf{R}^n) \subseteq L_2(\mathbf{R}^n), \quad \|f\|_2 \leq \sqrt{(2\pi)^n \|D\|_\infty} \|f\|_D. \tag{1.4}$$

Proof: Note first that

$$F/\sqrt{D} = G \in L_2(\mathbf{R}^n), \quad \sqrt{D} \in L_\infty(\mathbf{R}^n), \quad G\sqrt{D} = F \in L_2(\mathbf{R}^n).$$

By Plancherel's theorem the *Fourier transform* of f satisfies

$$F(t) \sim \frac{1}{(2\pi)^n} \int_{\mathbf{R}^n} f(x) \exp(-i(t,x))\, dx,$$

$$\int_{\mathbf{R}^n} |f(x)|^2\, dx = (2\pi)^n \int_{\mathbf{R}^n} |F(t)|^2\, dt \leq (2\pi)^n \|D\|_\infty \int_{\mathbf{R}^n} |F(t)|^2 / D(t) dt. \quad \square$$

Remark 1.1. *From the relations (1.2) and (1.4) it follows that convergence in the norm of $H_D(\mathbf{R}^n)$ always implies uniform convergence and convergence in the least square sense.*

First we consider *Paley-Wiener spaces* as harmonic Hilbert spaces. The defining function is the characteristic function $D = \chi_{[-b,b]}$ where $[-b,b]$ denotes the n-dimensional interval $[-b_1, b_1] \times ... \times [-b_n, b_n]$. We use the notation

$$PW_b(\mathbf{R}^n) = H_D(\mathbf{R}^n).$$

Any $f \in PW_b(\mathbf{R}^n)$ possesses the Fourier integral representation

$$f(x) = \int\limits_{[-b,b]} F(t) \exp(i(x,t)) dt, \quad F \in L_2(\mathbf{R}^n), \quad F(t) = 0 \ \ (t \notin [-b,b] \ ).$$

The inner product of $PW_b(\mathbf{R}^n)$ is given by $(f,g)_b := \int\limits_{[-b,b]} F(t)\overline{G(t)} dt$. The generating function of $PW_b(\mathbf{R}^n)$ is

$$d(x) = \int\limits_{[-b,b]} \exp(i(x,t)) dt = 2^n \prod_{j=1}^{n} \frac{\sin(b_j x_j)}{x_j} = 2^n \prod_{j=1}^{n} b_j \operatorname{sinc}(b_j x_j).$$

Define $h_{2b}(x) = \prod_{j=1}^n \operatorname{sinc}(b_j x_j)$. The translates $h_{2b}(\cdot - \pi\frac{k}{b})$, $k \in \mathbf{Z}^n$, form an orthonormal basis of $PW_b(\mathbf{R}^n)$. For any $f \in PW_b(\mathbf{R}^n)$ we have the *interpolation series representation*

$$f(x) = \sum_{k \in \mathbf{Z}^n} f(\pi\frac{k}{b}) h_{2b}(x - \pi\frac{k}{b}).$$

The simplest univariate Sobolev space $W^1(\mathbf{R}) = H_D(\mathbf{R})$ is obtained by choosing

$$D(t) = \frac{2}{1+t^2}\,.$$

The associated generating function is an exponential spline:

$$d(x) = 2\pi \exp(-|x|).$$

Choosing as defining function

$$D(t) = \exp(-|t|)$$

we get a harmonic Hilbert space of holomorphic functions $H^1(\mathbf{R})$ first considered by Paley and Wiener (see Katznelson [1976]). Its generating function is the holomorphic function

$$d(x) = \frac{2}{1+x^2}\,.$$

Assume that $H_D(\mathbf{R})$ is an univariate harmonic Hilbert space. The *anisotropic* method for constructing multivariate harmonic Hilbert spaces is defined by choosing as defining function

$$D_P(t_1, .., t_n) = D(t_1)..D(t_n).$$

The anisotropic multivariate harmonic Hilbert space is a tensor product space

$$H_{D_P}(\mathbf{R}^n) = H_D(\mathbf{R}) \otimes ... \otimes H_D(\mathbf{R}).$$

Its generating function is given by $d_P(x_1, .., x_n) = d(x_1)..d(x_n)$.

The anisotropic Sobolev space is denoted by

$$W^{(1,...,1)}(\mathbf{R}^n) = W^1(\mathbf{R}) \otimes ... \otimes W^1(\mathbf{R}).$$

while the anisotropic Hilbert space of holomorphic functions is denoted by

$$H^{(1,...,1)}(\mathbf{R}^n) = H^1(\mathbf{R}) \otimes ... \otimes H^1(\mathbf{R}).$$

Assume that $H_D(\mathbf{R})$ is an univariate harmonic Hilbert space. The *isotropic* method for constructing multivariate harmonic Hilbert spaces is defined by choosing as defining function

$$D_S(t_1, .., t_n) = (D(t_1)^{-1} + .. + D(t_n)^{-1})^{-n}. \tag{1.5}$$

Recall that $(\prod_{i=1}^n a_i)^{1/n} \leq \frac{1}{n}\sum_{i=1}^n a_i$ for positive numbers $a_1, ..., a_n$. Choosing $a_i = D(t_i)^{-1}$ we obtain

$$D_S(t_1, .., t_n) \leq \frac{1}{n^n} D_1(t_1)..D_n(t_n). \tag{1.6}$$

This shows that $H_{D_S}(\mathbf{R}^n)$ is an isotropic harmonic Hilbert space constructed from the univariate harmonic Hilbert space $H_D(\mathbf{R})$.

The isotropic Sobolev space $W^n(\mathbf{R}^n)$ is obtained in this way from the univariate Sobolev space $W^1(\mathbf{R})$. The isotropic harmonic Hilbert space of holomorphic functions obtained from $H^1(\mathbf{R})$ is denoted by $H^n(\mathbf{R}^n)$.

2. Approximation by Fourier Partial Integrals

The *Fourier partial integral* of $f \in A(\mathbf{R}^n)$ is defined by

$$S_b(f)(x) = \int_{\mathbf{R}^n} \chi_{[-b,b]}(t)F(t)\exp(i(x,t))dt.$$

S_b is a continuous projector on the Wiener algebra $A(\mathbf{R}^n)$ satisfying

$$\|S_b(f)\|_a \leq \|f\|_a .$$

Its range is denoted by $A_b(\mathbf{R}^n)$. S_b is called the *Fourier partial integral projector.* We have

$$\|f - S_b(f)\|_a = \int_{\mathbf{R}^n} (1 - \chi_{[-b,b]}(t))\,|F(t)|\,dt \to 0, \quad |b| \to \infty. \tag{2.1}$$

S_b is an *orthogonal projector* on $H_D(\mathbf{R}^n)$ with range $PW_b(\mathbf{R}^n)$. The qualitative convergence relation (2.1) can be improved:

Proposition 2.1. *Let $f \in H_D(\mathbf{R}^n)$. Then*

$$\|f - S_b(f)\|_a \leq \|d - S_b(d)\|_a^{\frac{1}{2}} \|f - S_b(f)\|_D . \tag{2.2}$$

Proof: We have

$$\|f - S_b(f)\|_a = \int_{\mathbf{R}^n} (1 - \chi_{[-b,b]}(t))\,|F(t)|\,\sqrt{D(t)}/\sqrt{D(t)}dt$$

$$\leq (\int_{\mathbf{R}^n} (1 - \chi_{[-b,b]}(t))\,|F(t)|^2\,/D(t)dt)^{\frac{1}{2}} (\int_{\mathbf{R}^n} (1 - \chi_{[-b,b]}(t))D(t)dt)^{\frac{1}{2}}$$

$$= \|d - S_b(d)\|_a^{\frac{1}{2}} \|f - S_b(f)\|_D .$$

□

Remark 2.1. *Suppose that $f \in A(\mathbf{R}^n)$ possesses a nonnegative Fourier transform F. Then we have*

$$\|f\|_\infty = \|f\|_a , \; \|f - S_b(f)\|_\infty = \|f - S_b(f)\|_a . \tag{2.3}$$

Thus we only establish estimates in the norm of the Wiener algebra $A(\mathbf{R}^n)$.

Note that

$$f(x) - S_b(f)(x) = (f, d(\cdot - x) - S_b(d(\cdot - x))_D.$$

Thus the linear functional defined by

$$L_x^b(f) := f(x) - S_b(f)(x)$$

is bounded on $H_D(\mathbf{R}^n)$ and we have in view of $S_b(d(\cdot - x)) = S_b(d)(\cdot - x))$

Proposition 2.2. *The norm of L_x^b is given by*

$$\|L_x^b\|_D = \|d - S_b(d)\|_a^{1/2}. \tag{2.4}$$

Remark 2.2. *Relations (2.2) and (2.3) imply*

$$\|f - S_b(f)\|_\infty \le \|d - S_b(d)\|_a^{1/2} \|f - S_b(f)\|_D \tag{2.5}$$

and by proposition 2.2 this estimate is best possible in $H_D(\mathbf{R}^n)$.

First we consider the anisotropic harmonic Hilbert space $H_{D_P}(\mathbf{R}^n)$. We assume for simplicity $d(0) = 1$ and $\underline{b} = b_j, \quad (j = 1, .., n)$. Note first that

$$1 - \chi_{[-b,b]}(t) \le \sum_{k=1}^{n} (1 - \chi_{[-\underline{b},\underline{b}]}(t_k)). \tag{2.6}$$

Then

$$\|d_P - S_b(d_P)\|_{D_P}^2 \le \sum_{k=1}^{n} \int_{\mathbf{R}^n} (1 - \chi_{[-\underline{b},\underline{b}]}(t_k)) D(t_1)..D(t_n)dt = n \left\|d - S_{\underline{b}}(d)\right\|_a.$$

Thus we have proved

Proposition 2.3. *Let $f \in H_{D_P}(\mathbf{R}^n)$. Then*

$$\|f - S_b(f)\|_\infty = o(\left\|d - S_{\underline{b}}(d)\right\|_a^{\frac{1}{2}}), \quad \underline{b} \to \infty. \tag{2.7}$$

This shows that in the case of anisotropic Hilbert space the error is determined by the univariate error.

We list two cases of anisotropic Hilbert spaces which are related to the univariate examples considered previously. For $D(t) = 2/(1 + t^2)$ we have

$$\|f - S_b(f)\|_\infty = o(\underline{b}^{-\frac{1}{2}}), \quad f \in W^{(1,..,1)}(\mathbf{R}^n) \tag{2.8}$$

while $D(t) = \exp(-|t|)$ implies

$$\|f - S_b(f)\|_\infty = o(\exp(-\frac{b}{2})), \quad f \in H^{(1,..,1)}(\mathbf{R}^n). \tag{2.9}$$

Next we consider the isotropic harmonic Hilbert space $H_{D_S}(\mathbf{R}^n)$. Taking into account (2.6) we can conclude

$$\|d_S - S_b(d_S)\|_a \le n \int_{\mathbf{R}^n} (1 - \chi_{[-\underline{b},\underline{b}]}(t_1)) D_S(t)dt. \tag{2.10}$$

The integral

$$J_1 = \int_{\mathbf{R}^n} (1 - \chi_{[-\underline{b},\underline{b}]}(t_1)) \prod_{j=2}^{n} ((1 - \chi_{[-\underline{b},\underline{b}]}(t_j)) + \chi_{[-\underline{b},\underline{b}]}(t_j)) D_S(t) dt \tag{2.11}$$

is a sum of integrals of the following type

$$J_{1,r} = \int_{\mathbf{R}^n} \prod_{s=1}^{r} (1 - \chi_{[-\underline{b},\underline{b}]}(t_s)) \prod_{i=r+1}^{n} \chi_{[-\underline{b},\underline{b}]}(t_i) D_S(t) dt, \quad (r = 1, .., n) \tag{2.12}$$

and analogous integrals which are obtained from the permutations of $t_2, .., t_n$. To obtain an asymptotic error estimate it is sufficient to derive bounds for $J_{1,r}$. Note that

$$D_S(t) \leq ((D(t_1)^{-1} + .. + D(t_r)^{-1})^{-r})^{\frac{n}{r}} \leq (D(t_1)..D(t_r))^{\frac{n}{r}}, \quad (r = 1, .., n). \tag{2.13}$$

Applying (2.13) to (2.12) we obtain

$$J_{1,r} \leq (2\underline{b})^{n-r} (\int_R (1 - \chi_{[-\underline{b},\underline{b}]}(t)) D(t)^{n/r} dt)^r. \tag{2.14}$$

This implies

Proposition 2.4. *Let $f \in H_{D_S}(\mathbf{R}^n)$. Then*

$$\|f - S_b(f)\|_\infty = o\left(\sup_{1 \leq r \leq n} \underline{b}^{\frac{n-r}{2}} \left(\int_{\mathbf{R}} (1 - \chi_{[-\underline{b},\underline{b}]}(t)) D(t)^{\frac{n}{r}} dt \right)^{\frac{r}{2}} \right). \tag{2.15}$$

As examples we list two cases of isotropic Hilbert spaces related to the univariate examples considered previously:

$$\|f - S_b(f)\|_\infty = o\left(\underline{b}^{-\frac{n}{2}}\right), \quad f \in W^n(\mathbf{R}^n), \tag{2.16}$$

$$\|f - S_b(f)\|_\infty = o\left(\underline{b}^{\frac{n-1}{2}} e^{-\frac{n}{2}\underline{b}}\right), \quad f \in H^n(\mathbf{R}^n). \tag{2.17}$$

3. Exponential-type Interpolation in Harmonic Hilbert Spaces

Interpolation is related to the process of periodization. Let $f \in A(\mathbf{R}^n)$, i. e.,

$$f(x) = \int_{\mathbf{R}^n} F(t) \exp(i(x,t)) dt, \quad F \in L_1(\mathbf{R}^n).$$

For $b = (b_1, .., b_n) \in]0, \infty[^n$ the $2b-$*periodization* of F is defined by

$$F_{2b}(t) := \sum_{k \in \mathbf{Z}^n} F(t + 2bk).$$

It is well known that $F_{2b} \in L_1([-b,b])$ and

$$f(\frac{\pi}{b}k) = \int_{[-b,b]} F_{2b}(t)\exp(i(\frac{\pi}{b}k,t))dt \quad (k \in \mathbf{Z}^n) . \tag{3.1}$$

This shows that the function of exponential type given by

$$T_b(f)(x) = \int_{\mathbf{R}^n} \chi_{[-b,b]}(t)F_{2b}(t)\exp(i(x,t))dt$$

interpolates f at the points $\frac{\pi}{b}k,\ k \in \mathbf{Z}^n$.

Clearly , T_b is a bounded projector on $A(\mathbf{R}^n)$ with range $A_b(\mathbf{R}^n)$ in view of

$$\|T_b(f)\|_a \le \|S_b(f)\|_a \le \|f\|_a .$$

Its kernel satisfies

$$\ker(T_b) = \left\{ f \in A(\mathbf{R}^n) : f(\frac{\pi}{b}k) = 0 \ (k \in \mathbf{Z}^n) \right\}.$$

Taking into account the uniqueness of the finite Fourier transform we obtain

Proposition 3.1. *Given $f \in A(\mathbf{R}^n)$ there is a unique function $g = T_b(f)$ in $A_b(\mathbf{R}^n)$ satisfying*

$$g(\frac{\pi}{b}k) = f(\frac{\pi}{b}k) \ (k \in \mathbf{Z}^n).$$

T_b is called the *exponential-type interpolation projector.* The relation between the projectors S_b and T_b is described by

$$T_bS_b = S_b, \quad S_bT_b = T_b.$$

T_b may be considered as a natural interpolatory approximation of the Fourier partial integral projector S_b due to the following

Proposition 3.2. *Let $f \in A(\mathbf{R}^n)$. Then*

$$\|S_b(f) - T_b(f)\|_a \le \|f - S_b(f)\|_a , \tag{3.2}$$

$$\|f - T_b(f)\|_a \le 2\,\|f - S_b(f)\|_a . \tag{3.3}$$

Proof: We have

$$\|S_b(f) - T_b(f)\|_a = \int_{[-b,b]} |F(t) - F_{2b}(t)|\,dt = \int_{[-b,b]} \left| \sum_{k\in\mathbf{Z}^n, k\neq 0} F(t+2bk) \right| dt$$

$$\le \int_{[-b,b]} \sum_{k\in\mathbf{Z}^n, k\neq 0} |F(t+2bk)|\,dt \ = \|f - S_b(f)\|_a . \qquad \square$$

Remark 3.1. *It follows from (3.3) the uniform approximation by exponetial-type interpolation is as good as uniform approximation by Fourier partial integrals.*

The approximation properties of proposition 3.2 can be converted into quantitative form for functions from a harmonic Hilbert space . By propositions 2.1 and 3.2 we obtain

Proposition 3.3. *Let* $f \in H_D(\mathbf{R}^n)$. *Then*

$$\|S_b(f) - T_b(f)\|_a \leq \|d - S_b(d)\|_a^{1/2} \|f - S_b(f)\|_D, \tag{3.4}$$

$$\|f - T_b(f)\|_a \leq 2\|d - S_b(d)\|_a^{1/2} \|f - S_b(f)\|_D. \tag{3.5}$$

In view of proposition 3.3 the error estimates for S_b as derived in section 2 carry over to T_b. In particular we have (see proposition 2.1)

Proposition 3.4. *Let* $f \in H_D(\mathbf{R}^n)$. *Then*

$$\|f - T_b(f)\|_\infty = o(\|d - S_b(d)\|_a^{\frac{1}{2}}). \tag{3.6}$$

For two cases of anisotropic Hilbert spaces related to the univariate examples considered previously we obtain

$$\|f - T_b(f)\|_\infty = o(\underline{b}^{-\frac{1}{2}}), \quad f \in W^{(1,..,1)}(\mathbf{R}^n), \tag{3.7}$$

$$\|f - T_b(f)\|_\infty = o(\exp(-\frac{\underline{b}}{2})), \quad f \in H^{(1,..,1)}(\mathbf{R}^n). \tag{3.8}$$

For the two cases of isotropic Hilbert spaces related to the univariate examples considered previously we obtain:

$$\|f - T_b(f)\|_\infty = o\left(\underline{b}^{-\frac{n}{2}}\right), \quad f \in W^n(\mathbf{R}^n), \tag{3.9}$$

$$\|f - T_b(f)\|_\infty = o\left(\underline{b}^{\frac{n-1}{2}} e^{-\frac{n}{2}\frac{b}{2}}\right), \quad f \in H^n(\mathbf{R}^n). \tag{3.10}$$

References

1. Babuška I., Über universal optimale Quadraturformeln Teil 1, Apl. mat. **13** (1968), 304-338.
2. Babuška I., Über universal optimale Quadraturformeln Teil 2, Apl. mat. **13** (1968), 368-404.
3. Chandrasekharan, K., *Classical Fourier transforms*, Springer,Berlin 1989.
4. Delvos F. J., Approximation by optimal periodic interpolation, Apl. mat. **35** (1990), 451-457.
5. Delvos F. J., Mean square approximation by optimal periodic interpolation, Apl. mat. **40** (1995), 267-283.
6. Delvos F. J., Interpolation in harmonic Hilbert spaces, Modelisation mathematique et Analyse numérique, (1996) to appear.
7. Garnir H. G., *Fonctions de variables réelles*, Gauthier Villars, Paris, 1965.
8. Katznelson Y., *An introduction to harmonic analysis*, Dover, New York 1976.
9. Prager M., Universally optimal approximation of functionals. Apl. mat. **24** (1979), 406-420.

Franz-Jürgen Delvos
FB Mathematik I ,Universität GH Siegen
Hölderlinstrasse 3
D-57068 Siegen , Germany

Interpolation by Continuous Function Spaces

Manfred von Golitschek

Abstract. Let $U \subset C(D)$ be an arbitrary n-dimensional linear subspace of continuous functions on a compact set $D \subset \mathbb{R}^s$. We construct interpolation operators $L : C(D) \to U$ which have an operator norm $\geq \sqrt{n} - 1$.

1. Introduction

There are famous interpolation operators with poor convergence properties in polynomial interpolation:

The Example of Runge (1901).
Let $P_n \in \mathcal{P}_n$ be the algebraic polynomial of degree $\leq n$ which interpolates the function $f(x) = 1/(1+25x^2)$, $-1 \leq x \leq 1$, at the equidistant points $(x_j)_{j=0}^n$, $x_j := -1+2j/n$, then the sequence $P_n(x)$ converges to $f(x)$ only for $|x| < 0.726\cdots$, but not for $0.726\cdots < |x| < 1$.

The Theorem of Faber.
For each sequence of partitions $\Delta_n : a \leq x_0^{(n)} < x_1^{(n)} < \cdots < x_n^{(n)} \leq b$, $n = 1, 2, \ldots$, of the interval $[a, b]$ there exists a function $f \in C[a, b]$ so that the sequence $P_n \in \mathcal{P}_n$ of polynomials, which interpolate f at the points Δ_n, does not converge uniformly on $[a, b]$ to f.

It is the purpose of this short note to present other interpolation problems with poor convergence, including interpolation by univariate spline functions, but also in multivariate interpolation.

In what follows, $D \subset \mathbb{R}^s$, $s = 1, 2, \ldots$, is a compact set, $C(D)$ is the set of all real-valued continuous functions on D, and $U \subset C(D)$ is an arbitrary linear space of dimension n, $n \in \mathbb{N}$. We write $\|f\|_\infty := \|f\|_{C(D)}$ for the uniform norm on D. We say that $T = \{t_j\}_{j=1}^n \subset D$ is an *interpolation set for* U if for each $(y_j)_{j=1}^n \in \mathbb{R}^n$ there exists a unique $u \in U$ so that $u(x_j) = y_j$, $j = 1, \ldots, n$. If

Multivariate Approximation and Splines
G. Nürnberger, J. W. Schmidt, and G. Walz (eds.), pp. 83–88.

ISBN 3-7643-5654-5.

$T = \{t_j\}_{j=1}^n \subset D$ is an interpolation set for U, then we call the pair (U, T) a *regular interpolation problem on* D, and the *interpolation operator* $L := L_{U,T} : C(D) \to U$ with $(Lf)(t_j) = f(t_j)$, $j = 1, \ldots, n$, is a linear projection onto U. Its operator norm (for the uniform norm $\|\cdot\|_\infty$ on D) is defined by

$$\|L\| := \sup_{f \in C(D), f \neq 0} \frac{\|Lf\|_\infty}{\|f\|_\infty}. \tag{1.1}$$

The size of the operator norm $\|L\|$ indicates if the operator L yields good or bad approximations Lf of f in U if compared with the error of best approximation $\text{dist}(f, U) := \min_{u \in U} \|f - u\|_\infty$:

(i) One has

$$\|f - Lf\|_\infty \leq (1 + \|L\|)\ \text{dist}(f, U). \tag{1.2}$$

That is, if the norm $\|L\|$ of the interpolation operator $L := L_{U,T}$ is small then the uniform error $\|f - Lf\|_\infty$ is only a little larger than the error of approximation $\text{dist}(f, U)$.

(ii) Conversely, let $\|L\|$ be large. For some $\varepsilon > 0$, let $f := f_\varepsilon \in C(D)$ satisfy

$$\|Lf_\varepsilon\|_\infty \ \geq\ \|L\| - \varepsilon, \quad \|f_\varepsilon\|_\infty = 1.$$

Then we have

$$\|f_\varepsilon - Lf_\varepsilon\|_\infty \ \geq\ \|Lf_\varepsilon\|_\infty - \|f_\varepsilon\|_\infty \ \geq\ \|L\| - \varepsilon - 1,$$

which is much larger than $\text{dist}(f_\varepsilon, U) \leq \|f_\varepsilon\|_\infty = 1$ if $\|L\|$ is large. Indeed,

$$\sup_{f \in C(D), f \notin U} \frac{\|f - Lf\|_\infty}{\text{dist}(f, U)} \ \geq\ \|L\| - 1. \tag{1.3}$$

(iii) Let $U_n \subset C(D)$ be a sequence of linear spaces of dimension $n = 1, 2, \ldots$, and let (U_n, T_n) be regular interpolation problems. If the supremum over the norms $\|L_n\|$ of the interpolation operators $L_n := L_{U_n,T_n}$ is $= \infty$, then the *Principle of Uniform Boundedness* states that there exists a function $f \in C(D)$ such that $\sup_n \|L_n f\|_\infty = \infty$.

Remark: Using (i), (iii) and the theorem of Weierstrass for algebraic polynomials, the theorem of Faber simply states that the supremum of the norms of interpolation operators of the regular interpolation problems $(\mathcal{P}_n, \Delta_n)$, $n \geq 1$, is $+\infty$.

2. Interpolation operators with large norm

In the setting of the last section, let $U \subset C(D)$ be an arbitrary n-dimensional space. We select a set $\Omega = \{t_j\}_{j=1}^{n+1}$ of $n+1$ distinct points of the compact set $D \subset \mathbb{R}^s$. We consider the sets

$$(2.1) \qquad T_k := \Omega \setminus \{t_k\} = \{t_j : j = 1, 2, \dots, n+1, \ j \neq k\}, \quad k = 1, \dots, n+1.$$

Moreover, we assume that

(2.2) all pairs (U, T_k), $k = 1, \dots, n+1$, are regular interpolation problems.

For a given function $f \in C(D)$ let $u^* := u_f^* \in U$ be the minimal solution of the least squares problem for U on Ω,

$$(2.3) \qquad A(f) := \sum_{j=1}^{n+1} (f(t_j) - u^*(t_j))^2 = \min_{u \in U} \sum_{j=1}^{n+1} (f(t_j) - u(t_j))^2.$$

The minimum u^* is characterized by the orthogonality relations

$$\sum_{j=1}^{n+1} (f(t_j) - u^*(t_j)) \ u(t_j) = 0, \quad u \in U,$$

which is equivalent to

$$(2.4) \qquad A(f) = \sum_{j=1}^{n+1} (f(t_j) - u^*(t_j))(f(t_j) - u(t_j)), \quad u \in U.$$

Let $L_k := L_{U,T_k} : C(D) \to U$ be the interpolation operator for the linear space U and the set T_k, that is, let $(L_k f)(t_j) = f(t_j)$, $j = 1, \dots, n+1$, $j \neq k$. If we insert $u := L_k f$ into (2.4), we get

$$(2.5) \qquad A(f) = (f(t_k) - u^*(t_k))(f(t_k) - (L_k f)(t_k)), \quad k = 1, \dots, n+1.$$

Theorem 2.1. *For each $f \in C(D)$ and each integer q, $1 \le q \le n$, at least q of the $n+1$ regular interpolation problems (2.2) satisfy (in the uniform norm on Ω)*

$$(2.6) \qquad \|f - L_k f\|_{C(\Omega)} \ge \sqrt{n+2-q} \ \|f - u^*\|_{C(\Omega)}.$$

Proof. Let $\Omega_q = \Omega_q(f)$ consist of q points of Ω so that

$$(2.7) \qquad \max_{t \in \Omega_q} |f(t) - u^*(t)| \le \min_{t \in \Omega \setminus \Omega_q} |f(t) - u^*(t)|.$$

For each $t = t_k \in \Omega_q$, (2.7) implies that $(n+2-q)(f(t)-u^*(t))^2 \le A(f)$, and (2.5) implies that

$$|f(t) - (L_k f)(t)| \;=\; \frac{A(f)}{|f(t) - u^*(t)|} \;\ge\; \sqrt{n+2-q}\,\sqrt{A(f)}.$$

Now, (2.6) follows for each k with $t_k \in \Omega_q$ since $\sqrt{A(f)} \ge \|f - u^*\|_{C(\Omega)}$. □

Lemma 2.2. *There exist functions $f \in C(D)$ which satisfy $u_f^* \equiv 0$ and*

$$\|f\|_{C(\Omega)} = \|f\|_{C(D)} > 0. \tag{2.8}$$

Proof. For fixed $y = (y_1, \ldots, y_{n+1}) \in \mathbb{R}^{n+1}$ let $v^* \in U$ be the minimal solution of the least squares problem

$$\min_{u \in U} \sum_{j=1}^{n+1} (y_j - u(t_j))^2. \tag{2.9}$$

We choose $y \in \mathbb{R}^{n+1}$ so that the minimum in (2.9) is positive and that $v^* \equiv 0$. We define the function $\Psi \in C(\mathbb{R}^s)$ by

$$\Psi(x) := \begin{cases} 1 - \|x\|_2, & \|x\|_2 \le 1 \\ 0, & \|x\|_2 > 1, \end{cases}$$

where $\|x\|_2$ denotes the Euclidean norm in $\mathbb{R}^s$. For each $N \in \mathbb{N}$, the function

$$f_N(x) \;:=\; \sum_{j=1}^{n+1} y_j \Psi(Nx - Nt_j)$$

is continuous on D. For large N, $f_N(t_j) = y_j$, $j = 1, \ldots, n+1$. Therefore, the minimal solution of (2.3) for the function f_N is $u_{f_N}^* \equiv 0$. Moreover, one has $\|f_N\|_{C(\Omega)} = \|f_N\|_{C(D)}$ for large N. Now, f_N has the properties of the lemma for all large N. □

Theorem 2.3. *For each $q \in \mathbb{N}$, $1 \le q \le n$, at least q of the $n+1$ regular interpolation problems $(U, \Omega \setminus \{t_k\})$, $k = 1, \ldots, n+1$, have an interpolation operator L_k with large norm:*

$$\|L_k\| := \sup_{f \in C(D), f \ne 0} \frac{\|L_k f\|_\infty}{\|f\|_\infty} \;\ge\; \sqrt{n+2-q} \;-\; 1. \tag{2.10}$$

Proof. We apply Theorem 2.1 to a function $f \in C(D)$ of Lemma 2.2. Its has the properties $u_f^* \equiv 0$ and (2.8). By Theorem 2.1, at least q of the $n+1$ regular interpolation problems $(U, \Omega \setminus \{t_k\})$ satisfy

$$\|f - L_k f\|_{C(\Omega)} \;\ge\; \sqrt{n+2-q}\; \|f - u^*\|_{C(\Omega)} \;=\; \sqrt{n+2-q}\; \|f\|_{C(\Omega)}.$$

Moreover,

$$\|L_k f\|_{C(D)} \geq \|L_k f\|_{C(\Omega)} \geq \|f - L_k f\|_{C(\Omega)} - \|f\|_{C(\Omega)}.$$

Therefore, and since $\|f\|_{C(\Omega)} = \|f\|_{C(D)}$, it follows that

$$\|L_k f\|_{C(D)} \geq \left(\sqrt{n+2-q} - 1\right) \|f\|_{C(D)},$$

which proves (2.10). □

3. Interpolation by polynomial splines

In this section we apply Theorem 2.3 to the polynomial spline spaces $U = \mathcal{S}_r(\Delta) \subset C^{r-2}[a,b]$ of order $r = 2, 3, \ldots$ and with simple knots

(3.1) $$\Delta : a = x_0 < x_1 < \cdots < x_n < x_{n+1} = b.$$

The dimension of $\mathcal{S}_r(\Delta)$ is $n + r$.
Let $\Omega = (t_j)_{j=1}^{n+r+1}$ be $n + r + 1$ distinct points in $[a,b]$ so that for $i = 1, 2, \ldots, n$

(3.2) at least $i + 1$ of the $t_j \in \Omega$ lie in the interval $[a, x_i)$,
at least $n + 2 - i$ of the $t_j \in \Omega$ lie in the interval $(x_i, b]$.

Then, each of the sets $T_k = \Omega \setminus \{t_k\}$, $k = 1, \ldots, n+r+1$, satisfies the Schoenberg - Whitney [1] condition; and all $(\mathcal{S}_r(\Delta), T_k)$ are regular interpolation problems. Therefore, Theorem 2.3 yields

Theorem 3.1. *Let $\Omega = \{t_j\}_{j=1}^{n+r+1} \subset [a,b]$ satisfy (3.2). For each $q \in \mathbb{N}$, $1 \leq q \leq n + r$, at least q of the $n + r + 1$ regular interpolation problems $(\mathcal{S}_r(\Delta), \Omega \setminus \{t_k\})$ have an interpolation operator L_k with large norm:*

(3.3) $$\|L_k\| \;\geq\; \sqrt{n+r+2-q} \;-\; 1.$$

Example 3.2 (Cubic Spline Interpolation, $r = 4$). For the knots Δ in (3.1) we define $\xi_0 := (x_1 - x_0)/2$, $\xi_1 := (x_2 - x_1)/2$, $\xi_n := (x_{n+1} - x_n)/2$, and take

$$\Omega := \{x_j\}_{j=0}^{n+1} \cup \{\xi_0\} \cup \{\xi_1\} \cup \{\xi_n\}.$$

Clearly, Ω has the property (3.2). By Theorem 3.1 for $q := n/2$ if n is even, for $q := (n+1)/2$ if n is odd, *for at least $n/2$ of the $n+2$ regular interpolation problems $(\mathcal{S}_r(\Delta), T_k)$,*

$$T_k \;:=\; \{x_j\}_{j=0, j\neq k}^{n+1} \cup \{\xi_0\} \cup \{\xi_1\} \cup \{\xi_n\}, \quad k = 0, \ldots, n+1,$$

the norm of the interpolation operator satisfies

$$\|L_k\| \geq \sqrt{\frac{n}{2}} - 1.$$

References

1. Schoenberg I. J. and Whitney A., On Pólya frequency functions III. The positivity of translation determinants with applications to the interpolation problem by spline curves. *Trans. Amer. Math. Soc.* **74**, (1953), 246–259.

Manfred von Golitschek
Institut für Angewandte Mathematik und Statistik
Universität Würzburg
97074 Würzburg
Germany
`goli@mathematik.uni-wuerzburg.de`

Discrete Characterization of Besov Spaces and Its Applications to Stochastics

Anna Kamont

Abstract. In this article, we present a review of discrete characterizations of Besov spaces and some applications of these results to regularity of realizations of random processes.

1. Introduction

This article presents a discrete characterization of Besov spaces on the d-dimensional cube $Q = [0,1]^d$. The characterization under considerstion uses only the values of the function at dyadic points.

It is well known that the Besov spaces are isomorphic to some sequence spaces, with the isomorphism given by the coefficients of a function in some spline or wavelet basis (see for example [8, 18]). However, to calculate these coefficients, we have to calculate inner products of the given function with some spline functions or wavelets. The calculation of these inner products is usually troublesome. This has been the motivation to obtain a characterization of Besov spaces which uses only finite linear combinations of the values of the function on dyadic mesh.

The discrete characterizations of the isotropic and anisotropic Besov spaces are presented in section 2. The characterization of this type was first proved by Ciesielski [5] for Hölder classes with exponent α, $0 < \alpha < 1$, on the interval $[0,1]$; the analogous result for Hölder classes on the d-dimensional cube was obtained by Bonic, Frampton and Tromba [1] (see also Ryll [20]). The extension of these results for Hölder classes in the uniform norm with the exponent $\alpha \geq 1$ is due to Ciesielski [6, 7] (univariate and multivariate case respectively). The results of this type for Besov spaces with the L^p parameter $p < \infty$ and the Hölder exponent $0 < \alpha < 1$ can be found in [11] (univariate case), [10] (isotropic multivariate case) and [15] (anisotropic multivariate case). Finally, the results for $p < \infty$ and $\alpha \geq 1$ are proved in [17].

Multivariate Approximation and Splines
G. Nürnberger, J. W. Schmidt, and G. Walz (eds.), pp. 89–98.

ISBN 3-7643-5654-5.

These characterizations are used to study regularity of realizations of random fields, e.g. fractional brownian motion, fractional Lévy fields on $\mathbb{R}^d$ and S^d, fractional anisotropic Wiener field (see [9, 10, 11, 16, 19]), brownian local time ([2, 3 ,4]), solutions of some stochastic differential equations ([12]). In section 3 we present the results on the regularity of Lévy and Wiener fields.

Let's start with some notation. For a function $f : Q \to \mathbb{R}$, $m \in \mathbb{N}$ and $\underline{h} \in \mathbb{R}^d$, the symbol $\Delta_{\underline{h}}^m f$ denotes the progressive difference of f of order m with the step $\underline{h}$, i.e.

$$\Delta_{\underline{h}}^m f(\underline{x}) = \sum_{i=0}^{m} (-1)^{i+j} \binom{m}{i} f(\underline{x} + i\underline{h}), \quad \underline{x} \in Q,$$

with the convention that $\Delta_{\underline{h}}^m f(\underline{x}) = 0$ if $\underline{x} + m\underline{h} \notin Q$. If $\underline{e}_i$ is the i-th unit vector in $\mathbb{R}^d$, and $h \in \mathbb{R}$, we use the abbreviation $\Delta_{h\underline{e}_i}^m f = \Delta_{h,i}^m f$. Moreover, for a subset of different directions $A = \{i_1, \ldots, i_k\} \subset \mathcal{D} = \{1, \ldots, d\}$ and $\underline{h} = (h_1, \ldots h_d)$, we denote by $\Delta_{\underline{h},A}^m f$ the mixed difference of f of order m with the step $\underline{h}$ and the set of directions A, i.e.

$$\Delta_{\underline{h},A}^m f = \Delta_{h_{i_1},i_1}^m \circ \ldots \circ \Delta_{h_{i_k},i_k}^m f.$$

By $\omega_{m,p}(f,t)$ we denote the isotropic modulus of smoothness of f of order m in the L^p norm, i.e.

$$\omega_{m,p}(f,t) = \sup_{\|\underline{h}\| \le t} \|\Delta_{\underline{h}}^m f\|_p, \tag{1}$$

where $\|\cdot\|_p$ denotes the standard L^p norm over Q, and $\|\cdot\|$ is the euclidean norm in $\mathbb{R}^d$; for $\underline{t} = (t_1, \ldots, t_d)$ with $t_i > 0$ and $A \subset \mathcal{D}$, we denote by $\omega_{m,A,p}(f,\underline{t})$ the mixed modulus of smoothness of f of order m with the set of directions A in the L^p norm, i.e.

$$\omega_{m,A,p}(f,\underline{t}) = \sup_{\underline{h}} \|\Delta_{\underline{h},A}^m f\|_p, \tag{2}$$

with the supremum taken over $\underline{h} = (h_1, \ldots, h_d)$ with $|h_i| \le t_i$, $i = 1, \ldots, d$.

Definition 1.1. *Let $1 \le p, q \le \infty$, $m \in \mathbb{N}$, $0 < \alpha < m$. Let for $f \in L^p(Q)$*

$$\|f\|_{p,q}^{\alpha} = \|f\|_p + \left(\int_0^{1/m} \left(\frac{\omega_{m,p}(f,t)}{t^{\alpha}} \right)^q \frac{dt}{t} \right)^{1/q}, \tag{3}$$

with the integral replaced by suitable supremum in case $q = \infty$, and

$$B_{p,q}^{\alpha}(Q) = \{f \in L^p(Q) : \ \|f\|_{p,q}^{\alpha} < \infty\}, \tag{4}$$

(with $L^\infty(Q)$ replaced by $C(Q)$ for $p = \infty$). For $q = \infty$ put

$$B_{p,\infty}^{\alpha,0}(Q) = \{f \in B_{p,\infty}^{\alpha}(Q) : \ \omega_{m,p}(f,t) = o(t^{\alpha}) \ \text{as} \ t \to 0\}. \tag{5}$$

The space $B_{p,q}^{\alpha}(Q)$ is called the isotropic Besov space with parameters p, q and α.

Before the definition of the anisotropic Besov space, we introduce some notation. For a vector $\underline{a} = (a_1, \ldots, a_d)$ and $A \subset \mathcal{D}$ denote $\underline{a}(A) = (\tilde{a}_1, \ldots, \tilde{a}_d)$ with $\tilde{a}_i = a_i$ for $i \in A$ and $\tilde{a}_i = 0$ for $i \notin A$; for two vectors $\underline{a} = (a_1, \ldots, a_d)$ and $\underline{b} = (b_1, \ldots, b_d)$ put $\underline{a}^{\underline{b}} = \prod_{i=1}^d a_i^{b_i}$, $\underline{a} \cdot \underline{b} = \sum_{i=1}^d a_i b_i$ and $|\underline{a}| = \sum_{i=1}^d |a_i|$; moreover, put $\underline{1} = (1, \ldots, 1) \in \mathbb{N}^d$.

Definition 1.2. *Let $1 \leq p, q \leq \infty$, $m \in \mathbb{N}$, $\underline{\alpha} = (\alpha_1, \ldots, \alpha_d)$, with $0 < \alpha_i < m$. Let for $f \in L^p(Q)$*

$$\|f\|_{p,q}^{\underline{\alpha}} = \|f\|_p + \sum_{\emptyset \neq A \subset \mathcal{D}} \left(\int_{[0,1/m]^d} \left(\frac{\omega_{m,A,p}(f, \underline{t})}{\underline{t}^{\underline{\alpha}(A)}} \right)^q \frac{d\underline{t}}{\underline{t}^{\underline{1}(A)}} \right)^{1/q}, \tag{6}$$

with the integral replaced by suitable supremum in case $q = \infty$, and

$$B_{p,q}^{\underline{\alpha}}(Q) = \{ f \in L^p(Q) : \ \|f\|_{p,q}^{\underline{\alpha}} < \infty \}, \tag{7}$$

(with $L^\infty(Q)$ replaced by $C(Q)$ for $p = \infty$). For $q = \infty$ put

$$\begin{aligned} B_{p,\infty}^{\underline{\alpha},0}(Q) = &\{ f \in B_{p,\infty}^{\underline{\alpha}}(Q) : \\ &\forall_{\emptyset \neq A \subset \mathcal{D}} \ \omega_{m,A,p}(f, \underline{t}) = o(\underline{t}^{\underline{\alpha}(A)}) \ \text{as} \ \min_{i \in A} t_i \to 0 \}. \end{aligned} \tag{8}$$

The space $B_{p,q}^{\underline{\alpha}}(Q)$ is called the anisotropic Besov space with parameters p, q and $\underline{\alpha}$.

2. Discrete characterization of Besov spaces

In this section we present the discrete characterization of both isotropic and anisotropic Besov spaces. The proofs of these results can be found in [17] and are not presented here. The main tools used in the proofs are: (i) the characterization of Besov spaces in terms of the projections on spline subspaces, and (ii) Rabienkij's construction of interpolating splines. The constants obtained in these estimates are independent of p. This and formula (9) below enables us to extend the results of subsections 2.1 and 2.2 to Besov spaces in some Orlicz norms; these results are formulated in subsection 2.3.

2.1. The isotropic Besov spaces

Let's start with the description of the characterization of the isotropic Besov spaces. For $\mu \in \mathbb{N}$, let $\mathcal{T}_\mu = \{0, \frac{1}{2^\mu}, \ldots, \frac{2^\mu - 1}{2^\mu}, 1\}$ be the dyadic partition of the interval $[0, 1]$ with the step $\frac{1}{2^\mu}$, and let

$$\mathcal{T}_\mu^d = \mathcal{T}_\mu \times \ldots \mathcal{T}_\mu \subset Q.$$

Moreover, let for $m \in \mathbb{N}$ and $i \in \mathcal{D}$

$$\mathcal{T}_\mu^d(m,i) = \Big\{ \underline{x} \in \mathcal{T}_\mu^d : \underline{x} + \frac{m}{2^\mu} \underline{e}_i \in \mathcal{T}_\mu^d \Big\}.$$

For $f \in C(Q)$ let for $1 \le p < \infty$

$$\Lambda_{\mu,p}^{(m,i)}(f) = \Big(\frac{1}{2^{d\mu}} \sum_{\underline{x} \in \mathcal{T}_\mu^d(m,i)} |\Delta^m_{\frac{1}{2^\mu},i} f(\underline{x})|^p \Big)^{1/p},$$

and for $p = \infty$

$$\Lambda_{\mu,\infty}^{(m,i)}(f) = \sup_{\underline{x} \in \mathcal{T}_\mu^d(m,i)} |\Delta^m_{\frac{1}{2^\mu},i} f(\underline{x})|.$$

It follows from Sobolev's embedding theorem that $B^\alpha_{p,q}(Q) \subset C(Q)$ for $\alpha > \frac{d}{p}$. For such parameters p and α, we have the following characterization of the space $B^\alpha_{p,q}(Q)$:

Theorem 2.1. *Let $1 \le p,q \le \infty$, $m \in \mathbb{N}$ and $\frac{d}{p} < \alpha < m$. Let $\nu \in \mathbb{N}$ be such that $2^\nu > m$. Then $f \in B^\alpha_{p,q}(Q)$ if and only if $f \in C(Q)$ and*

$$\Big(\frac{1}{2^{d\nu}} \sum_{\underline{x} \in \mathcal{T}_\nu^d} |f(\underline{x})|^p \Big)^{1/p} + \sum_{i=1}^d \Big(\sum_{\mu=\nu}^\infty (2^{\alpha\mu} \cdot \Lambda_{\mu,p}^{(m,i)}(f))^q \Big)^{1/q} < \infty,$$

with the sums replaced by suprema in case $q = \infty$. The above formula defines an equivalent norm in the space $B^\alpha_{p,q}(Q)$. Moreover,

$$f \in B^{\alpha,0}_{p,\infty}(Q) \quad \textit{iff} \quad \lim_{\mu \to \infty} \big(\max_{i \in \mathcal{D}} 2^{\alpha\mu} \cdot \Lambda_{\mu,p}^{(m,i)}(f) \big) = 0.$$

2.2. The anisotropic Besov spaces

Let's introduce the quantites needed for the discrete characterization of the anisotropic Besov spaces. For $\underline{\mu} = (\mu_1, \dots \mu_d) \in \mathbb{N}^d$, let

$$\mathcal{T}_{\underline{\mu}} = \mathcal{T}_{\mu_1} \times \ldots \times \mathcal{T}_{\mu_d} \quad \text{and} \quad \underline{t}_{\underline{\mu}} = \Big(\frac{1}{2^{\mu_1}}, \dots, \frac{1}{2^{\mu_d}} \Big).$$

For $m \in \mathbb{N}$ and $A \subset \mathcal{D}$, let

$$\mathcal{T}_{\underline{\mu}}(m,A) = \Big\{ \underline{x} \in \mathcal{T}_{\underline{\mu}} : \underline{x} + m \underline{t}_{\underline{\mu}}(A) \in \mathcal{T}_{\underline{\mu}} \Big\}.$$

Now, for $f \in C(Q)$ and $1 \le p < \infty$ define

$$\Omega_{\underline{\mu},p}^{(m,A)}(f) = \Big(\frac{1}{2^{\underline{\mu} \cdot \underline{1}}} \sum_{\underline{x} \in \mathcal{T}_{\underline{\mu}}(m,A)} |\Delta^m_{\underline{t}_{\underline{\mu}},A} f(\underline{x})|^p \Big)^{1/p},$$

and for $p = \infty$

$$\Omega^{(m,A)}_{\underline{\mu},\infty}(f) = \sup_{\underline{x}\in\mathcal{T}_{\underline{\mu}}(m,A)} |\Delta^m_{\underline{t}_{\underline{\mu}},A} f(\underline{x})|.$$

In addition, for $\nu \in \mathbb{N}$ and $A \subset \mathcal{D}$ let

$$\mathbb{N}^d(\nu, A) = \{\underline{\mu} = (\mu_1, \ldots, \mu_d) \in \mathbb{N}^d : \ \mu_i \geq \nu \ \text{for} \ i \in A, \ \ \mu_i = \nu \ \text{for} \ i \notin A\}.$$

Now, from Sobolev's embedding theorem, for p and $\underline{\alpha} = (\alpha_1, \ldots, \alpha_d)$ such that $\alpha_i > \frac{1}{p}$ for all $i = 1, \ldots, d$, we have $B^{\underline{\alpha}}_{p,q}(Q) \subset C(Q)$; for this range of parameters the following characterization holds:

Theorem 2.2. *Let $1 \leq p, q \leq \infty$, $m \in \mathbb{N}$ and $\underline{\alpha} = (\alpha_1, \ldots, \alpha_d)$ with $\frac{1}{p} < \alpha_i < m$, $i = 1, \ldots, d$. Let $\nu \in \mathbb{N}$ be such that $2^\nu > m$. Then $f \in B^{\underline{\alpha}}_{p,q}(Q)$ if and only if $f \in C(Q)$ and*

$$\Big(\frac{1}{2^{d\nu}} \sum_{\underline{x}\in\mathcal{T}^d_\nu} |f(\underline{x})|^p\Big)^{1/p} + \sum_{\emptyset\neq A\subset\mathcal{D}} \Big(\sum_{\underline{\mu}\in\mathbb{N}^d(\nu,A)} (2^{\underline{\mu}(A)\cdot\underline{\alpha}(A)} \cdot \Omega^{(m,A)}_{\underline{\mu},p}(f))^q\Big)^{1/q} < \infty,$$

with the sums replaced by the appropriate suprema in case $q = \infty$. The above formula defines an equivalent norm in $B^{\underline{\alpha}}_{p,q}(Q)$. Moreover, $f \in B^{\underline{\alpha},0}_{p,\infty}(Q)$ iff

$$\forall_{\emptyset\neq A\subset\mathcal{D}} \ \ 2^{\underline{\mu}(A)\cdot\underline{\alpha}(A)} \Omega^{(m,A)}_{\underline{\mu},p}(f) = o(1) \quad \textit{as} \quad |\underline{\mu}(A)| \to \infty, \quad \underline{\mu} \in \mathbb{N}^d(\nu, A).$$

2.3. Besov spaces in Orlicz norms

Now, we describe Besov-type spaces in Orlicz norms. Let M be an N-function; denote by $L_M(Q)$ the corresponding Orlicz space on Q, and by $\|\cdot\|_M$ the appropriate norm. For $f \in L_M(Q)$, the isotropic modulus of smoothness of order m, $\omega_{m,M}(f,t)$, and the anisotropic moduli of smoothness of order m with the set of directions A, $\omega_{m,A,M}(f,\underline{t})$, in the Orlicz norm $\|\cdot\|_M$, are defined by the formulae analogous to (1) and (2); then, for $1 \leq q \leq \infty$, $0 < \alpha < m$ and $\underline{\alpha} = (\alpha_1, \ldots, \alpha_d)$ with $0 < \alpha_i < m$, the respective norms $\|\cdot\|^\alpha_{M,q}$ and $\|\cdot\|^{\underline{\alpha}}_{M,q}$ are defined by the formulae analogous to (3) and (6), and finally the isotropic and anisotropic Besov spaces in Orlicz norm $B^\alpha_{M,q}(Q)$, $B^{\alpha,0}_{M,\infty}(Q)$, $B^{\underline{\alpha}}_{M,q}(Q)$ and $B^{\underline{\alpha},0}_{M,\infty}(Q)$ are defined as in (4), (5), (7) and (8), respectively.

We are interested in a particular scale of Orlicz spaces. Namely, for $\gamma > 0$, let $M_\gamma(u)$ be the N-function equivalent to $\exp(|u|^\gamma)$; the corresponding spaces $L_{M_\gamma}(Q)$ form a scale of intermediate spaces between the $L^\infty(Q)$ space and the $L^p(Q)$ spaces with $p < \infty$. It is known that (see [13, 14])

$$\|f\|_{M_\gamma} \sim \sup_{p\geq 1} \frac{\|f\|_p}{p^{1/\gamma}}. \tag{9}$$

Thus

$$\omega_{m,M_\gamma}(f,t) \sim \sup_{p\geq 1} \frac{\omega_{m,p}(f,t)}{p^{1/\gamma}}$$

and

$$\omega_{m,A,M_\gamma}(f,\underline{t}) \sim \sup_{p\geq 1} \frac{\omega_{m,A,p}(f,\underline{t})}{p^{1/\gamma}}.$$

Note that for any $\gamma > 0$, $1 \leq q \leq \infty$, $\alpha > 0$ and $\underline{\alpha} = (\alpha_1, \dots, \alpha_d)$ with $\alpha_i > 0$ we have $B^\alpha_{M_\gamma,q}(Q) \subset C(Q)$ and $B^{\underline{\alpha}}_{M_\gamma,q}(Q) \subset C(Q)$. To describe the discrete characterization of these spaces, define for $f \in C(Q)$, $\gamma > 0$, $i \in \mathcal{D}$ and $\mu \in \mathbb{N}$

$$\Lambda^{(m,i)}_{\mu,M_\gamma}(f) = \sup_{p\geq 1} \frac{\Lambda^{(m,i)}_{\mu,p}(f)}{p^{1/\gamma}}.$$

Similarly, let for $\underline{\mu} \in \mathbb{N}^d$ and $A \subset \mathcal{D}$

$$\Omega^{(m,A)}_{\underline{\mu},M_\gamma}(f) = \sup_{p\geq 1} \frac{\Omega^{(m,A)}_{\underline{\mu},p}(f)}{p^{1/\gamma}}.$$

Now, we have the following characterizations of the isotropic and anisotropic Besov spaces in the Orlicz norms.

Theorem 2.3. *Let $1 \leq q \leq \infty$, $\gamma > 0$, $m \in \mathbb{N}$ and $0 < \alpha < m$. Let $\nu \in \mathbb{N}$ be such that $2^\nu > m$. Then $f \in B^\alpha_{M_\gamma,q}(Q)$ if and only if $f \in C(Q)$ and*

$$\sup_{\underline{x}\in \mathcal{T}^d_\nu} |f(\underline{x})| + \sum_{i=1}^{d} \Big(\sum_{\mu=\nu}^{\infty} (2^{\alpha\mu} \cdot \Lambda^{(m,i)}_{\mu,M_\gamma}(f))^q \Big)^{1/q} < \infty,$$

with the sums replaced by suprema in case $q = \infty$. The above formula defines an equivalent norm in the space $B^\alpha_{M_\gamma,q}(Q)$, and

$$f \in B^{\alpha,0}_{M_\gamma,\infty}(Q) \quad \textit{iff} \quad \lim_{\mu\to\infty} \big(\max_{i\in\mathcal{D}} 2^{\alpha\mu} \cdot \Lambda^{(m,i)}_{\mu,M_\gamma}(f) \big) = 0.$$

Theorem 2.4. *Let $1 \leq q \leq \infty$, $\gamma > 0$, $m \in \mathbb{N}$ and $\underline{\alpha} = (\alpha_1, \dots, \alpha_d)$ with $0 < \alpha_i < m$, $i = 1, \dots, d$. Let $\nu \in \mathbb{N}$ be such that $2^\nu > m$. Then $f \in B^{\underline{\alpha}}_{M_\gamma,q}(Q)$ if and only if $f \in C(Q)$ and*

$$\sup_{\underline{x}\in \mathcal{T}^d_\nu} |f(\underline{x})| + \sum_{\emptyset\neq A\subset\mathcal{D}} \Big(\sum_{\underline{\mu}\in\mathbb{N}^d(\nu,A)} (2^{\underline{\mu}(A)\cdot\underline{\alpha}(A)} \cdot \Omega^{(m,A)}_{\underline{\mu},M_\gamma}(f))^q \Big)^{1/q} < \infty,$$

with the sums replaced by the appropriate suprema in case $q = \infty$. Moreover, the above formula defines an equivalent norm in $B^{\underline{\alpha}}_{M_\gamma,q}(Q)$, and $f \in B^{\underline{\alpha},0}_{M_\gamma,\infty}(Q)$ iff

$$\forall_{\emptyset\neq A\subset\mathcal{D}} \;\; 2^{\underline{\mu}(A)\cdot\underline{\alpha}(A)} \Omega^{(m,A)}_{\underline{\mu},M_\gamma}(f) = o(1) \quad \textit{as} \quad |\underline{\mu}(A)| \to \infty, \quad \underline{\mu} \in \mathbb{N}^d(\nu, A).$$

3. Applications to the regularity of realizations of stochastic processes

We give two examples of the results concerning the regularity of realizations of random fields, obtained by means of the discrete characterization of Besov spaces. The fields under consideration are the fractional Lévy field and the anisotropic fractional Wiener field (with d-dimensional parameter). These fields are multidimensional analogues of the fractional brownian motion with 1-dimensional parameter. The detailed proofs can be found in [9] and [16] respectively; see also [11] for the 1-dimensional version.

Example 3.1 – the fractional Lévy field.

The fractional Lévy field with parameter β, $0 < \beta < 2$, is a gaussian random field $\{L^{(\beta)}(\underline{t}) : \underline{t} \in \mathbb{R}^d\}$, with continuous realizations, $Pr\{L^{(\beta)}(\underline{0}) = 0\} = 1$, $EL^{(\beta)}(\underline{t}) = 0$ for $\underline{t} \in \mathbb{R}^d$ and the covariance

$$EL^{(\beta)}(\underline{t})L^{(\beta)}(\underline{s}) = \frac{\|\underline{t}\|^\beta + \|\underline{s}\|^\beta - \|\underline{t}-\underline{s}\|^\beta}{2},$$

where $\|\cdot\|$ is the euclidean norm on $\mathbb{R}^d$.

Note that for $\underline{x} \in T^d_\mu(2,i)$, $\Delta^2_{\frac{1}{2^\mu},i} L^{(\beta)}(\underline{x})$ is a centered gaussian random variable, with the variance $\sim \frac{1}{2^{\beta\mu}}$; moreover, one can check that for any pair $\underline{x}, \underline{y} \in T^d_\mu(2,i)$

$$\left| E\, \Delta^2_{\frac{1}{2^\mu},i} L^{(\beta)}(\underline{x}) \cdot \Delta^2_{\frac{1}{2^\mu},i} L^{(\beta)}(\underline{y}) \right| \le \frac{C}{2^{\beta\mu}} \frac{1}{1 + \sum_{j=1}^d (2^\mu |x_j - y_j|)^{4-\beta}},$$

with some constant C independent of μ. Then, it is proved with the help of these estimates that for any $p < \infty$ and $i = 1, \ldots, d$

$$Pr\Big\{0 < \underline{\lim}_{\mu\to\infty} 2^{\mu\beta/2} \Lambda^{(2,i)}_{\mu,p}(L^{(\beta)}\,|_Q) \le \overline{\lim}_{\mu\to\infty} 2^{\mu\beta/2} \Lambda^{(2,i)}_{\mu,p}(L^{(\beta)}\,|_Q) < \infty\Big\} = 1,$$

and (for $\gamma = 2$)

$$Pr\Big\{0 < \underline{\lim}_{\mu\to\infty} 2^{\mu\beta/2} \Lambda^{(2,i)}_{\mu,M_2}(L^{(\beta)}\,|_Q) \le \overline{\lim}_{\mu\to\infty} 2^{\mu\beta/2} \Lambda^{(2,i)}_{\mu,M_2}(L^{(\beta)}\,|_Q) < \infty\Big\} = 1.$$

These estimates and Theorems 2.1 and 2.3 imply that for $0 < \frac{d}{p} < \frac{\beta}{2}$

$$Pr\Big\{L^{(\beta)}\,|_Q \in B^{\beta/2}_{p,\infty}(Q)\Big\} = 1, \quad Pr\Big\{L^{(\beta)}\,|_Q \in B^{\beta/2,0}_{p,\infty}(Q)\Big\} = 0,$$

$$Pr\Big\{L^{(\beta)}\,|_Q \in B^{\beta/2}_{M_2,\infty}(Q)\Big\} = 1, \quad Pr\Big\{L^{(\beta)}\,|_Q \in B^{\beta/2,0}_{M_2,\infty}(Q)\Big\} = 0.$$

On the other hand, for $p = \infty$ we have

$$Pr\Big\{0 < \underline{\lim}_{\mu\to\infty} \frac{2^{\mu\beta/2}}{\sqrt{\mu}} \Lambda^{(2,i)}_{\mu,\infty}(L^{(\beta)}\mid_Q) \le \overline{\lim}_{\mu\to\infty} \frac{2^{\mu\beta/2}}{\sqrt{\mu}} \Lambda^{(2,i)}_{\mu,\infty}(L^{(\beta)}\mid_Q) < \infty\Big\} = 1,$$

which gives

$$Pr\Big\{\omega_{1,\infty}(L^{(\beta)}\mid_Q, t) = O\big(\sqrt{t^\beta \log(1/t)}\big),\ t \to 0\Big\} = 1,$$

$$Pr\Big\{\omega_{1,\infty}(L^{(\beta)}\mid_Q, t) = o\big(\sqrt{t^\beta \log(1/t)}\big),\ t \to 0\Big\} = 0.$$

Example 3.2 – the fractional anisotropic Wiener field.

The fractional anisotropic Wiener field with parameter $\underline{\beta} = (\beta_1, \dots, \beta_d)$, $0 < \beta_i < 2$, is a gaussian random field $\{W^{(\underline{\beta})}(\underline{t}) : \underline{t} \in \mathbb{R}^d\}$, with continuous realizations, $Pr\{W^{(\underline{\beta})}(\underline{0}) = 0\} = 1$, $EW^{(\underline{\beta})}(\underline{t}) = 0$ for any $\underline{t} \in \mathbb{R}^d$ and the covariance kernel being the tensor product of the covariance kernels of the fractional brownian motion with the parameter β_i, i.e.

$$EW^{(\underline{\beta})}(\underline{t})W^{(\underline{\beta})}(\underline{s}) = \prod_{i=1}^{d} K_{\beta_i}(t_i, s_i),$$

where $K_\beta(t,s) = \frac{1}{2}(|t|^\beta + |s|^\beta - |t-s|^\beta)$. For the realizations of $W^{(\underline{\beta})}$, the anisotropic regularity is studied. Note that for $\underline{\mu} \in \mathbb{N}^d(\nu, A)$ and $\underline{x} \in \mathcal{T}_{\underline{\mu}}(2, A)$ the second order mixed differences $\Delta^2_{\underline{t}_\mu, A} W^{(\underline{\beta})}(\underline{x})$ are centered gaussian random variables, with variance $\sim \underline{t_\mu}^{\underline{\beta}(A)}$, and for $\underline{x}, \underline{y} \in \mathcal{T}_{\underline{\mu}}(2, A)$ the following estimate holds

$$|E\Delta^2_{\underline{t}_\mu, A} W^{(\underline{\beta})}(\underline{x}) \cdot \Delta^2_{\underline{t}_\mu, A} W^{(\underline{\beta})}(\underline{y})| \le C \underline{t_\mu}^{\underline{\beta}(A)} \prod_{i\in A} \frac{1}{1 + (2^{\mu_i}|x_i - y_i|)^{4-\beta_i}}.$$

Using these estimates we prove that the terms $\Omega^{(m,A)}_{\underline{\mu},p}(W^{(\underline{\beta})}\mid_Q)$ for $p < \infty$ and $\Omega^{(m,A)}_{\underline{\mu},M_2}(W^{(\underline{\beta})}\mid_Q)$ fulfill a.s. the conditions of Theorems 2.2. and 2.4 with the parameter $\underline{\beta}/2$, so we get for p such that $0 < \frac{1}{p} < \frac{\beta_i}{2}$, $i = 1, \dots, d$

$$Pr\{W^{(\underline{\beta})}\mid_Q \in B^{\underline{\beta}/2}_{p,\infty}(Q)\} = 1, \quad Pr\{W^{(\underline{\beta})}\mid_Q \in B^{\underline{\beta}/2,0}_{p,\infty}(Q)\} = 0,$$

$$Pr\{W^{(\underline{\beta})}\mid_Q \in B^{\underline{\beta}/2}_{M_2,\infty}(Q)\} = 1, \quad Pr\{W^{(\underline{\beta})}\mid_Q \in B^{\underline{\beta}/2,0}_{M_2,\infty}(Q)\} = 0.$$

Acknowledgements. This work was supported by KBN grant 2 P301 019 06 and the author's participation in the conference was possible thanks to the financial support from the organizers.

References

1. Bonic R., Frampton J., and Tromba A., Λ- manifolds, J. Funct. Anal. **3**, (1969), 310–320.
2. Boufoussi B., Régularite du temps local brownien dans les espaces de Besov-Orlicz, Studia Math. **118**, (1996), 145–156.
3. Boufoussi B. and Kamont A., Temps local brownien et espaces de Besov anisotropiques, to appear in Stoch. and Stoch. Rep.
4. Boufoussi B. and Roynette B., Le temps local appartient p.s. à $B^{1/2}_{p,\infty}$, C. R. Acad. Sci. Paris **316**, série I, (1993), 843–848.
5. Ciesielski Z., On the isomorphism of the space H_α and m, Bull. Acad. Pol. Serie des Sc. Math., Ast. et Ph. **8** (4), (1960), 217–222.
6. Ciesielski Z., Approximation by splines and its application to Lipschitz classes and to stochastic processes, in *Teoria priblizenii funkcii, Proc. of Conference in Kaluga 1975*, Nauka, Moscow, 1977, 397–404.
7. Ciesielski Z., Properties of realizations of random fields, in *Mathematical Statistics nad Probability, Proc. of Sixth Int. Conf. Wisla 1978*, Springer, Lecture Notes in Statistics 2, 1980, 97–110.
8. Ciesielski Z. and Figiel T., Spline bases in classical function spaces on compact C^∞ manifolds, part II, Studia Math. **76**, (1983), 95–136.
9. Ciesielski Z. and Kamont A., Lévy's fractional Brownian random field and function spaces, Acta Sci. Math. (Szeged) **60**, (1995), 99–118.
10. Ciesielski Z. and Kamont A., On the fractional Lévy's field on the sphere S^d, East J. Approx. **1**, (1995), 111–123.
11. Ciesielski Z., Kerkyacharian G., and Roynette B., Quelques espaces fonctionnels associés à des processus gaussiens, Studia Math. **107**, (1993), 171–204.
12. Deaconu M. and Roynette B., Besov regularity for the solutions of Walsh equation, preprint Institut Elie Cartan 95/6.
13. Fernique X., Régularité de processus gaussiens, Invent. Math. **12** (1971), 304–320.
14. Johnson W. B., Schechtman G. and Zinn J., Best constants in moment inequalities for linear combinations of independent and exchangeable random variables, Ann. Probab. **13**, (1985), 234–253.
15. Kamont A., Isomorphism of some anisotropic Besov and sequence spaces, Studia Math. **110**, (1994), 169–189.

16. Kamont A., On the fractional anisotropic Wiener field, Probab. and Math. Stat. **16**, (1996), 85–98.

17. Kamont A., A discrete characterization of Besov spaces, to appear in Approx. Theory Appl.

18. Meyer Y., *Wavelets and operators*, Cambridge Univ. Press, 1992.

19. Roynette B., Mouvement brownien et espaces de Besov, Stoch. and Stoch. Rep. **43**, (1993), 221–260.

20. Ryll J., Schauder bases for the space of continuous functions on an n-dimensional cube, Comment. Math. **27**, (1973), 201–213.

Anna Kamont
Instytut Matematyczny PAN
ul. Abrahama 18
81 – 825 Sopot
Poland
A.Kamont@impan.gda.pl

One-Sided Approximation and Interpolation Operators Generating Hyperbolic Sigma-Pi Neural Networks

Burkhard Lenze

Abstract. In this paper, we show how to design three-layer feedforward neural networks with hyperbolic sigma-pi units in the hidden layer in order to act as one-sided approximation and interpolation devices for regular gridded data. We obtain the concrete networks in real-time using a one-shot learning scheme based on special approximation operators which are generated by sampling the given discrete information on a regular grid. In this context, it is essential that we do not require any smoothness conditions regarding the underlying data function f. At the end of the paper we briefly discuss an application of our strategy.

1. Introduction

As it is well-known there is a quite intimate connection between approximation and interpolation theory and neural network design. Without any claim of completeness we mention the papers [2,1,4,8,9]. In the following, we again use some techniques from approximation and interpolation theory in order to construct three-layer feedforward networks for constrained discrete data processing. The general idea is to sample the given discrete information on a regular grid and to use this information to initialize approximation, resp. interpolation, operators which approximate the data from below and from above, resp., interpolate the data. Moreover, the resulting operators can be easily implemented in the framework of a three-layer feedforward neural network with hyperbolic sigma-pi units in the hidden layer (cf. [4,7,9] for details concerning such networks). Summing up, we obtain three-layer feedforward sigma-pi neural networks of hyperbolic type for one-sided approximation and interpolation of regular gridded data. At the end of this introduction, let

Multivariate Approximation and Splines
G. Nürnberger, J. W. Schmidt, and G. Walz (eds.), pp. 99–112.

ISBN 3-7643-5654-5.

us add a few words about the organization of the paper. We start with some notational preliminaries and introduce the concept of sigmoidal functions for designing neural networks. Then, we define our two basic operators which yield one-sided approximations or interpolations with respect to given regular discrete information. Finally, we apply the operators in the context of a concrete example.

2. Notation and Results

We essentially use the same notation as in [7]. Let $n \in \mathbb{N}$ be given and $\mathbf{a} = (a_1, a_2, \ldots, a_n), \mathbf{b} = (b_1, b_2, \ldots, b_n) \in \mathbb{R}^n$, with $\mathbf{a} < \mathbf{b}$ (i.e., $a_k < b_k$, $1 \leq k \leq n$) the endpoints of the interval $[\mathbf{a}, \mathbf{b}] \subset \mathbb{R}^n$,

$$[\mathbf{a}, \mathbf{b}] := \{\mathbf{x} \in \mathbb{R}^n \mid a_k \leq x_k \leq b_k \ , \ 1 \leq k \leq n\} \ . \tag{2.1}$$

By means of a standard translation argument we may assume that the point $\mathbf{a} \in \mathbb{R}^n$ is always equal to the origin, i.e., without loss of generality we have $\mathbf{a} = \mathbf{0}$. We choose $J_1, J_2, \ldots, J_n \in \mathbb{N}$ and define $\mathbf{h} \in \mathbb{R}^n$ componentwise by

$$h_k := \frac{b_k}{J_k} \ , \ \ 1 \leq k \leq n \ . \tag{2.2}$$

We introduce on the interval $[\mathbf{0}, \mathbf{b}]$ the regular grid with grid points

$$\mathbf{h_j} := (h_1 j_1, h_2 j_2, \ldots, h_n j_n) \in [\mathbf{0}, \mathbf{b}] \ , \ \ \mathbf{0} \leq \mathbf{j} \leq \mathbf{J} \ , \tag{2.3}$$

where $\mathbf{j} := (j_1, j_2, \ldots, j_n)$ and $\mathbf{J} := (J_1, J_2, \ldots, J_n)$. We set $\mathbf{h} = \mathbf{h_e}$ with $\mathbf{e} := (1, 1, \ldots, 1)$.

At each grid point $\mathbf{h_j}$, $\mathbf{0} \leq \mathbf{j} \leq \mathbf{J}$, there may be given a real value $f(\mathbf{h_j}) \in \mathbb{R}$ which we may assume to come from some underlying function $f : [\mathbf{0}, \mathbf{b}] \to \mathbb{R}$. Our problem is to design a neural network which is able to model the given discrete information $(\mathbf{h_j}, f(\mathbf{h_j}))$, $\mathbf{0} \leq \mathbf{j} \leq \mathbf{J}$, under the constraint of one-sidedness. To solve this problem we proceed in two steps: In the first step, we define the so-called hyperbolic cardinal translation-type operators which comprise the data set from below and from above and can be interpreted as feedforward neural networks. In the second step, we show that these operators give rise to proper approximation and/or interpolation operators depending on a suitable choice of their free parameters.

We first of all need some further definitions which in part may already be found in [4]. For arbitrary $\mathbf{c}, \mathbf{d} \in \mathbb{R}^n$, with $\mathbf{c} < \mathbf{d}$, let

$$Cor[\mathbf{c}, \mathbf{d}] := \{\mathbf{x} \in \mathbb{R}^n \mid x_k = c_k \ \text{ or } \ x_k = d_k \ , \ 1 \leq k \leq n\} \tag{2.4}$$

be the set of corners of the interval $[\mathbf{c}, \mathbf{d}]$. Moreover, let

$$\gamma(\mathbf{x}, \mathbf{c}) := \#\{k \in \{1, 2, \ldots, n\} \mid x_k = c_k\} \ , \ \ \mathbf{x} \in Cor[\mathbf{c}, \mathbf{d}] \ . \tag{2.5}$$

In (2.5), # denotes the number of distinct elements of the set under consideration and $n - \gamma(\mathbf{x}, \mathbf{c})$ is nothing else but the well-known Hamming distance from $\mathbf{x}$ to $\mathbf{c}$. Now, for a given function $f : [\mathbf{c}, \mathbf{d}] \to \mathbb{R}$ the so-called corresponding interval function or iterated difference Δ_f of f on $[\mathbf{c}, \mathbf{d}]$ is defined by

$$\Delta_f[\mathbf{c}, \mathbf{d}] := \sum_{\mathbf{x} \in Cor[\mathbf{c}, \mathbf{d}]} (-1)^{\gamma(\mathbf{x}, \mathbf{c})} f(\mathbf{x}) \ . \tag{2.6}$$

Moreover, based on the linear functional Δ we introduce some kind of measure for the positive, resp. negative, local discrete variation of f, namely,

$$\Delta_{P_f}[\mathbf{c}, \mathbf{d}] := \begin{cases} \Delta_f[\mathbf{c}, \mathbf{d}] & \text{for} \quad \Delta_f[\mathbf{c}, \mathbf{d}] > 0, \\ 0 & \text{for} \quad \Delta_f[\mathbf{c}, \mathbf{d}] \leq 0, \end{cases} \tag{2.7}$$

$$\Delta_{N_f}[\mathbf{c}, \mathbf{d}] := \begin{cases} -\Delta_f[\mathbf{c}, \mathbf{d}] & \text{for} \quad \Delta_f[\mathbf{c}, \mathbf{d}] < 0, \\ 0 & \text{for} \quad \Delta_f[\mathbf{c}, \mathbf{d}] \geq 0. \end{cases} \tag{2.8}$$

Note that with the above definitions we have the identities

$$\Delta_f[\mathbf{c}, \mathbf{d}] = \Delta_{P_f}[\mathbf{c}, \mathbf{d}] - \Delta_{N_f}[\mathbf{c}, \mathbf{d}] \tag{2.9}$$

and

$$|\Delta_f[\mathbf{c}, \mathbf{d}]| = \Delta_{P_f}[\mathbf{c}, \mathbf{d}] + \Delta_{N_f}[\mathbf{c}, \mathbf{d}] \ . \tag{2.10}$$

In the last of these basic definitions we introduce the notion of a sigmoidal function.

Definition 2.1. *A bounded measurable function $\sigma : \mathbb{R} \to \mathbb{R}$ is called a sigmoidal function, if*

$$\lim_{\xi \to -\infty} \sigma(\xi) = 0 \quad \textit{and} \quad \lim_{\xi \to \infty} \sigma(\xi) = 1. \tag{2.11}$$

Examples of sigmoidal functions are the unit step function $\mathbf{1} : \mathbb{R} \to \{0, 1\}$,

$$\mathbf{1}(\xi) := \begin{cases} 0, & \xi < 0, \\ 1, & \xi \geq 0, \end{cases} \tag{2.12}$$

or scaled integrated B-splines as discussed in Section 3. Aside from these general sigmoidal functions we also need some special functions of this type to define our one-sided hyperbolic-type operators. We let σ_ℓ, resp. σ_u, denote sigmoidal functions satisfying

$$\sigma_\ell(\xi) \leq \mathbf{1}(\xi), \quad \text{resp.} \quad \mathbf{1}(\xi) \leq \sigma_u(\xi), \tag{2.13}$$

for all $\xi \in \mathbb{R}$. We are now prepared to introduce the hyperbolic one-sided approximation operators $\Omega_\ell^{(\mathbf{h},\beta)}$ and $\Omega_u^{(\mathbf{h},\beta)}$. For σ_ℓ and σ_u sigmoidal functions with property (2.13), $\beta > 0$ a free dilation parameter, $\mathbf{e} := (1, 1, \ldots, 1) \in \mathbb{Z}^n$, $\mathbf{h} \in \mathbb{R}^n$ given by (2.2), and $f(\mathbf{h_j}) \in \mathbb{R}$, $\mathbf{0} \leq \mathbf{j} \leq \mathbf{J}$, the given discrete data set induced by some underlying function $f : [\mathbf{0}, \mathbf{b}] \to \mathbb{R}$, the operators $\Omega_\ell^{(\mathbf{h},\beta)}$ and $\Omega_u^{(\mathbf{h},\beta)}$ are defined for all $\mathbf{x} \in \mathbb{R}^n$ as

$$\Omega_\ell^{(\mathbf{h},\beta)}(f)(\mathbf{x}) := \\ 2^{1-n} \sum_{\substack{\mathbf{j} \in \mathbb{Z}^n \\ -\mathbf{e} \leq \mathbf{j} \leq \mathbf{J}}} \sigma_\ell \left(\beta \prod_{k=1}^n \left(\frac{x_k}{h_k} - j_k - \frac{1}{2} \right) \right) \Delta_{P_f} \left[\mathbf{h_j}, \mathbf{h_{j+e}} \right] \\ - \sigma_u \left(\beta \prod_{k=1}^n \left(\frac{x_k}{h_k} - j_k - \frac{1}{2} \right) \right) \Delta_{N_f} \left[\mathbf{h_j}, \mathbf{h_{j+e}} \right] \tag{2.14}$$

and

$$\Omega_u^{(\mathbf{h},\beta)}(f)(\mathbf{x}) := \\ 2^{1-n} \sum_{\substack{\mathbf{j} \in \mathbb{Z}^n \\ -\mathbf{e} \leq \mathbf{j} \leq \mathbf{J}}} \sigma_u \left(\beta \prod_{k=1}^n \left(\frac{x_k}{h_k} - j_k - \frac{1}{2} \right) \right) \Delta_{P_f} \left[\mathbf{h_j}, \mathbf{h_{j+e}} \right] \\ - \sigma_\ell \left(\beta \prod_{k=1}^n \left(\frac{x_k}{h_k} - j_k - \frac{1}{2} \right) \right) \Delta_{N_f} \left[\mathbf{h_j}, \mathbf{h_{j+e}} \right] , \tag{2.15}$$

where we agree to set

$$f(\mathbf{h_j}) := 0 , \quad \text{for } \mathbf{h_j} \notin [\mathbf{0}, \mathbf{b}] , \tag{2.16}$$

in order to obtain the compact notation of the finite sums appearing in (2.14) and (2.15). Obviously, the operators are basically induced by the given sigmoidal functions σ_ℓ and σ_u evaluated at componentwise products of shifted and scaled arguments, so-called hyperbolic-type arguments, and the interval functions Δ_{P_f} and Δ_{N_f} which make use of all underlying discrete information. Moreover, let us note already here that the difference of the operators has a very compact representation which is always nonnegative, namely,

$$\Omega_u^{(\mathbf{h},\beta)}(f)(\mathbf{x}) - \Omega_\ell^{(\mathbf{h},\beta)}(f)(\mathbf{x}) \\ = 2^{1-n} \sum_{\substack{\mathbf{j} \in \mathbb{Z}^n \\ -\mathbf{e} \leq \mathbf{j} \leq \mathbf{J}}} (\sigma_u - \sigma_\ell) \left(\beta \prod_{k=1}^n \left(\frac{x_k}{h_k} - j_k - \frac{1}{2} \right) \right) \left| \Delta_f \left[\mathbf{h_j}, \mathbf{h_{j+e}} \right] \right| \\ \geq 0 \quad \text{for all } \mathbf{x} \in \mathbb{R}^n. \tag{2.17}$$

At the moment, it is not clear why we call these operators "one-sided approximation operators" (with respect to the sampled discrete information). This terminology will be justified by Theorem 2.1 which is based on the following lemma.

Lemma 2.1. *Using the notations and definitions given above we have for all grid points* $\mathbf{h_i} \in [\mathbf{0}, \mathbf{b}]$, $\mathbf{0} \le \mathbf{i} \le \mathbf{J}$,

$$f(\mathbf{h_i}) = 2^{1-n} \sum_{\substack{\mathbf{j} \in \mathbb{Z}^n \\ -\mathbf{e} \le \mathbf{j} \le \mathbf{J}}} \mathbf{1}\left(\prod_{k=1}^{n}\left(i_k - j_k - \frac{1}{2}\right)\right) \Delta_f\left[\mathbf{h_j}, \mathbf{h_{j+e}}\right] . \tag{2.18}$$

Proof: The proof may essentially be found in [3], Lemmas 3.1 and 3.2. □

With the above lemma we now can formulate our first basic result.

Theorem 2.1. *Let us assume that* f, σ_ℓ, σ_u *and* $\mathbf{h} = \mathbf{h_e}$ *are given as introduced above. Then for all grid points* $\mathbf{h_i} \in [\mathbf{0}, \mathbf{b}]$, $\mathbf{0} \le \mathbf{i} \le \mathbf{J}$, *and for all* $\beta > 0$ *we have*

$$\Omega_\ell^{(\mathbf{h},\beta)}(f)(\mathbf{h_i}) \le f(\mathbf{h_i}) \le \Omega_u^{(\mathbf{h},\beta)}(f)(\mathbf{h_i}) , \tag{2.19}$$

i.e., the operators yield one-sided approximations for the function f *on the regular grid* $\mathbf{h_i}$, $\mathbf{0} \le \mathbf{i} \le \mathbf{J}$.

Proof: Let $\mathbf{h_i} \in [\mathbf{0}, \mathbf{b}]$ and $\beta > 0$ be given arbitrarily. Moreover, we only consider the first inequality because the second one may be proved in a completely similar way. Using the one-sided properties of σ_ℓ and σ_u with respect to $\mathbf{1}$ and the nonnegativity of Δ_{P_f} and Δ_{N_f} we obtain by means of Lemma 2.1

$$\begin{aligned}
&\Omega_\ell^{(\mathbf{h},\beta)}(f)(\mathbf{h_i}) \\
&= 2^{1-n} \sum_{\substack{\mathbf{j} \in \mathbb{Z}^n \\ -\mathbf{e} \le \mathbf{j} \le \mathbf{J}}} \sigma_\ell\left(\beta \prod_{k=1}^{n}\left(\frac{i_k h_k}{h_k} - j_k - \frac{1}{2}\right)\right) \Delta_{P_f}\left[\mathbf{h_j}, \mathbf{h_{j+e}}\right] \\
&\qquad - \sigma_u\left(\beta \prod_{k=1}^{n}\left(\frac{i_k h_k}{h_k} - j_k - \frac{1}{2}\right)\right) \Delta_{N_f}\left[\mathbf{h_j}, \mathbf{h_{j+e}}\right] \\
&\le 2^{1-n} \sum_{\substack{\mathbf{j} \in \mathbb{Z}^n \\ -\mathbf{e} \le \mathbf{j} \le \mathbf{J}}} \mathbf{1}\left(\beta \prod_{k=1}^{n}\left(i_k - j_k - \frac{1}{2}\right)\right) \Delta_{P_f}\left[\mathbf{h_j}, \mathbf{h_{j+e}}\right] \qquad (2.20) \\
&\qquad - \mathbf{1}\left(\beta \prod_{k=1}^{n}\left(i_k - j_k - \frac{1}{2}\right)\right) \Delta_{N_f}\left[\mathbf{h_j}, \mathbf{h_{j+e}}\right] \\
&= 2^{1-n} \sum_{\substack{\mathbf{j} \in \mathbb{Z}^n \\ -\mathbf{e} \le \mathbf{j} \le \mathbf{J}}} \mathbf{1}\left(\beta \prod_{k=1}^{n}\left(i_k - j_k - \frac{1}{2}\right)\right) \Delta_f\left[\mathbf{h_j}, \mathbf{h_{j+e}}\right] \\
&= f(\mathbf{h_i}) .
\end{aligned}$$

□

One of the immediate consequences of the above one-sided approximation theorem is the fact that these operators can – in general – not be linear. It is an easy exercise to show that one-sidedness on a discrete set and linearity necessarily result in interpolation on the given set. However, our operators are of interpolation type only in special situations, but not in general (cf. Theorem 2.4). For those interested in linear operators we note that for special sigmoidal functions the arithmetic mean of Ω_ℓ and Ω_u results in linear approximation and interpolation operators without the one-sidedness property which have been intensively studied in [5,6,7]. In the following, however, we are interested in the properties of the one-sided approximation operators Ω_ℓ and Ω_u. We start with a simple theorem which fixes their main mapping properties.

Theorem 2.2. *Let us assume that f, σ_ℓ, σ_u, and $\mathbf{h} = \mathbf{h_e}$ are given as introduced above and $\beta > 0$ is given arbitrarily.*
(a) The operators $\Omega_\ell^{(\mathbf{h},\beta)}$ and $\Omega_u^{(\mathbf{h},\beta)}$ are ordered in the usual sense, i.e., we have $\Omega_\ell^{(\mathbf{h},\beta)}(f) \leq \Omega_u^{(\mathbf{h},\beta)}(f)$.
(b) The operators $\Omega_\ell^{(\mathbf{h},\beta)}$ and $\Omega_u^{(\mathbf{h},\beta)}$ are pairwise antisymmetric, i.e., we have $\Omega_\ell^{(\mathbf{h},\beta)}(f) = -\Omega_u^{(\mathbf{h},\beta)}(-f)$.
(c) The operators $\Omega_\ell^{(\mathbf{h},\beta)}$ and $\Omega_u^{(\mathbf{h},\beta)}$ are absolutely homogeneous, i.e., for all $M \geq 0$ we have $\Omega_\ell^{(\mathbf{h},\beta)}(Mf) = M\Omega_\ell^{(\mathbf{h},\beta)}(f)$ and $\Omega_u^{(\mathbf{h},\beta)}(Mf) = M\Omega_u^{(\mathbf{h},\beta)}(f)$.

Proof: The easy proof is a direct consequence of the defintion of the operators and is left to the reader. □

After these basic results we are now interested in the approximation properties of the operators. Here, we obtain the following completely satisfying result.

Theorem 2.3. *Let us assume that f, σ_ℓ, σ_u, and $\mathbf{h} = \mathbf{h_e}$ are given as introduced above. Then for all $\mathbf{x} \in [\mathbf{0}, \mathbf{b}]$ with*

$$\prod_{k=1}^{n} \left(\frac{x_k}{h_k} - j_k - \frac{1}{2} \right) \neq 0\,, \quad \mathbf{j} \in \mathbb{Z}^n, \tag{2.21}$$

we have

$$\lim_{\beta\to\infty} \left(\Omega_u^{(\mathbf{h},\beta)}(f)(\mathbf{x}) - \Omega_\ell^{(\mathbf{h},\beta)}(f)(\mathbf{x}) \right) = 0\,. \tag{2.22}$$

On the other hand, for all $\mathbf{x} \in [\mathbf{0}, \mathbf{b}]$ with

$$\prod_{k=1}^{n} \left(\frac{x_k}{h_k} - j_k - \frac{1}{2} \right) = 0\,, \tag{2.23}$$

for some $\mathbf{j} \in \mathbb{Z}^n$ we have

$$\lim_{\beta\to\infty} \left(\Omega_u^{(\mathbf{h},\beta)}(f)(\mathbf{x}) - \Omega_\ell^{(\mathbf{h},\beta)}(f)(\mathbf{x}) \right) = \tag{2.24}$$

$$= 2^{1-n}(\sigma_u - \sigma_\ell)(0) \sum_{\substack{\mathbf{j}\in\mathbb{Z}^n \\ \prod_{k=1}^n\left(\frac{x_k}{h_k}-j_k-\frac{1}{2}\right)=0}} \left|\Delta_f\left[\mathbf{h_j}, \mathbf{h_{j+e}}\right]\right| .$$

Especially, for all $\epsilon > 0$ there exists a compact subset $K \subset [\mathbf{0}, \mathbf{b}]$ with the property that the Lebesgue measure of $[\mathbf{0}, \mathbf{b}] \setminus K$ is less than ϵ such that we have

$$\lim_{\beta\to\infty}\left(\Omega_u^{(\mathbf{h},\beta)}(f) - \Omega_\ell^{(\mathbf{h},\beta)}(f)\right) = 0 \quad \text{uniformly on } K . \tag{2.25}$$

Proof: Let $\mathbf{x} \in [\mathbf{0}, \mathbf{b}]$ be given with

$$\prod_{k=1}^n \left(\frac{x_k}{h_k} - j_k - \frac{1}{2}\right) \neq 0 , \quad \mathbf{j} \in \mathbb{Z}^n. \tag{2.26}$$

Since by means of (2.11) the sigmoidal functions satisfy

$$\lim_{\beta\to\infty} (\sigma_u - \sigma_\ell)(\beta\xi) = 0 \quad \text{for } \xi \in \mathbb{R} \setminus \{0\} , \tag{2.27}$$

representation (2.17) implies

$$\begin{aligned} &\lim_{\beta\to\infty}\left(\Omega_u^{(\mathbf{h},\beta)}(f)(\mathbf{x}) - \Omega_\ell^{(\mathbf{h},\beta)}(f)(\mathbf{x})\right) \\ &= 2^{1-n} \sum_{\substack{\mathbf{j}\in\mathbb{Z}^n \\ -\mathbf{e}\leq\mathbf{j}\leq\mathbf{J}}} \lim_{\beta\to\infty} (\sigma_u - \sigma_\ell)\left(\beta \prod_{k=1}^n \left(\frac{x_k}{h_k} - j_k - \frac{1}{2}\right)\right) \left|\Delta_f\left[\mathbf{h_j}, \mathbf{h_{j+e}}\right]\right| \\ &= 0 . \end{aligned} \tag{2.28}$$

Now, let $\mathbf{x} \in [\mathbf{0}, \mathbf{b}]$ be a vector satisfying

$$\prod_{k=1}^n \left(\frac{x_k}{h_k} - j_k - \frac{1}{2}\right) = 0 \tag{2.29}$$

for some $\mathbf{j} \in \mathbb{Z}^n$. In this case, we conclude as follows

$$\begin{aligned} &\lim_{\beta\to\infty}\left(\Omega_u^{(\mathbf{h},\beta)}(f)(\mathbf{x}) - \Omega_\ell^{(\mathbf{h},\beta)}(f)(\mathbf{x})\right) \\ &= 2^{1-n} \sum_{\substack{\mathbf{j}\in\mathbb{Z}^n \\ \prod_{k=1}^n\left(\frac{x_k}{h_k}-j_k-\frac{1}{2}\right)=0}} \lim_{\beta\to\infty} (\sigma_u - \sigma_\ell)\left(\beta \prod_{k=1}^n \left(\frac{x_k}{h_k} - j_k - \frac{1}{2}\right)\right) \left|\Delta_f\left[\mathbf{h_j}, \mathbf{h_{j+e}}\right]\right| \\ &\quad + 2^{1-n} \sum_{\substack{\mathbf{j}\in\mathbb{Z}^n \\ \prod_{k=1}^n\left(\frac{x_k}{h_k}-j_k-\frac{1}{2}\right)\neq 0}} \lim_{\beta\to\infty} (\sigma_u - \sigma_\ell)\left(\beta \prod_{k=1}^n \left(\frac{x_k}{h_k} - j_k - \frac{1}{2}\right)\right) \left|\Delta_f\left[\mathbf{h_j}, \mathbf{h_{j+e}}\right]\right| \\ &= 2^{1-n}(\sigma_u - \sigma_\ell)(0) \sum_{\substack{\mathbf{j}\in\mathbb{Z}^n \\ \prod_{k=1}^n\left(\frac{x_k}{h_k}-j_k-\frac{1}{2}\right)=0}} \left|\Delta_f\left[\mathbf{h_j}, \mathbf{h_{j+e}}\right]\right| . \end{aligned} \tag{2.30}$$

The final statement of the theorem is a direct consequence of Egorov's Theorem (cf. [11], p. 57, for details) since by means of the first part of the proof we know that $\left(\Omega_u^{(\mathbf{h},\beta)}(f) - \Omega_\ell^{(\mathbf{h},\beta)}(f)\right)$ converges to zero for $\beta \to \infty$ almost everywhere on $[\mathbf{0}, \mathbf{b}]$. □

Remarks.
(1) The detailed pointwise result of Theorem 2.3 obviously implies closely related L_p-type results. We omit the details.
(2) With a view towards applications, the main consequence of Theorem 2.3 is to note that in case of approximate reconstruction of f on a finer grid one should choose the finer grid an odd number of times finer than the initial grid in order to avoid the more delicate asymptotic behaviour given in (2.24).
Now, we take a look at the interpolation properties of our operators. In detail, we obtain the following theorem.

Theorem 2.4. *Let us assume that f, σ_ℓ, σ_u and $\mathbf{h} = \mathbf{h_e}$ are given as introduced above. Moreover, let $M > 0$ be a real constant and let σ_ℓ and σ_u satisfy the identity*

$$\sigma_\ell(\xi) = \mathbf{1}(\xi) = \sigma_u(\xi) \tag{2.31}$$

for all $|\xi| \geq M$. Then for all grid points $\mathbf{h_i} \in [\mathbf{0}, \mathbf{b}]$, $\mathbf{0} \leq \mathbf{i} \leq \mathbf{J}$, and for all $\beta \geq 2^n M$ we have

$$\Omega_\ell^{(\mathbf{h},\beta)}(f)(\mathbf{h_i}) \;=\; f(\mathbf{h_i}) \;=\; \Omega_u^{(\mathbf{h},\beta)}(f)(\mathbf{h_i})\,, \tag{2.32}$$

i.e., the operators interpolate the function f on the regular grid $\mathbf{h_i}$, $\mathbf{0} \leq \mathbf{i} \leq \mathbf{J}$.

Proof: Let $\mathbf{h_i} \in [\mathbf{0}, \mathbf{b}]$ and $\beta \geq 2^n M$ be given arbitrarily. Moreover, we only consider the first equality because the second one may be proved in a completely similar way. Using supposition (2.31) on σ_ℓ and σ_u with respect to $\mathbf{1}$ and the fact that

$$\left|\prod_{k=1}^{n}\left(i_k - j_k - \frac{1}{2}\right)\right| \geq 2^{-n} \tag{2.33}$$

for all $\mathbf{i}, \mathbf{j} \in \mathbb{Z}^n$ we obtain

$$\begin{aligned}
&\Omega_\ell^{(\mathbf{h},\beta)}(f)(\mathbf{h_i}) \\
&\quad = 2^{1-n} \sum_{\substack{\mathbf{j}\in\mathbb{Z}^n \\ -\mathbf{e}\leq\mathbf{j}\leq\mathbf{J}}} \sigma_\ell\left(\beta\prod_{k=1}^{n}\left(\frac{i_k h_k}{h_k} - j_k - \frac{1}{2}\right)\right)\Delta_{P_f}\left[\mathbf{h_j}, \mathbf{h_{j+e}}\right] \\
&\qquad\qquad - \sigma_u\left(\beta\prod_{k=1}^{n}\left(\frac{i_k h_k}{h_k} - j_k - \frac{1}{2}\right)\right)\Delta_{N_f}\left[\mathbf{h_j}, \mathbf{h_{j+e}}\right]
\end{aligned}$$

$$= 2^{1-n} \sum_{\substack{\mathbf{j} \in \mathbb{Z}^n \\ -\mathbf{e} \le \mathbf{j} \le \mathbf{J}}} \sigma_\ell \left(\beta \prod_{k=1}^{n} \left(i_k - j_k - \frac{1}{2} \right) \right) \Delta_{P_f} \left[\mathbf{h_j}, \mathbf{h_{j+e}} \right] \tag{2.34}$$

$$- \sigma_u \left(\beta \prod_{k=1}^{n} \left(i_k - j_k - \frac{1}{2} \right) \right) \Delta_{N_f} \left[\mathbf{h_j}, \mathbf{h_{j+e}} \right]$$

$$= 2^{1-n} \sum_{\substack{\mathbf{j} \in \mathbb{Z}^n \\ -\mathbf{e} \le \mathbf{j} \le \mathbf{J}}} \mathbf{1} \left(\beta \prod_{k=1}^{n} \left(i_k - j_k - \frac{1}{2} \right) \right) \Delta_{P_f} \left[\mathbf{h_j}, \mathbf{h_{j+e}} \right]$$

$$- \mathbf{1} \left(\beta \prod_{k=1}^{n} \left(i_k - j_k - \frac{1}{2} \right) \right) \Delta_{N_f} \left[\mathbf{h_j}, \mathbf{h_{j+e}} \right]$$

$$= 2^{1-n} \sum_{\substack{\mathbf{j} \in \mathbb{Z}^n \\ -\mathbf{e} \le \mathbf{j} \le \mathbf{J}}} \mathbf{1} \left(\beta \prod_{k=1}^{n} \left(i_k - j_k - \frac{1}{2} \right) \right) \Delta_f \left[\mathbf{h_j}, \mathbf{h_{j+e}} \right] = f(\mathbf{h_i}) .$$

The last equality again follows from Lemma 2.1. □

3. Application

As already stated, we will now apply our one-sided network operators to a concrete problem. Therefore, we first of all have to answer the question of which sigmoidal functions we should use in order to generate our operators, resp., networks. Because of the essential conditions (2.11) and (2.13) a nice way to get sigmoidal functions of the requisite type is to start with the well-known B-splines (cf. [10], for example) and integrate them. We consider the first few cases. In detail, we define B_0, B_1 and B_2 as

$$B_0(\xi) := \begin{cases} \frac{1}{2}, & |\xi| \le 1 , \\ 0, & |\xi| > 1 , \end{cases} \tag{3.1}$$

$$B_1(\xi) := \begin{cases} 1 - |\xi|, & |\xi| \le 1 , \\ 0, & |\xi| > 1 , \end{cases} \tag{3.2}$$

$$B_2(\xi) := \begin{cases} (\xi + 1)^2, & -1 \le \xi \le -\frac{1}{3} , \\ -2\xi^2 + \frac{2}{3}, & -\frac{1}{3} \le \xi \le \frac{1}{3} , \\ (\xi - 1)^2, & \frac{1}{3} \le \xi \le 1 , \\ 0, & |\xi| > 1 . \end{cases} \tag{3.3}$$

Integrating – for example – B_1 leads to

$$\sigma(\xi) := \begin{cases} 0, & \xi \le -1 , \\ \frac{1}{2}\xi^2 + \xi + \frac{1}{2}, & -1 < \xi \le 0 , \\ -\frac{1}{2}\xi^2 + \xi + \frac{1}{2}, & 0 < \xi \le 1 , \\ 1, & \xi > 1 , \end{cases} \tag{3.4}$$

which is obviously a sigmoidal function. Moreover, this sigmoidal function is monotone increasing, continuously differentiable, and identical with the unit step function $\mathbf{1}$ outside the interval $[-1,1]$. Therefore, proper choices for our lower and upper sigmoidal functions are special translates of σ, namely,

$$\sigma_\ell(\xi) := \sigma(\xi - 1) \quad \text{and} \quad \sigma_u(\xi) := \sigma(\xi + 1)\,. \tag{3.5}$$

In the following, we will consider the special case $n = 2$ and use only the scaling factor $\beta = 4$ (non-interpolation case, smoothing effect). In detail, our operators are

$$\Omega_{\ell|u}^{(\mathbf{h},\beta)}(f)(\mathbf{x}) := \frac{1}{2} \sum_{\substack{\mathbf{j}\in\mathbb{Z}^2 \\ -\mathbf{e}\leq\mathbf{j}\leq\mathbf{J}}} \sigma_{\ell|u}\left(\beta \prod_{k=1}^{2}\left(\frac{x_k}{h_k} - j_k - \frac{1}{2}\right)\right) \Delta_{P_f}\left[\mathbf{h}_\mathbf{j}, \mathbf{h}_{\mathbf{j}+\mathbf{e}}\right] \tag{3.6}$$

$$- \sigma_{u|\ell}\left(\beta \prod_{k=1}^{2}\left(\frac{x_k}{h_k} - j_k - \frac{1}{2}\right)\right) \Delta_{N_f}\left[\mathbf{h}_\mathbf{j}, \mathbf{h}_{\mathbf{j}+\mathbf{e}}\right]$$

with $\mathbf{e} := (1,1) \in \mathbb{Z}^2$, $\beta = 4$, and $\mathbf{x} \in \mathbb{R}^2$.
As a test function for our operators we take a look at a special problem containing different types of edged surfaces. More precisely, using the three standard norms (with an abused notation for the maximum norm)

$$\|(x_1, x_2)\|_0 := \max\{|x_1|, |x_2|\}\,, \tag{3.7}$$

$$\|(x_1, x_2)\|_1 := |x_1| + |x_2|\,, \tag{3.8}$$

$$\|(x_1, x_2)\|_2 := \sqrt{x_1^2 + x_2^2}\,, \tag{3.9}$$

we introduce the test function $f : [0,1]^2 \to \mathbb{R}$ defined as

$$f(x_1, x_2) := \sum_{r=0}^{2}\sum_{s=0}^{2} B_r(7\,\|(x_1, x_2) - (0.2 + 0.3s, 0.2 + 0.3r)\|_s)\,. \tag{3.10}$$

We set $[\mathbf{0}, \mathbf{b}] := [0,1] \times [0,1]$, $\beta = 4$, $\mathbf{J} := (60, 60)$, and $\mathbf{h} := (\frac{1}{60}, \frac{1}{60}) = \frac{1}{60}\mathbf{e}$, and obtain – for example – for the difference operator

$$\left(\Omega_u^{(\frac{1}{60}\mathbf{e},4)} - \Omega_\ell^{(\frac{1}{60}\mathbf{e},4)}\right)(f)(\mathbf{x}) \tag{3.11}$$

$$= \frac{1}{2} \sum_{\substack{-1\leq j_1\leq 60 \\ -1\leq j_2\leq 60}} (\sigma_u - \sigma_\ell)\left(4 \prod_{k=1}^{2}\left(\frac{x_k}{\frac{1}{60}} - j_k - \frac{1}{2}\right)\right) \left|\Delta_f\left[\frac{1}{60}\mathbf{j}, \frac{1}{60}(\mathbf{j}+\mathbf{e})\right]\right|.$$

The plots of the test function f, its one-sided approximations, and its error surface according to the parameters fixed above are shown in Figures 1–4, where we have used the parametrization q ,

$$q : [0,60]^2 \to [0,1]^2\,,\quad (u,v) \mapsto \left(\frac{u}{60}, \frac{v}{60}\right). \tag{3.12}$$

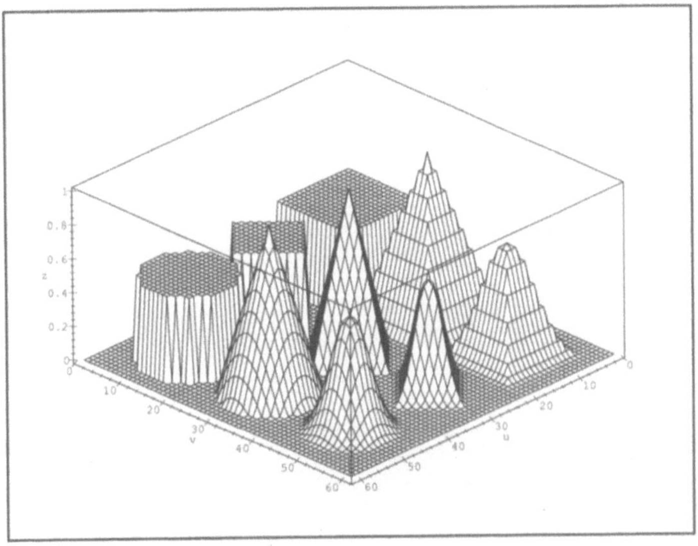

plot of $f(q(u,v))$

Figure 1

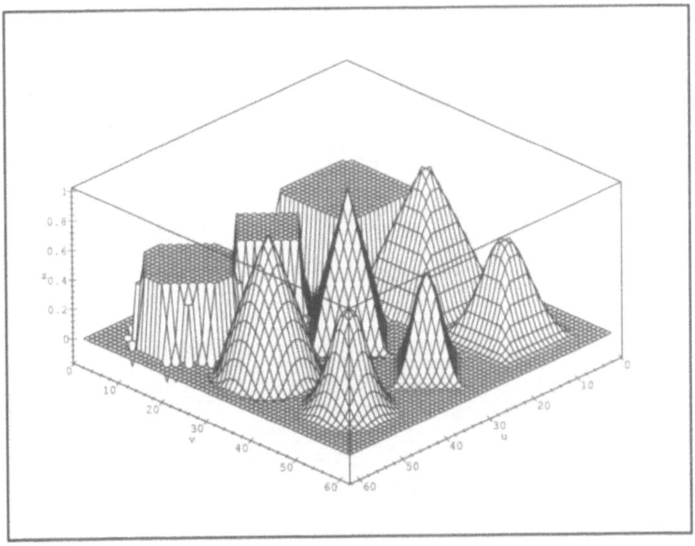

plot of $\Omega_\ell^{(\frac{1}{60}\mathbf{e},4)}(f)(q(u,v))$

Figure 2

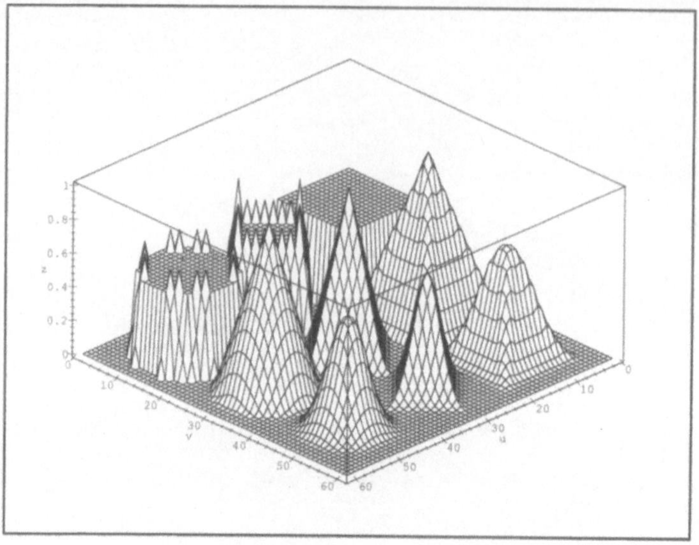

plot of $\Omega_u^{(\frac{1}{60}\mathbf{e},4)}(f)(q(u,v))$

Figure 3

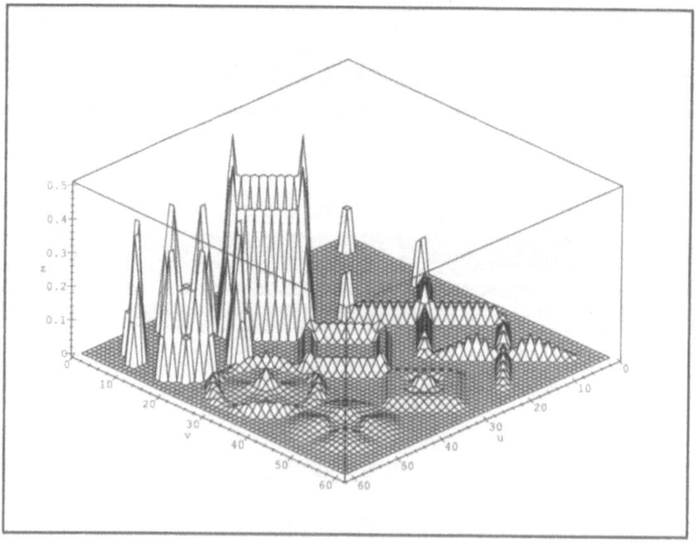

plot of $(\Omega_u^{(\frac{1}{60}\mathbf{e},4)} - \Omega_\ell^{(\frac{1}{60}\mathbf{e},4)})(f)(q(u,v))$

Figure 4

Figures 1–3 show that there are almost no differences between the one-sided approximations and the original function f in regions where f is smooth. Near sharp edges of f , however, the operators act as smoothing operators which necessarily imply an increase of local errors and induce Gibbs-like phenomena. A more detailed look at the error function plotted in Figure 4 in fact shows that the most significant differences occur at those parts of f with maximal lack of smoothness. In general, a proper strategy of generating one-sided approximations of the discussed type should choose the smoothness of σ_ℓ and σ_u according to the smoothness of the surface or image which is to be modelled.

4. Conclusions

In this paper, we presented a real-time scheme for generating one-sided approximation operators for discrete regular gridded data. The operators themselves could be implemented as three-layer feedforward sigma-pi neural networks with manageable complexity (for details connecting these type of operators and neural networks comp. [7]). Finally, an example showed that the operators work best on those parts of the given discrete information which approximately possess the same smoothness as the constrained modelling operators.

References

1. Chui C. K. and Li X., Approximation by ridge functions and neural networks with one hidden layer, J. Appr. Theory **70** (1992), 131–141.
2. Hornik K., Stinchcombe M., and White H., Multilayer feedforward networks are universal approximators, Neural Networks **2** (1989), 359–366.
3. Lenze B., On multidimensional Lebesgue-Stieltjes convolution operators, in: *Multivariate Approximation Theory IV*, C.K. Chui, W. Schempp, and K. Zeller (eds.), ISNM 90, Birkhäuser Verlag, Basel, 1989, 225–232.
4. Lenze B., Constructive multivariate approximation with sigmoidal functions and applications to neural networks, in: *Numerical Methods of Approximation Theory*, D. Braess and L.L. Schumaker (eds.), ISNM 105, Birkhäuser Verlag, Basel, 1992, 155–175.
5. Lenze B., Quantitative approximation results for sigma-pi-type neural network operators, in: *Multivariate Approximations: From CAGD to Wavelets*, K. Jetter and F. Utreras (eds.), World Scientific, Singapore, 1993, 193–209.
6. Lenze B., Local behaviour of neural network operators –Approximation and Interpolation–, Analysis **13** (1993), 377–387.
7. Lenze B., How to make sigma-pi neural networks perform perfectly on regular training sets, Neural Networks **7** (1994), 1285–1293.

8. Mhaskar H. N. and Micchelli C. A., Degree of approximation by neural and translation networks with a single hidden layer, Adv. in Appl. Math. **16** (1995), 151–183.
9. Pinkus A., TDI-subspaces of $C(\mathbb{R}^d)$ and some density problems from neural networks, J. Appr. Theory **85** (1996), 269–287.
10. Schumaker L. L., *Spline Functions: Basic Theory*, John Wiley & Sons, New York, 1981.
11. Wheeden R. L. and Zygmund A., *Measure and Integral*, Marcel Dekker Inc., New York, 1977.

Burkhard Lenze
Fachbereich Informatik
Fachhochschule Dortmund
Postfach 105018
D-44047 Dortmund
Germany
lenze@fh-dortmund.de

Unconstrained Minimization of Quadratic Splines and Applications

Wu Li

Abstract. In this paper, we give a review on unconstrained minimization of quadratic splines and some related application problems. We also point out new research directions. In particular, we will discuss construction of merit functions from the augmented Lagrangian function and the regularized gap function, Newton directions for singular Hessians, finite termination of Newton methods for minimization of convex quadratic splines, implementation of **QPspline** method (a Newton method for strictly convex quadratic programs) by using hot-starts and matrix updating, and estimates of the distance from a point to the intersection of closed convex subsets.

1. Introduction

Recently, there is a growing interest in converting constrained optimization problems to unconstrained minimization problems. Examples include merit functions for complementarity problems and variational inequality problems [10,12,13,15,16, 23,32,40,43,44,49,50,51,52,53], unconstrained reformulations of quadratic programming problems [6,9,18,19,28,30,31,34,35,36], and unconstrained formulations of optimal truss topology design problems [2,3]. Such unconstrained reformulations usually provide new perspectives on the structure of the original optimization problems and lead to novel methods for solving the original problems. We are particularly interested in those reformulations that can be modeled as the unconstrained minimization of a quadratic spline f:

$$\min_{x\in\mathbb{R}^n} f(x). \tag{1}$$

A continuously differentiable function $f(x)$ on $\mathbb{R}^n$ is called a *quadratic spline* if there exist finitely many quadratic functions $f_1, f_2, \dots, f_m$ on $\mathbb{R}^n$ such that

$$f(x) \in \{f_1(x), f_2(x), \dots, f_m(x)\} \quad \text{for } x \in \mathbb{R}^n. \tag{2}$$

Multivariate Approximation and Splines
G. Nürnberger, J. W. Schmidt, and G. Walz (eds.), pp. 113–128.

ISBN 3-7643-5654-5.

Or equivalently, $f(x)$ is said to be a quadratic spline if f is a continuously differentiable function and there exist polyhedral subsets $W_1, \ldots, W_m$ of $\mathbb{R}^n$ such that $\mathbb{R}^n = \bigcup_{i=1}^m W_i$ and f is a quadratic function on each W_i (cf. [Lemma 1, 48]). That is, quadratic splines are differentiable piecewise quadratic functions.

In the next three sections, we will discuss three research topics related to unconstrained minimization of a quadratic spline: (i) constrained minimization problems that can be converted to the unconstrained minimization problem (1); (ii) numerical algorithms for solving (1) that take advantage of the structure of the original problem; (iii) error bounds that are useful for convergence analysis of algorithms for solving (1).

2. Unconstrained Reformulations

Consider a differentiable nonlinear minimization problem:

$$\min_{x \in X} f(x), \tag{3}$$

where X is a closed convex subset of $\mathbb{R}^n$ and $f(x)$ is a differentiable function on $\mathbb{R}^n$. The classical penalty function approach is to find a function $f_\alpha(x)$ depending on a penalty parameter α such that (3) can be converted to the unconstrained minimization of $f_\alpha(x)$. One major difficult issue in the penalty function theory is the determination of the penalty parameter α. Even in the simple case that $f(x)$ is a strictly convex quadratic function and $X = \{x \in \mathbb{R}^n : l \leq x \leq u\}$ is a "box" in $\mathbb{R}^n$, this is not easy [18,19]. Moreover, in general, $f_\alpha(x)$ is much more complicated than $f(x)$ and the characteristics of (3) such as convexity and quadratic nature would get lost in the conversion. Some penalty functions such as the ℓ_1 penalty function lead to simple yet nondifferentiable $f_\alpha(x)$ [6]. Our study on unconstrained reformulations of quadratic programming problems shows that it is possible to obtain unconstrained reformulations of certain quadratic programming problems (including quadratic programming problems with simple bound constraints and strictly convex quadratic programming problems) such that the new unconstrained minimization problems are differentiable and keep certain characteristics of the original problems (including convexity and quadratic nature). What is more important is that new philosophies emerged from our study on constructing merit functions for quadratic programming problems: merit functions for (3) can be constructed from the augmented Lagrangian for (3) and from the regularized gap function introduced originally for variational inequality problems.

2.1. Augmented Lagrangian

Consider the following quadratic programming problem:

$$\min_x \frac{1}{2} x^T Q x - d^T x \quad \text{subject to} \quad l \leq Ax \leq u, \tag{4}$$

where Q is an $n \times n$ symmetric matrix, $d \in \mathbb{R}^n$ is a vector of n components, A is an $m \times n$ matrix, $l, u \in \mathbb{R}^m$, and the superscript "T" denotes the transpose of a vector (or matrix). Here we assume that (4) has a solution and some components of l and u may be $-\infty$ and ∞, respectively.

The corresponding augmented Lagrangian function $L_\alpha(x, y)$ introduced independently by Hestenes [14] and Powell [45] for equality constraints and by Rockafellar [46,47] for inequality constraints can be written in the following unified form:

$$L_\alpha(x,y) := \frac{1}{2}x^TQx - d^Tx + \frac{\alpha}{2}\left\|\left(\frac{1}{\alpha}(Ax-u)+y\right)_+\right\|^2 + \frac{\alpha}{2}\left\|\left(\frac{1}{\alpha}(l-Ax)-y\right)_+\right\|^2 - \frac{\alpha}{2}\|y\|^2, \tag{5}$$

where y is the Lagrange multiplier corresponding to two-sided inequality constraints, α is a penalty parameter, $\|\cdot\|$ denotes the Euclidean norm, and $(z)_+$ is the vector whose i-th component is $\max\{0, z_i\}$. The idea of using one Lagrange multiplier for two-sided inequality constraints was first introduced by Bertsekas [5]. For quadratic programming problem (4), the Lagrange multiplier y satisfies the following equation:

$$Qx - d + A^Ty = 0. \tag{6}$$

By using (6) to eliminate either x or y in $L_\alpha(x, y)$, we can get primal or dual merit functions for (4).

2.2. Primal Merit Functions

When $m = n$ and $A = I$ is the $n \times n$ identity matrix, (4) reduces to a quadratic programming problem with simple bound constraints (or box constraints):

$$\min_x \frac{1}{2}x^TQx - d^Tx \quad \text{subject to} \quad l \le x \le u. \tag{7}$$

Moreover, (6) gives the following explicit relation between y and x:

$$y = d - Qx. \tag{8}$$

The Method of Multipliers solves a sequence of unconstrained minimization of $L_{\alpha_k}(x, y^k)$ with respect to x while updating the penalty parameter α_k and the multiplier y^k after each unconstrained minimization iteration [4]. For quadratic programming problem (7), instead of updating y^k as in the Method of Multipliers, we can consider the following unconstrained minimization of a quadratic spline:

$$\min_{x\in\mathbb{R}^n} L_\alpha(x, d - Qx). \tag{9}$$

It was proved [Theorems 10 and 12, 31] that if $0 < 2\alpha\|Q\| < 1$, then *a global (or local) solution of (9) is a global (or local) solution of (7) and vice versa.*

2.3. Dual Merit Functions

On the other hand, if Q is an $n \times n$ nonsingular matrix, then (6) gives an explicit representation of x in terms of y:

$$x = Q^{-1}(d - A^T y). \tag{10}$$

By using (10) to eliminate x in $L_\alpha(x, y)$, we obtain a function of y and can consider the following dual maximization problem:

$$\max_{y \in \mathbb{R}^m} \; L_\alpha \left[Q^{-1}(d - A^T y), y\right]. \tag{11}$$

This is the dual merit function first introduced by Li and Swetits. When Q is positive definite and $\alpha > \|AQ^{-1}A^T\|$, it was proved [35,36] that $L_\alpha \left[Q^{-1}(d - A^T y), y\right]$ *is a concave quadratic spline; moreover,* $\bar{x}$ *is a solution of (4) if and only if* $\bar{x} = Q^{-1}(d - A^T \bar{y})$, *where* $\bar{y}$ *is a solution of (11).* See Theorem 2.6 of [36] for details, where the result was stated for $-L_\alpha \left[Q^{-1}(d - A^T y), y\right]$.

This suggests that $L_\alpha \left[Q^{-1}(d - A^T y), y\right]$ should be considered as a dual merit function for (4) with a nonsingular Q.

2.4. Reduced Dual Merit Functions

Another related problem is the quadratic programming problem with the standard constraints:

$$\min_x \; \frac{1}{2} x^T Q x - d^T x \quad \text{subject to} \quad Ax = b, \; l \le x \le u. \tag{12}$$

Replacing A by $\begin{pmatrix} A \\ I \end{pmatrix}$, l by $\begin{pmatrix} b \\ l \end{pmatrix}$, and u by $\begin{pmatrix} b \\ u \end{pmatrix}$ in (4), we get (12). Theoretically, one can consider (12) as a special form of (4). However, the special form of constraints in (12) makes the problem simpler. For example, when $Q = I$ is the $n \times n$ identity matrix, we have

$$\frac{1}{2} x^T Q x - d^T x = \frac{1}{2} x^T x - d^T x = \frac{1}{2}\|x - d\|^2 - \frac{1}{2}\|d\|^2.$$

In this case, (12) is reduced to the following least distance problem:

$$\min_x \; \frac{1}{2}\|x - d\|^2 \quad \text{subject to} \quad Ax = b, \; l \le x \le u. \tag{13}$$

The least distance problem is also a special form of constrained interpolation problems in approximation theory (cf. [8,9] and references therein) and can be written as an unconstrained minimization of a convex quadratic spline $\Phi(y)$ whose gradient is $A(d - A^T y)_l^u - b$, where z_l^u is the vector whose i-th component is $\max\{l_i, \min\{u_i, z_i\}\}$ (the median of $\{l_i, z_i, u_i\}$), *i.e.*, z_l^u is the componentwise truncation of z by l from below and by u from above. Note that finding a global minimizer of $\Phi(y)$ is equivalent to solving the following piecewise affine equation:

$$A(d - A^T y)_l^u = b. \tag{14}$$

It was proved [34,28] that $\bar{x}$ *is a solution of (13) if and only if* $\bar{x} = (d - A^T \bar{y})_l^u$, *where* $\bar{y}$ *is a solution of (14).*

2.5. Regularized Gap Function

Our merit functions are derived from the augmented Lagrangian by variable elimination and are completely different from the classical penalty functions. They are also different from the merit functions for complementarity problems or variational inequality problems, which were based on NCP-functions and the regularized gap function [10,12,13,15,16, 23,40,44,49,50,51, 52,53]. Mangasarian and Solodov [43] constructed a merit function for nonlinear complementarity problems by using an implicit Lagrangian, which is closely related to the regularized gap function [16,44]. It turns out that the merit function $L_\alpha(x, d - Qx)$ for (7) can also be expressed in terms of the regularized gap function (cf. (17)).

Note that the stationary points for (3) are solutions of the following variational inequality problem:

$$(y - x)^T \nabla f(x) \geq 0 \quad \text{for} \quad y \in X, \tag{15}$$

where ∇f denotes the gradient of f. Consider the regularized gap function for the variational inequality (15) [15,16]:

$$G_\alpha(x) = \max_{y \in X} \left\{ (x - y)^T \nabla f(x) - \frac{1}{2\alpha} \|x - y\|^2 \right\}, \tag{16}$$

where $\alpha > 0$ is a regularization parameter. Then it was proved [32] that

$$f(x) - G_\alpha(x) = L_\alpha(x, d - Qx), \tag{17}$$

provided $X = \{x \in \mathbb{R}^n : l \leq x \leq u\}$, $f(x) = \frac{1}{2}x^T Qx - d^T x$, $m = n$, and $A = I$ is the $n \times n$ identity matrix. That is, when the nonlinear minimization problem (3) reduces to the quadratic programming problem (7), we can also use the regularized gap function to construct a merit function. Note that the implicit Lagrangian is equal to $\left(G_\alpha(x) - G_{\frac{1}{\alpha}}(x)\right)$ [16,44], which is different from $(f(x) - G_\alpha(x))$.

2.6. Future Research Problems

One major question is how much of the theory on unconstrained reformulation of quadratic programs can be applied to constrained minimization problems. In the case of quadratic programming problems, we still have many unanswered questions. Here are some open problems related to unconstrained reformulations of constrained minimization problems:

(i) If Q is nonsingular (but not positive definite), is there any relation between (4) and (11)?

(ii) Can we always reformulate (12) as a minimization problem with only equality constraints? Can we always eliminate simple bound constraints for constrained minimization problems?

(iii) For any quadratic programming problem, is it possible to construct a differentiable piecewise quadratic merit function whose local (or global) minimizers are local (or global) solutions of the original problem?

(iv) For constrained minimization problem (3), can we always use $f(x) - G_\alpha(x)$ as a merit function? Based on (17), what should be a dual form of G_α that corresponds to $L_\alpha(Q^{-1}(d - A^T y), y)$?

(v) How can we extend the theory on merit functions for quadratic programming problems to constrained minimization problems?

3. Numerical Methods

Unconstrained reformulations of quadratic programming problems given in the last section have been very useful in developing new algorithms for solving quadratic programming problems. In particular, there are Newton-type methods, matrix-splitting methods, row-action methods, conjugate-gradient methods, and other descent methods for solving convex quadratic programs (4) by using unconstrained reformulations (9), (11), or (14) [28,30,34,35,36]. Here we first illustrate why **QPspline** method, a Newton-type method for solving (9), is actually an active-set method that allows swapping of many working active indices in one iteration and terminates in finitely many iterations. For certain quadratic programming problems, **QPspline** method outperforms QPROG in IMSL library [17,45,22] (an active-set method that allows only adding or deleting one index in the working active set and requires the dual feasibility) (cf. [36]). Then we discuss issues related to numerical methods for solving (4): Newton directions for singular Hessians, finite termination with singular Hessians, large-scale problems, and hybrid methods.

3.1. Newton Methods as Active-Set Methods

To get a rough idea of how a Newton method (called **QPspline** method in [36]) for finding a maximizer of $L_\alpha\left[Q^{-1}(d - A^T y), y\right]$ relates to an active-set method for solving (4) that allows swapping of many working active indices in one iteration, we consider the one-sided version of (4):

$$\min_x \ \frac{1}{2} x^T Q x - d^T x \quad \text{subject to} \quad Ax \le u. \tag{18}$$

Then the corresponding Lagrangian reduces to the following form:

$$L_\alpha(x, y) = \frac{1}{2} x^T Q x - d^T x + \frac{\alpha}{2} \left\| \left(\frac{1}{\alpha}(Ax - u) + y \right)_+ \right\|^2 - \frac{\alpha}{2} \|y\|^2. \tag{19}$$

Let $x(y) := Q^{-1}(d - A^T y)$. Then (11) becomes

$$\max_{y \in \mathbb{R}^m} \ \frac{1}{2} x(y)^T Q x(y) - d^T x(y) + \frac{\alpha}{2} \left\| \left(\frac{1}{\alpha}(Ax(y) - u) + y \right)_+ \right\|^2 - \frac{\alpha}{2} \|y\|^2. \tag{20}$$

Let us consider a nonsmooth Newton method for solving (20). The optimality condition for (20) can be simplified to the following system of piecewise affine equations:

$$-\alpha\left(I-\frac{1}{\alpha}AQ^{-1}A^T\right)\left[y-\left(\frac{1}{\alpha}(Ax(y)-u)+y\right)_+\right]=0, \tag{21}$$

which is equivalent to

$$y-\left(\frac{1}{\alpha}(Ax(y)-u)+y\right)_+=0, \tag{22}$$

since $\alpha > \|AQ^{-1}A^T\|$ and $\left(I-\frac{1}{\alpha}AQ^{-1}A^T\right)$ is positive definite.

Note that if $\bar{y}$ is a solution of (22), then $x(\bar{y})$ is a solution of (18). The corresponding active indices are

$$\{i : \bar{y}_i > 0\} = \left\{i : \left(\frac{1}{\alpha}(Ax(\bar{y})-u)+\bar{y}\right)_i > 0\right\},$$

since $(Ax(\bar{y})-u)_i = 0$ if $\bar{y}_i > 0$ and $(Ax(\bar{y})-u)_i \le 0$ if $\bar{y}_i = 0$. Therefore, for a given iterate y^k, it is natural to choose the working active index set as

$$J_k := \left\{i : \frac{1}{\alpha}(Ax(y^k)-u)_i + y_i^k > 0\right\}. \tag{23}$$

However, a nonsmooth Newton direction for (22) is a solution z^k of the following system of linear equations:

$$\begin{aligned} z_i &= -\left(y^k-\left(\frac{1}{\alpha}(Ax(y^k)-u)+y^k\right)_+\right)_i \quad \text{for } i \notin J_k, \\ A_iQ^{-1}A^Tz &= \left(y^k-\left(\frac{1}{\alpha}(Ax(y^k)-u)+y^k\right)_+\right)_i \quad \text{for } i \in J_k, \end{aligned} \tag{24}$$

where A_i denotes the i-th row of A. For the moment, we assume that (24) is consistent (which is true if $\{A_i : i \in J_k\}$ are linearly independent). Then (y^k+z^k) is actually a solution of the following strictly convex quadratic program with equality constraints:

$$\min_x \frac{1}{2}x^TQx - d^Tx \quad \text{subject to} \quad A_ix = u_i \quad \text{for } i \in J_k.$$

Therefore, if $\{A_i : i \in J_k\}$ are linearly independent and J_k is an active index set, then (y^k+z^k) solves (20) and $x(y^k+z^k)$ is the solution of (18). In general, if J_k is not an active index set, then we apply a line search to $L_\alpha(x(y^k+tz^k), y^k+tz^k)$

(a concave quadratic function of one variable t) for a stepsize t_k and get the next iterate $y^{k+1} = y^k + t_k z^k$. Due to concavity of $L_\alpha(x(y), y)$, we can use a Newton method with exact line search to generate a sequence of iterates y^k such that J_k is an active index set when k is large enough. Thus, **QPspline** method finds a solution of (18) in finitely many iterations. See [36] for details. The reformulation (20) provides an intelligent way of swapping many working active indices in each iteration. Note that the exact line search finds a solution of the following monotone piecewise affine equation of one variable t:

$$\left(z^k - \frac{1}{\alpha}AQ^{-1}A^T z^k\right)^T \left(y^k + tz^k - \left(\frac{1}{\alpha}(Ax(y^k + tz^k) - u) + y^k + tz^k\right)_+\right) = 0,$$

which is very easy to solve (cf. [34,35]).

3.2. Newton Directions for Singular Hessians

Two subtle issues involved are the following: (i) inconsistency of (24) and (ii) multiple solutions of (24). However, by clever algebraic manipulations, we can resolve these two problems without much difficulty [36]. In terms of active-set methods, we do not require any feasibility of y^k and/or $x(y^k)$, since our method is a global Newton method for solving (20) (the maximization of a concave quadratic spline). Also, theoretically, the finite termination is independent of degeneracy [36]. *However, for a general convex quadratic spline, how should we compute Newton directions if matrices in generalized Hessians are singular?*

3.3. Finite Termination with Singular Hessians

Note that the gradient of $L_\alpha(x(y), y)$ is a piecewise affine function, which is a semismooth function. The coefficient matrix of the system (24) corresponds to a member of the generalized Jacobian of the piecewise affine function $y - \left(\frac{1}{\alpha}(Ax(y) - u) + y\right)_+$ at y^k. Therefore, **QPspline** method does belong to the class of Newton methods for semismooth piecewise affine equations [48]. The reason for using (24) is that it relates **QPspline** method to an active-set method for solving (18) and it is essential for finite termination of **QPspline** method. This leads to the following question:

> *Given a convex quadratic spline $f(x)$ on $\mathbb{R}^n$ that has a minimizer, is it possible to design a Newton method that finds a minimizer of f in finitely many steps?*

This is a difficult question to answer since members in the generalized Hessians of f are only positive semidefinite. If one uses some regularization technique, then finite termination might not happen. It would be interesting to give a structural description of those convex quadratic splines that have some internal "active indices" (such as J_k) and then to modify **QPspline** method mentioned above in such a way that it could find minimizers of those special convex quadratic splines in finitely many iterations.

3.4. Matrix Updating for Newton Directions

From (24) we know that the Newton direction involves a factorization of the principle submatrix $A_{J_k}Q^{-1}A_{J_k}^T$ of $AQ^{-1}A^T$ corresponding to the index set J_k. In implementation of **QPspline** method, we must decide how to update a factorization of $A_{J_k}Q^{-1}A_{J_k}^T$ in the case that the difference between J_{k+1} and J_k is relatively small. This is very important in the case of a large number of active constraints at a solution. We do not want to factorize a large submatrix in each iteration when the algorithm tries to locate a few missing active constraints in the last few iterations. *An effective matrix updating technique should greatly improve the performance of* **QPspline** method.

3.5. Large-Scale Problems

However, for a large-scale problem, finding a Newton direction means solving a large system of linear equations. In such a case, matrix factorizations or inversions are not very practical. For example, if Q and A are large sparse matrices, then solving (24) directly might not be practical. Note that (24) is essentially equivalent to finding $\{z_i^k : i \in J_k\}$. The underlying linear system to be solved can be reformulated as follows:

$$A_{J_k}Q^{-1}A_{J_k}^T v = b^k, \tag{25}$$

where b^k is defined in such a way that a solution of (25) gives the J_k-components of a solution z^k of (24):

$$b^k = \left(y^k - \left(\frac{1}{\alpha}(Ax(y^k) - u) + y^k\right)_+\right)_{J_k} + A_{J_k}Q^{-1}A_{J_k^c}^T \left(y^k - \left(\frac{1}{\alpha}(Ax(y^k) - u) + y^k\right)_+\right)_{J_k^c}.$$

Here $J_k^c = \{i : i \notin J_k\}$ and A_J (or w_J) denotes the matrix (or vector) obtained by deleting the rows (or components) of A (or w) whose indices are not in J. We can find an approximate solution of the system by using an iterative method to solve the following expanded system:

$$Qw - A_{J_k}^T v = 0 \quad \text{and} \quad A_{J_k}w = b^k. \tag{26}$$

Note that (26) is an asymmetric and positive semidefinite linear system that keeps the sparsity of original matrices. Once we get a solution (v^k, w^k) of (26), the vector v^k is a solution of (25) and it will yield a solution z^k of (24). Moreover, if (25) has a unique solution (*i.e.*, A_{J_k} is row independent), then (26) is a nonsingular linear system and also has a unique solution. In this sense, (26) and (25) are completely equivalent. Therefore, the reformulation (20) does not alter the sparsity structure of the original problem in the sense that we can explore the sparsity pattern by working with (26). *One future research topic is to have an implementation of* **QPspline** method *by exploring the sparsity of Q and A in computing Newton directions.*

3.6. Hybrid Numerical Methods

Note that **QPspline** method can start at any given point y^k. This allows one to combine a descent method for (20) with **QPspline** method to get a hybrid method for solving the original quadratic problem (18). For example, we can use a special descent method (such as conjugate gradient methods proposed in [30]) to find a relatively accurate approximate solution y^k (as a hot-start) and then switch to **QPspline** method if the descent method does not make much "progress". Of course, there are many choices for such a hybrid method and their effectiveness has to be tested by extensive numerical experiments.

4. Error Bounds and Convergence Analysis

Estimates of the distance from an approximate solution to the solution set of an optimization problem are indispensable in convergence analysis of linearly constrained optimization problems. We are interested in error bounds for minimization of convex quadratic splines, which will be useful for convergence analysis of numerical algorithms for finding a minimizer of a convex quadratic spline. Here we discuss general global error bounds for constrained minimization problems and point out future research directions.

4.1. Convex Inequalities and Minimization Problems

Consider the following constrained minimization problem:

$$\min_{x \in K} f(x), \tag{27}$$

where K is a closed convex subset of $\mathbb{R}^n$ and f is a convex function on $\mathbb{R}^n$. Let $f_{\min}$ be the minimum objective function value of (27) and $S = \{x \in K : f(x) = f_{\min}\}$. Then for an approximate solution x of (27), one type of error bounds is to estimate $\operatorname{dist}(x, S)$ in terms of $(f(x) - f_{\min})_+$ and $\operatorname{dist}(x, K)$. Here $\operatorname{dist}(x, S)$ (or $\operatorname{dist}(x, K)$) denotes the distance from x to the set S (or K). In general, one might expect the following type of error bounds:

$$\begin{aligned} \operatorname{dist}(x, S) \le \lambda\Big(& (f(x) - f_{\min})_+ + \operatorname{dist}(x, K) \\ & + (f(x) - f_{\min})_+^r + \operatorname{dist}(x, K)^r \Big) \quad \text{for } x \in \mathbb{R}^n, \end{aligned} \tag{28}$$

where λ is some positive constant and $0 < r \le 1$. This is a general form for global error bounds. For $r = 1$, (28) becomes

$$\operatorname{dist}(x, S) \le \gamma\,((f(x) - f_{\min})_+ + \operatorname{dist}(x, K)) \quad \text{for } x \in \mathbb{R}^n, \tag{29}$$

where $\gamma = (\lambda + 1)$ is a positive constant. Note that (29) implies that (27) has weak sharp minima:

$$\operatorname{dist}(x, S) \le \gamma(f(x) - f_{\min})_+ = \lambda(f(x) - f_{\min}) \quad \text{for } x \in K. \tag{30}$$

This type of estimates is very important in convergence analysis of numerical algorithms for solving (27). For example, (30) implies the finite termination of proximal point methods for solving (27) [11].

We want to point out that error bounds for constrained convex inequalities can be considered as a special case of (28). In fact, consider the following constrained convex inequalities [27]:

$$g_i(x) \le 0 \quad \text{for } 1 \le i \le m \quad \text{and} \quad x \in K, \tag{31}$$

where g_i are convex functions on $\mathbb{R}^n$.

First note that (31) can be rewritten as (27) with

$$f(x) := \max\{g_1(x), \ldots, g_m(x), 0\}.$$

Here we add an artificial function $g_0(x) \equiv 0$ to get the equivalent formulation. In this case, $f_{\min} = 0$ and (28) reduces to the following global error bound for constrained convex inequalities:

$$\begin{aligned} \operatorname{dist}(x, C) \le \lambda \Bigg(\left(\max_{1 \le i \le m} g_i(x) \right)_+ + \operatorname{dist}(x, K) \\ + \left(\max_{1 \le i \le m} g_i(x) \right)_+^r + \operatorname{dist}(x, K)^r \Bigg) \quad \text{for } x \in \mathbb{R}^n, \end{aligned} \tag{32}$$

where C is the solution set of (31).

4.2. Error Bounds for Convex Inequalities

A special case of (31) is a system of convex inequalities:

$$g_i(x) \le 0 \quad \text{for } 1 \le i \le m. \tag{33}$$

Here $K = \mathbb{R}^n$ and $\operatorname{dist}(x, K) = 0$. In this case, global error bounds of type (29) become

$$\operatorname{dist}(x, C) \le \gamma \left(\max_{1 \le i \le m} g_i(x) \right)_+ \quad \text{for } x \in \mathbb{R}^n, \tag{34}$$

where C is the solution set of (33). Global error bound (34) holds under the Slater condition and asymptotic constraint qualifications [1,7,24,25,26,27,37,39, 41,42]. However, in many applications, (33) does not satisfy the Slater condition. Characterizations for global error bounds (34) are given in terms of directional derivatives in normal directions [27,26], but these conditions are not easy to verify. However, when g_i are either affine or convex quadratic functions on $\mathbb{R}^n$, Pang and Wang [54] showed that

$$\operatorname{dist}(x, C) \le \lambda \left(\left(\max_{1 \le i \le m} g_i(x) \right)_+ + \left(\max_{1 \le i \le m} g_i(x) \right)_+^r \right) \quad \text{for } x \in \mathbb{R}^n, \tag{35}$$

where $r = \frac{1}{2^{m+1}}$. Actually, they introduced a concept called degree of singularity for (33) that is an integer s between 1 and $(m+1)$ and proved that (35) holds for $r = \frac{1}{2^s}$. When (33) satisfies the weak Slater condition, Luo and Luo [38] established the following Hoffman-type error bound:

$$\text{dist}(x, C) \leq \gamma \left(\max_{1 \leq i \leq m} g_i(x) \right)_+ \quad \text{for } x \in \mathbb{R}^n. \tag{36}$$

Later, Li [33] proved that (36) holds if and only if convex quadratic inequalities (33) satisfy Abadie's constraint qualification. Note that (36) is Hoffman's error bound [20] for approximate solutions of linear inequalities when g_i are affine functions.

4.3. Convex Piecewise Quadratic Minimization Problems

In convergence analysis of numerical algorithms for minimization of convex piecewise quadratic functions, one usually has to use error bounds for convex quadratic inequalities or convex piecewise quadratic inequalities. In fact, when K is a polyhedral set and f is a convex quadratic function, (29) is equivalent to (30) [Theorem 6, 33]. This type of estimates is very important in convergence analysis of numerical algorithms for solving (27). For example, (30) implies the finite termination of proximal point methods for solving (27) [11]. In the case that X is a polyhedral set and $f(x)$ is a quadratic function with some very "weak" convexity condition, Luo and Pang [39] proved that (28) holds with $r = \frac{1}{2}$. When K is a polyhedral set and f is a convex function that is quadratic on each of finitely many polyhedral sets $W_1, \dots, W_m$ with $\mathbb{R}^n = \bigcup_{i=1}^m W_i$, then [Corollary 2.8, 29]

$$\text{dist}(x, S) \leq \lambda \left((f(x) - f_{\min}) + \sqrt{(f(x) - f_{\min})} \right) \quad \text{for } x \in K. \tag{37}$$

The estimate (37) is essential in proving the finite termination of **QPspline** method for solving (4) (with a positive definite Q) [36], in analysis of linear convergence of descent methods [28] and conjugate gradient methods [30] for unconstrained minimization of convex quadratic splines, and in establishing rate of convergence for proximal point algorithms [29].

One related problem is to estimate the distance from a point x to the intersection of finitely many closed convex subsets $C_1, \dots, C_m$ of $\mathbb{R}^n$ in the following form:

$$\text{dist}\left(x, \bigcap_{1 \leq i \leq m} C_i \right) \leq \lambda \cdot \max_{1 \leq i \leq m} \left(\text{dist}(x, C_i) + \text{dist}(x, C_i)^r \right) \quad \text{for } x \in \mathbb{R}^n. \tag{38}$$

Some local estimates of type (38) were given by Lewis and Pang in [27]. For $r = 1$, (38) was established by Hoffmann [21] under the assumption that $C_1, \dots, C_{m-1}$ are bounded and $\bigcap_{i=1}^m C_i$ is contained in the interior of $\bigcap_{i=1}^{m-1} C_i$. Note that by choosing $C_1 := \{x \in \mathbb{R}^n : f(x) - f_{\min} \leq 0\}$ and $C_2 := K$, we may derive error bounds of type (28) from (38).

References

[1] Auslender A. and Crouzeix J.-P., Global regularity theorems, Math. Oper. Res. **13** (1988), 243–253.

[2] Bendsøe M. P. and Ben-Tal A., A new method for optimal truss topology design, SIAM J. Optim. **3** (1993), 322–345.

[3] Bendsøe M. P., Ben-Tal A., and Zowe J., Optimization methods for truss geometry and topology design, Structural Optim. **7** (1994), 141–159.

[4] Bertsekas D. P., Approximation procedures based on the method of multipliers, J. Optim. Theory Appl. **23** (1977), 487–510.

[5] Bertsekas D. P., *Constrained Optimization and Lagrange Multiplier Methods*, Academic Press, New York, 1982.

[6] Coleman T. F. and Hulbert L. A., A globally and superlinearly convergent algorithm for convex quadratic programs with simple bounds, SIAM J. Optim. **3** (1993), 298–321.

[7] Deng S., Global error bounds for convex inequalities in Banach spaces, Preprint, Department of Mathematical Sciences, Northern Illinois University, De Kalb, IL, October 1995, revised July 1996.

[8] Deutsch F., Interpolation from a convex subset of Hilbert space: a survey of some recent results, in *Approximation Theory, Wavelets and Applications* (S. P. Singh, ed.), Kluwer Academic Publishers, Boston, 1995, pp. 95–105.

[9] Deutsch F., Li W., and Ward J., A dual approach to constrained interpolation from a convex subset of Hilbert space, May, 1996, to appear in J. Approx. Theory.

[10] Facchinei F. and Soares J., A new merit function for nonlinear complementarity problems and a related algorithm, Technical Report, Dipartimento di Informatica e Sistemistica, Università di Roma "La Sapienza", Rome, Italy, Dec. 1994.

[11] Ferris M. C., Finite termination of the proximal point algorithm, Math. Programming **50** (1991), 359–366.

[12] Fischer A., A special Newton-type optimization method, Optimization **24** (1992), 269–284.

[13] Fischer A., An NCP-function and its use for the solution of complementarity problems, to appear in *Recent Advances in Nonsmooth Optimization*, D.-Z. Du, L. Qi, and R. S. Womersley (eds.), World Scientific Publishers, Singapore, 1995.

[14] Hestenes M. R., Multiplier and gradient methods, J. Optim. Theory Appl. **4** (1969), 303–320.

[15] Fukushima M., Equivalent differentiable optimization problems and descent methods for asymmetric variational inequality problems, Math. Programming **53** (1992), 99–110.

[16] Fukushima M., Merit functions for variational inequality and complementarity problems, Technical Report, June, 1995, Graduate School of Information Science, Nara Institute of Science and Technology, 8916-5 Takayama, Ikoma, Nara 630-01, Japan.

[17] Goldfarb D. and Idnani A., A numerically stable dual method for solving strictly convex quadratic programs, Math. Programming **27** (1983), 1–33.

[18] Grippo L. and Lucidi S., A differentiable exact penalty function for bound constrained quadratic programming problems, Optimization **22** (1991), 557–578.

[19] Grippo L. and Lucidi S., On the solution of a class of quadratic programs using a differentiable exact penalty function, in *System Modelling and Optimization (Leipzig, 1989)*, H. J. Sebastian and K. Tammer eds. Lecture Notes in Control and Inform. Sci., No. 143, Springer, Berlin, 1990, pp. 764–773.

[20] Hoffman A. J., On approximate solutions of systems of linear inequalities, J. Res. Natl. Bur. Standards **49** (1952), 263–265.

[21] Hoffmann A., The distance to the intersection of two convex sets expressed by the distances to each of them, Math. Nachr., **157** (1992), 81–98.

[22] *IMSL 32-Bit Fortran Numerical Libraries*, Visual Numerics, Inc., Houston, 1994.

[23] Kanzow C., Nonlinear complementarity as unconstrained optimization, to appear in J. Optim. Theory Appl.

[24] Klatte D., Lipschitz stability and Hoffman's error bounds for convex inequality systems, Preprint, October 1995, revised June 1996, to appear in *Parametric Optimization and Related Topics IV*, the Proceedings of the Conference (Enschede 1995).

[25] Klatte D., Hoffman's error bound for systems of convex inequalities, Preprint, April 1996, revised July 1996, to appear in *Mathematical Programming with Data Perturbations*, A. V. Fiacco, ed., Marcel Dekker Publ.

[26] Klatte D. and Li W., Asymptotic constraint qualifications and global error bounds for convex inequalities, submitted to Mathematical Programming, October, 1996.

[27] Lewis A. S. and Pang J.-S., Error bounds for convex inequality systems, Preprint, Department of Mathematical Sciences, The Johns Hopkins University, Baltimore, Maryland 21218-2689, USA, June 1996.

[28] Li W., Linearly convergent descent methods for unconstrained minimization of a convex quadratic spline, J. Optim. Theory Appl. **86** (1995), 145–172.

[29] Li W., Error bounds for piecewise quadratic programs and applications, SIAM J. Control Optim. **33** (1995), 1510–1529.

[30] Li W., A conjugate gradient method for unconstrained minimization of strictly convex quadratic splines, Math. Programming **72** (1996), 17–32.

[31] Li W., Differentiable piecewise quadratic exact penalty functions for quadratic programs with simple bound constraints, SIAM J. Optim. **6** (1996), 299–315.

[32] Li W., A merit function and a Newton-type method for symmetric linear complementarity problems, to appear in *Proceeding of International Conference on Complementarity Problems* (held at the Johns Hopkins University, November, 1995), edited by M. Ferris and J.-S. Pang.

[33] Li W., Abadie's constraint qualification, metric regularity, and error bounds for differentiable convex inequalities, to appear in SIAM J. Optim.

[34] Li W., Pardalos P., and Han C.-G., Gauss-Seidel method for least distance problems, J. Optim. Theory Appl. **75** (1992), 487–500.

[35] Li W. and Swetits J., A Newton method for convex regression, data smoothing, and quadratic programming with bounded constraints, SIAM J. Optim. **3** (1993), 466–488.

[36] Li W. and Swetits J., A new algorithm for strictly convex quadratic programs, to appear in SIAM J. Optim.

[37] Li W. and Singer I., Global error bounds for convex multifunctions and applications, Preprint, Department of Mathematics and Statistics, Old Dominion University, Norfolk, VA 23529, USA, June 1996.

[38] Luo X.-D. and Luo Z.-Q., Extension of Hoffman's error bound to polynomial systems, SIAM J. Optim. **4** (1994), 383–392.

[39] Luo Z.-Q. and Pang J.-S., Error bounds for analytical systems and their applications, Math. Programming **67** (1994), 1–28.

[40] Luo Z.-Q. and Tseng P., A new class of merit functions for the nonlinear complementarity problem, Preprint, Department of Mathematics, University of Washington, Seattle, WA 98195, Oct. 1995.

[41] Mangasarian O. L., A condition number for differentiable convex inequalities, Math. Oper. Res. **10** (1985), 175–179.

[42] Mangasarian O. L., Error bounds for nondifferentiable convex inequalities under a strong Slater constraint qualification, Mathematical Programming Technical Report 96–04, Computer Science Department, University of Wisconsin–Madison, July 1996.

[43] Mangasarian O. L. and Solodov M. V., Nonlinear complementarity as unconstrained and constrained minimization, Math. Programming **62** (1993), 277–297.

[44] Peng J. M., Equivalence of variational inequality problems to unconstrained optimization, Technical Report, State Key Laboratory of Scientific and Engineering Computing, Academia Sinica, Beijing, China, December, 1994.

[45] Powell M. J. D., A method for nonlinear constraints in minimization problems, in *Optimization*, Edited by R. Fletcher, Academic Press, New York, 1969, pp. 283–298.

[46] Rockafellar R. T., New applications of duality in convex programming, in *Proceedings of the Fourth Conference on Probability Theory*, Brasov, Romania, 1971, pp. 73–81.

[47] Rockafellar R. T., The multiplier method of Hestenes and Powell applied to convex programming, J. Optim. Theory Appl. **12** (1973), 555–562.

[48] Sun J., On piecewise quadratic Newton and trust region problems, to appear in Math. Programming.

[49] Taji K. and Fukushima M., A new merit function and a successive quadratic programming algorithm for variational inequality problems, to appear in SIAM J. Optim.

[50] Tseng P., Yamashita N., and Fukushima M., Equivalence of complementarity problems to differentiable minimization: A unified approach, to appear in SIAM J. Optim.

[51] Yamashita N. and Fukushima M., On stationary points of the implicit Lagrangian for nonlinear complementarity problems, J. Optim. Theory Appl. **84** (1995), 653–663.

[52] Yamashita N. and Fukushima M., Equivalent unconstrained minimization and global error bounds for variational inequality problems, Graduate School of Information Science, Nara Institute of Science and Technology, 8916-5 Takayama, Ikoma, Nara 630-01, Japan.

[53] Yamashita N. and Fukushima M., A new NCP-function and its properties, Graduate School of Information Science, Nara Institute of Science and Technology, 8916-5 Takayama, Ikoma, Nara 630-01, Japan.

[54] Wang T. and Pang J.-S., Global error bounds for convex quadratic inequality systems, Optimization **31** (1994) 1–12.

Wu Li
Department of Mathematics and Statistics
Old Dominion University
Norfolk, Virginia 23529-0077
U.S.A.
`wuli@math.odu.edu`

Interpolation by Translates of a Basis Function

Will Light

Abstract. Interpolation using radial basis functions has become an established and useful technique in multi-dimensional data fitting. It is a special case of interpolation by translates of a basis function. All the methods used in practice have a strong connection with a variational principle. In this paper we provide a straightforward approach to the variational principle, and make some useful practical deductions from the theory.

1. Introduction

The theory of minimal norm interpolants is a very powerful one, which we intend to describe briefly in this introduction. This theory has its beginnings in a Hilbert space X. An essential assumption is that this Hilbert space should be embeddable in a suitable space of continuous functions. For our purposes, we will assume that X is embeddable in $C(\mathbb{R}^n)$. The precise meaning of this is that $X \subset C(\mathbb{R}^n)$ and that the natural linear mapping from X to $C(\mathbb{R}^n)$ induced by this containment is continuous. Here the topology on $C(\mathbb{R}^n)$ is that of convergence on compact sets. Thus a family of seminorms indexed by the compact sets K in $\mathbb{R}^n$ is defined by

$$\rho_K(f) = \sup_{x \in K} |f(x)|, \qquad f \in C(\mathbb{R}^n).$$

Why is this embedding property important? Well, it has the effect that every continuous, linear functional on $C(\mathbb{R}^n)$ is also a continuous linear functional on X. This is a simple fact: for $x \in \mathbb{R}^n$, define the mapping $\gamma : X \to \mathbb{C}$ by $\gamma(f) = f(x)$ for $f \in X$. This mapping is a composition of two continuous mappings. One of these is the map from $C(\mathbb{R}^n)$ to $\mathbb{C}$ defined by the same formula as γ, and the other is the embedding mapping from X into $C(\mathbb{R}^n)$.

Let $a_1, \ldots, a_m$ be a set of points in $\mathbb{R}^n$. Fix $f \in X$ and consider the set

$$G = \{v \in X : v(a_j) = f(a_j),\ j = 1, 2, \ldots, m\}.$$

Multivariate Approximation and Splines
G. Nürnberger, J. W. Schmidt, and G. Walz (eds.), pp. 129–140.

ISBN 3-7643-5654-5.

This set is a closed, convex subset of X. As such it has a unique point of minimal norm. We write this element as Uf, and refer to it as the *minimal norm interpolant to f on $a_1, \ldots, a_m$*. The theory of minimal norm interpolants is well-developed, and has an added richness when the Hilbert space is embedded in a space of continuous functions. The best reference for this sort of material is Golomb and Weinberger [4], but we would also refer to the survey in [9], since it is much more closely aligned with our present development. The element Uf is specified by the fact that it lies in the intersection of a finite number of hyperplanes in X. In general, the task of determining an element of minimal norm in such a large set could be difficult. However, the Riesz representation theorem allows us to find elements $q_1, \ldots, q_m \in X$ such that $f(a_j) = (f, q_j)$ for $j = 1, \ldots, m$. We call q_j the *representative* of a_j in X. The element Uf is completely characterised by the conditions that $Uf \in G$ and $Uf \perp G_0$, where G_0 is the set

$$G_0 = \{v \in X : v(a_j) = (v, q_j) = 0 \text{ for } j = 1, \ldots, m\}.$$

It is now easy to see that Uf has the form $Uf = \sum_{j=1}^{m} \lambda_j q_j$, where $\lambda_1, \ldots, \lambda_m \in \mathbb{C}$. Indeed, any element in $G_0^{\perp}$ has this general form. (Note that we are assuming X is a Hilbert space over the complex field.) The $\lambda_1, \ldots, \lambda_m$ are determined by the interpolation equations

$$f(a_s) = (Uf)(a_s) = \sum_{j=1}^{m} \lambda_j q_j(a_s), \qquad s = 1, \ldots, m.$$

This shows that the minimal norm interpolant is always in the finite-dimensional subspace spanned by $q_1, \ldots, q_m$, which makes its calculation rather less daunting. Error estimation is another very powerful aspect of the theory. Suppose $x \in \mathbb{R}^n$, and $q_0 \in X$ is such that $v(x) = (v, q_0)$ for all $v \in X$. Then because $f - Uf$ belongs to G_0, we can write

$$|f(x) - (Uf)(x)| = (f - Uf, q_0) \leq \|q_0|_{G_0}\| \, \|f - Uf\|.$$

Here,

$$\|q_0|_{G_0}\| = \sup\left\{ \frac{|(v, q_0)|}{\|v\|} : v \in G_0 \right\}.$$

It is perhaps hardly surprising that this theory is so clean. This is really a consequence of the assumption that X is a Hilbert space which is embeddable in the space $C(\mathbb{R}^n)$. However, if we wish to have a theory with practical applications then three essential problems must be surmounted:

(i) The Hilbert space X must be constructed in such a way that it is embeddable in $C(\mathbb{R}^n)$.
(ii) The representative of a point evaluation functional at any point x in $\mathbb{R}^n$ must be obtainable in an explicit form.
(iii) The error estimate must have a computable form.

There are two common approaches to the construction of a suitable space X. The first is to define an inner product on a subspace of $C(\mathbb{R}^n)$. This subspace is then shown to be complete, or alternatively, its completion is shown to be in $C(\mathbb{R}^n)$. This approach has been followed successfully by Madych and Nelson (see [10] and [11]) and Schaback (see [13] for references). The second approach is to construct an inner product. The space X is then defined to be the set of all objects for which the inner product makes sense and is finite. One then has the task of showing that X is embeddable in $C(\mathbb{R}^n)$. This approach was followed by Duchon in his variational treatment of polyharmonic splines [2].

We intend in the next section to show how the Duchon programme works out in quite a general situation. Then in Section 3, we will provide some examples of applications, plus some indication of the relationship of our work to that of other authors. The full theory involves considerable technical machinery. Our approach here will be to outline the results, and the thread of proof. The interested reader is referred to [9] for the details.

2. Spaces of Distributions

We are going to talk about spaces of tempered distributions. Consequently, $\mathcal{S}$ will be the space of rapidly decreasing functions, endowed with the usual topology. A tempered distribution is then a continuous (complex-valued) linear functional on $\mathcal{S}$; that is, a member of $\mathcal{S}'$. The advantage of using tempered distributions is that the Fourier transform is an isomorphism from $\mathcal{S}'$ to $\mathcal{S}'$. Our reference for the theory of distributions is either Rudin [12] or Hörmander [7]. Our first space is defined as follows.

Definition 1. *Let $w \in C(\mathbb{R}^n \setminus \{0\})$ be such that $w(x) > 0$ if $x \neq 0$. Then*

$$Y = \{f \in \mathcal{S}' : \widehat{f} \in L^1_{\text{loc}}(\mathbb{R}^n) \text{ and } \textstyle\int_{\mathbb{R}^n} |\widehat{f}|^2 w < \infty \}.$$

A norm on Y is defined by

$$\|f\|_Y = \left(\int_{\mathbb{R}^n} |\widehat{f}|^2 w \right)^{1/2}.$$

The function w will sometimes be referred to as the 'weight function' for Y. Clearly, w can have a singularity only at zero, and so $1/w \in C(\mathbb{R}^n \setminus \{0\})$. The reader will not come to much harm if she thinks of the space Y as consisting of functions. If we were to do this for a moment, what sort of functions f would be in Y? The significant element is the behaviour of $\widehat{f}$ at zero, where there must be a 'big' enough zero in $\widehat{f}$ to counteract the singularity of the weight function at zero. Of course, if w does not have a singularity at zero, then Y will be quite a big space. In addition, $\widehat{f}$ must have some sort of decay properties at infinity so as to obtain the correct integrability against w. If one makes these considerations precise, Y can be recognised by a convenient isometric isomorphism.

Theorem 1. *Let $w \in C(\mathbb{R}^n \setminus \{0\})$ be such that $w(x) > 0$ for $x \neq 0$. Suppose $1/w \in L^1_{\text{loc}}(\mathbb{R}^n)$ and there exists $\mu \in \mathbb{R}$ such that $\{w(x)\}^{-1} = \mathcal{O}(|x|^\mu)$ as $|x| \to \infty$. Then the mapping $I : Y \to L^2(\mathbb{R}^n)$ defined by $If = \sqrt{w}\widehat{f}$ is an isometric isomorphism of Y onto $L^2(\mathbb{R}^n)$.*

One can get a better feeling for the character of the space Y by exploring the various sets of functions which are dense in Y. There are a lot of interesting results which can be proved, but we simply give the one which fits immediately into the present exposition.

Theorem 2. *Let $w \in C(\mathbb{R}^n \setminus \{0\})$ be such that $w(x) > 0$ if $x \neq 0$. Then the set $\mathcal{S} \cap Y$ is dense in Y.*

Note that in line with our above discussion, it is generally the case that $\mathcal{S}$ is not a subset of Y. One has to impose extra conditions on w to ensure this.

Theorem 3. *Let $w \in C(\mathbb{R}^n \setminus \{0\})$ be such that $w(x) > 0$ if $x \neq 0$. Suppose $w \in L^1_{\text{loc}}(\mathbb{R}^n)$, and that there is a $\mu \in \mathbb{R}$ such that $w(x) = \mathcal{O}(|x|^\mu)$ as $x \to \infty$. Then $\mathcal{S} \subset Y$.*

Proof: There exist constants $C, R > 0$ such that for all $|x| > R$, $w(x) \leq C|x|^\mu$. Take $\phi \in \mathcal{S}$. Then

$$\int_{\mathbb{R}^n} \widehat{\phi}^2 w = \int_{|x| \leq R} \widehat{\phi}^2 w + \int_{|x| > R} \widehat{\phi}^2 w.$$

Since $w \in L^1_{\text{loc}}(\mathbb{R}^n)$, the first integral on the right-hand side of the above equality is finite. Since $\phi \in \mathcal{S}$, $\widehat{\phi}$ is also a rapidly decreasing function. Since w has polynomial growth for $|x| > R$, the second integral on the right-hand side of the above equality is also finite. Hence $\phi \in Y$. □

Our second space of distributions is very similar in nature to Y.

Definition 2. *Let $w \in C(\mathbb{R}^n \setminus \{0\})$ be such that $w(x) > 0$ if $x \neq 0$. Let $\alpha \in \mathbb{Z}^n_+$. Then*

$$Y_\alpha = \{f \in \mathcal{S}' : \widehat{D^\alpha f} \in L^1_{\text{loc}}(\mathbb{R}^n) \text{ and } \textstyle\int_{\mathbb{R}^n} |\widehat{D^\alpha f}|^2 w < \infty\}.$$

In Y_α, a seminorm $|\cdot|_\alpha$ may be defined by

$$|f|_\alpha = \left\{ \int_{\mathbb{R}^n} |\widehat{D^\alpha f}|^2 w \right\}^{1/2}, \qquad f \in Y_\alpha.$$

Theorem 4. *The mapping D^α maps Y_α onto Y, and $\|D^\alpha f\|_Y = |f|_\alpha$ for all $f \in Y_\alpha$.*

This Theorem gives a seemingly obvious connection between the two spaces Y and Y_α. Most of the proof is straightforward, but the fact that D^α maps Y onto Y_α is quite a sophisticated result, as far as we are able to tell. To understand this, take

g in Y. If the distributional partial differential equation $D^\alpha f = g$ has a tempered solution, then it is easy to check that this tempered solution is in Y_α, and so the mapping D^α is onto as required. However, the problem of the existence of a tempered solution to the above equation seems to be a difficult one. Hörmander [6], established that every partial differential equation with constant coefficients, whose right-hand side is tempered, has a tempered solution.

It is also straightforward to prove that Y_α is complete, providing that by completeness we mean only that every Cauchy sequence in Y_α has a limit in Y_α. The uniqueness of limits is not possible, since Y_α is usually only endowed with a seminorm.

We now can construct the main space for our theory.

Definition 3. *Let $w \in C(\mathbb{R}^n \setminus \{0\})$ be such that $w(x) > 0$ if $x \neq 0$. Let $k \in \mathbb{Z}_+$. Then $X = \bigcap_{\substack{|\alpha|=k \\ \alpha \in \mathbb{Z}_+^n}} Y_\alpha$. A seminorm on X is given by*

$$|f| = \left(\sum_{|\alpha|=k} c_\alpha \int_{\mathbb{R}^n} |\widehat{D^\alpha f}|^2 w \right)^{1/2},$$

where the numbers c_α, $\alpha \in \mathbb{Z}_+^n$, $|\alpha| = k$, are defined by the identity $|x|^{2k} = \sum_{|\alpha|=k} c_\alpha x^{2\alpha}$, $x \in \mathbb{R}^n$.

The somewhat peculiar choice of norm on X, involving the binomial coefficients c_α, is made so that if w has spherical symmetry, then the seminorm on X has spherical symmetry. It also has the useful effect that

$$\begin{aligned} |f|^2 &= \sum_{|\alpha|=k} c_\alpha \int_{\mathbb{R}^n} w(x) |(\widehat{D^\alpha f})(x)|^2 \, dx \\ &= \int_{\mathbb{R}^n} w(x) \sum_{|\alpha|=k} c_\alpha x^{2\alpha} |\widehat{f}(x)|^2 \, dx \\ &= \int_{\mathbb{R}^n} w(x) |x|^{2k} |\widehat{f}(x)|^2 \, dx. \end{aligned}$$

We will make use of this fact frequently. However, the reader should be on her guard here. If one knows that f is in X, then one cao evaluate the seminorm of f using the above equations. However, one cannot write

$$X = \left\{ f \in \mathcal{S}' : \int_{\mathbb{R}^n} w(x) |x|^{2k} |\widehat{f}(x)|^2 \, dx < \infty \right\}.$$

One really cannot omit the specification of the integrability properties of f. So the space above might be presumed to imply that $\widehat{f}$ is in $L^1_{\text{loc}}(\mathbb{R}^n)$. In this case, a different (and smaller) space than X would be implied. The following result should help the reader to understand X a little better.

Theorem 5. *Let $w \in C(\mathbb{R}^n \setminus \{0\})$ be such that $w(x) > 0$ if $x \neq 0$. The seminorm in X has kernel π_{k-1}. If $1/w \in L^1_{\mathrm{loc}}(\mathbb{R}^n)$ and there exists $\mu \in \mathbb{R}$ such that $\{w(x)\}^{-1} = \mathcal{O}(|x|^{\mu})$ as $|x| \to \infty$, then X is complete. Furthermore, the set $\{f \in X : \widehat{f} \in C_0^\infty(\mathbb{R}^n)\}$ is dense in X.*

Now recall our agenda as set in Section 1. We wish to show that X is embeddable in the space $C(\mathbb{R}^n)$. Also our theory began with a Hilbert space, rather than a semi-Hilbert space. Associated with the seminorm as given in Definition 3 is the semi-inner product:

$$\langle f, g\rangle = \sum_{|\alpha|=k} c_\alpha \int_{\mathbb{R}^n} w\, \widehat{D^\alpha f}\, \overline{\widehat{D^\alpha g}}, \qquad f, g \in X.$$

Because the kernel of the seminorm is π_{k-1}, we can try to define an inner product on X by choosing points $a_1, \ldots, a_\ell$ in $\mathbb{R}^n$ which are unisolvent with respect to π_{k-1} and then set

$$(f, g) = \sum_{s=1}^{\ell} f(a_s)\overline{g(a_s)} + \langle f, g\rangle, \qquad f, g \in X. \tag{1}$$

The difficulty is that we cannot make sense of such a definition until we know that point evaluations are valid in X. So the above procedure is a bit delicate. However, things can be arranged so that everything proceeds in a legal fashion. The result is a generalisation of the Sobolev embedding theorem for our particular spaces.

Theorem 6. *Let $k \in \mathbb{Z}_+^n$ and $\mu \in \mathbb{R}$ be such that $k + \mu - n/2 > 0$. Let $\{a_1, \ldots, a_\ell\}$ be a set in $\mathbb{R}^n$ which is unisolvent with respect to π_{k-1}. Define*

$$(f, g) = \sum_{s=1}^{\ell} f(a_s)\overline{g(a_s)} + \langle f, g\rangle, \qquad f, g \in X.$$

Then $(\cdot, \cdot)$ defines an inner product on X. Furthermore, given $x \in \mathbb{R}^n$, there is a constant $C > 0$ such that $|f(x)| \le C\sqrt{(f, f)}$ for all $f \in X$.

The result quoted here is rather weaker than that given in [9]. One can show that X is embedded in $C^{(\beta)}(\mathbb{R}^n)$ for an appropriate value of β, providing $k+\mu-n/2 > |\beta|$, but this level of generality is probably not helpful in the present exposition.

3. Representatives for Point Evaluations

The only part of our programme declared in Section 1 which remains to be completed is the determination of the representative for a general point evaluation functional in $\mathbb{R}^n$. Let's try to develop the appropriate element in X. During our exposition, there will be a number of places where mathematical subtleties are glossed over, so as to get a smooth exposition! It will help to adopt the notation

V_α for the polynomial $V_\alpha(x) = (ix)^\alpha$, $\alpha \in \mathbb{Z}_+^n$. A standard result from the Fourier transform theory is that $\widehat{D^\alpha f} = V_\alpha \widehat{f}$. Fix $x \in \mathbb{R}^n$. At first we want to dispose of the point evaluations in the inner product as specified in Equation (1). We therefore suppose that g in X is such that $g(a_1) = \cdots = g(a_\ell) = 0$. Let us also suppose $\widehat{g}$ is in $C_0^\infty(\mathbb{R}^n)$. Now we waot an element r_x in X such that

$$\begin{aligned} g(x) &= (g, r_x) \\ &= \langle g, r_x \rangle \\ &= \sum_{|\alpha|=k} c_\alpha \int_{\mathbb{R}^n} \overline{(D^\alpha r_x)\widehat{\ }(y)} \widehat{(D^\alpha g)}(y) w(y)\, dy. \end{aligned}$$

We cao regard the last integral above as defining the action of a distribution on a test function. It will help to denote by $[f, \psi]$ the action of the distribution f on the test function $\psi \in \mathcal{S}$. This allows us to write

$$\begin{aligned} [T_x\delta, g] = g(x) &= \sum_{|\alpha|=k} c_\alpha \left[w\, \overline{(D^\alpha r_x)\widehat{\ }}, \widehat{D^\alpha g} \right] \\ &= \sum_{|\alpha|=k} c_\alpha \left[w \overline{V_\alpha}\, \overline{\widehat{r_x}}, V_\alpha \widehat{g} \right] \\ &= \left[w \sum_{|\alpha|=k} c_\alpha \overline{V_\alpha} V_\alpha \overline{\widehat{r_x}}, \widehat{g} \right] \\ &= \left[w|\cdot|^{2k} \overline{\widehat{r_x}}, \widehat{g} \right]. \end{aligned}$$

One way to make the above equality hold is as follows. Firstly, define e_x so that $e_x(y) = e^{ixy}$, for all $y \in \mathbb{R}^n$. Then observe that $(e_x 1)\widehat{\ } = T_x \widehat{1} = T_x \delta$. We wish to obtain

$$\left[w|\cdot|^{2k} \overline{\widehat{r_x}}, \widehat{g} \right] = [T_x\delta, g] = [(e_x.1)\widehat{\ }, g] = [e_x, \widehat{g}].$$

Thus we should demaod that as distributions,

$$w|\cdot|^{2k} \overline{\widehat{r_x}} = e_x.$$

Defining the distribution $\phi \in \mathcal{S}$ by the equation $r_x = T_x \overline{\phi}$ gives

$$e_x = w|\cdot|^{2k} \overline{(T_x\overline{\phi})\widehat{\ }} = w|\cdot|^{2k} \overline{e_{-x}\widehat{\overline{\phi}}} = w|\cdot|^{2k} e_x \widehat{\phi}.$$

It is then tempting to make the unknown distribution $\widehat{\phi}$ the subject of this formula:

$$\widehat{\phi} = \frac{1}{w|\cdot|^{2k}}.$$

There are however problems with this approach. We will have to make sense of the quotient $1/(w|\cdot|^{2k})$ as a distribution. It might be possible to interpret $1/w$ successfully, but in general, the singularity in $1/|\cdot|^{2k}$ is difficult to handle. There are a number of ways of dealing with this difficulty. Duchon [2] appeals to the theory of pseudo-functions as defined in Schwartz [14]. This in turn would involve us in the notion of the finite part of a divergent integral as defined by Hadamard [5]. Other approaches via analytic continuation or the theory of homogeneous distributions are also possible, see [7]. However, one can avoid these technicalities by a rather elegant trick, which is found in [9]. Another way to proceed is to eschew the opportunity to carry out the whole rearrangement. Instead, write

$$|\cdot|^{2k}\widehat{\phi} = \frac{1}{w}. \tag{2}$$

If the right-hand side of Equation (2) can be interpreted as a tempered distribution, then the result from Hörmaoder, which we have already seen, guarantees a tempered solution to Equation (2). Of course, there are many solutions to Equation (2), and any two of these differ by a solution of the homogeneous equation:

$$|\cdot|^{2k}\widehat{\phi} = 0.$$

These solutions are distributions supported at the origin, and at first sight, one might conclude that any polynomial q in π_{2k-1} was a solution. However, the requirement that $D^\alpha q$ lies in $L^1_{\text{loc}}(\mathbb{R}^n)$ forces q to lie in π_{k-1}. Thus ϕ is completely determined by Equation (2), up to a polynomial of degree $k-1$. This indeterminacy arises because we have only arranged that $(g, r_x) = g(x)$ on the subspace of X given by $\{g \in X : g(a_j) = 0,\ j = 1, \ldots, \ell\}$. Requiring that r_x have the correct action on the whole space X takes up the freedom and gives us the following result.

Theorem 7. *Let $w \in C(\mathbb{R}^n \setminus \{0\})$ be such that $w(x) > 0$ if $x \neq 0$. Suppose also $1/w \in L^1_{\text{loc}}(\mathbb{R}^n)$, and $\mu \in \mathbb{R}$ is such that $\{w(x)\}^{-1} = \mathcal{O}(|x|^{-2\mu})$ as $|x| \to \infty$. Let X be defined as in Section 2, with inner product $(\cdot,\cdot) : X \times X \to \mathbb{C}$. Let $a_1, \ldots, a_\ell \in \mathbb{R}^n$ be a unisolvent set of points with respect to π_{k-1}, and let $p_1, \ldots, p_\ell \in \pi_{k-1}$ be such that $p_j(a_s) = 1$ if $j = s$ and is zero otherwise. Let ϕ be a solution of Equation (2). Let $x \in \mathbb{R}^n$. The unique element $r_x \in X$ which satisfies $(f, r_x) = f(x)$ for all $f \in X$ is given by*

$$\begin{aligned}
&\phi(y-x) - \sum_{j=1}^{\ell} p_j(x)\phi(y-a_j) \\
&\quad - \sum_{s=1}^{\ell}\Big(\phi(a_s - x) - \sum_{j=1}^{\ell} p_j(x)\phi(a_s - a_j)\Big)p_s(y) + \sum_{s=1}^{\ell} p_s(x)p_s(y).
\end{aligned}$$

4. The Interpolant

We know from Section 1 that the interpolant to any f in X at $a_1, \dots, a_m$ is of the form $\sum_{j=1}^m \lambda_j r_{a_j}$. If $a_1, \dots, a_m$ contains a unisolvent set of points, then it is easy to see from Theorem 7 that each function r_{a_j} can be expressed as a linear combination of translates of the function ϕ plus a polynomial. This shows immediately that

$$Uf = \sum_{j=1}^{m} \beta_j \phi(\cdot - a_j) + p,$$

where p lies in π_{k-1}. This equation explains the title of this paper. The function ϕ is the basis function, whose translates are being used to form the interpolant. The polynomial term appears so that the variational theory remains consistent. However, if the original form of the interpolant in terms of the functions r_{a_j} is used, then the interpolation equations which define $\lambda_1, \dots, \lambda_m$ are

$$\sum_{j=1}^{m} \lambda_j r_{a_j}(a_s) = f(a_s), \qquad s = 1, \dots, m.$$

The related interpolation matrix whose (s, j) element is $r_{a_j}(a_s)$ is now easily seen to be positive definite. This should be of some advantage in computational work.

Furthermore, quite explicit error estimates are available, as promised in Section 1. We give an example.

Theorem 8. *Let w be a weight function satisfying the assumptions of Theorem 7. Suppose $a_1, \dots, a_m \in \mathbb{R}^n$ are so ordered that $a_1, \dots, a_\ell$ is a unisolvent set with respect to π_{k-1}. Let $p_1, \dots, p_\ell \in \pi_{k-1}$ be such that $p_s(a_j) = 1$ if $s = j$ and is zero otherwise. Let X be defined as at the beginning of Section 2, with seminorm $|\cdot|$. Suppose $k + \mu - n/2 > 0$. Let ϕ be a solution of Equation (2). Let $f \in X$, and let Uf be the minimal norm interpolant to f on $a_1, \dots, a_m$. For any $x \in \mathbb{R}^n$,*

$$\big|f(x) - (Uf)(x)\big| \le \rho(x)|f|,$$

where ρ is the power function, given by

$$\rho(x) = \Big(\phi(0) - \sum_{j=1}^{\ell} p_j(x)\big(\phi(x-a_j) + \phi(a_j - x)\big) + \sum_{j=1}^{\ell}\sum_{s=1}^{\ell} p_j(x)p_s(x)\phi(a_s - a_j)\Big)^{1/2}.$$

From this error estimate it is possible to deduce asymptotic error estimates. Weak assumptions are needed on the set over which the error bound is taken. Without going into these, the type of result obtained is given in the following Metatheorem.

Metatheorem 1. *Let Ω be a bounded subset of $\mathbb{R}^n$ which is connected and has non-empty interior. Let $k+\mu-n/2>0$, and let J denote the largest integer less than $2k+2\mu-n$. For each $h>0$, let $\mathcal{A}_h$ be a finite, π_{k-1}-unisolvent subset of Ω with $\sup_{t\in\overline{\Omega}}\inf_{a\in\mathcal{A}_h}|t-a|\le h$. For each $f\in X$, let $U_hf\in X$ be the minimal norm interpolant to f on $\mathcal{A}_h$. There exists a constant $C>0$, independent of h, such that for all $f\in X$,*

$$\|f-U_hf\|_{\infty,\Omega}\le Ch^{\min\{k,J/2\}}|f|.$$

The proof of this result has as an important ingredient a Taylor series argument involving the function ϕ. This observation helps to explain why the parameter J appears in the Metatheorem. It is the amount of continuity present in ϕ, so that $\phi\in C^{(J)}(\mathbb{R}^n)$.

5. Conditional Positive Definiteness

A glance at the literature on radial basis functions will rapidly convince the reader that conditional positive definiteness is an important idea in the theory. We want in this final Section to tie our approach to this fundamental concept. Since we are working with distributions, it is natural to use a distributional notion of conditional positive definiteness. The appropriate notion comes from Gelfand and Vilenkin [3].

Definition 4. *A distribution $f\in\mathcal{S}'$ is said to be conditionally positive definite of order $s>0$ if the inequality $\left[p\overline{p}\widehat{f},\psi\overline{\psi}\right]\ge 0$ holds for all $\psi\in\mathcal{S}$ and for all homogeneous polynomials p of degree s.*

Theorem 9. *Let $w\in C(\mathbb{R}^n\setminus\{0\})$ be such that $w(x)>0$ for $x\neq 0$. Let $1/w\in L^1_{\mathrm{loc}}(\mathbb{R}^n)$, and suppose there is a $\mu\in\mathbb{R}$ such that $\{w(x)\}^{-1}=\mathcal{O}(|x|^{-2\mu})$ as $|x|\to\infty$. Let ϕ be a distribution whose Fourier transform is given by $\widehat{\phi}=(w|\cdot|^{2k})^{-1}$. Then ϕ is conditionally positive definite of order k.*

Notes. The above Theorem and its proof will give the reader a very instructive glimpse into the technicalities needed to give correct arguments and statements in this area. As mentioned in Section 3, the statement that $(w|\cdot|^{2k})^{-1}$ is a distribution is cavalier, and should be made precise. Such attention to detail is not a trifle: the poor specification of $\widehat{\phi}$ will manifest itself in the following proof.

Proof: Let V_α be defined as at the beginning of Section 3. Let p be any homogeneous polynomial of degree k. We may write $p=\sum_{|\alpha|=k}a_\alpha V_\alpha$, where $a_\alpha\in\mathbb{C}$, $\alpha\in\mathbb{Z}^n_+$ and $|\alpha|=k$. Observe now that $\left[p\overline{p}\widehat{\phi},\psi\overline{\psi}\right]=\left[\widehat{\phi},p\overline{p}\psi\overline{\psi}\right]$, for any $\psi\in\mathcal{S}$. Consider first for $\gamma\in\mathbb{Z}^n_+$ and $|\gamma|<2k$,

$$D^\gamma(p\psi)=\sum_{|\alpha|=k}a_\alpha D^\gamma(V_\alpha\psi).$$

Using the Leibniz theorem, there exist coefficients b_δ such that

$$D^\gamma(p\psi) = \sum_{|\alpha|=k} a_\alpha \sum_{0\le\delta\le\gamma} b_\delta V_{\alpha-\delta} D^{(\gamma-\delta)}\psi.$$

It follows that if $|\gamma| < k$, then in the above sum, $|\alpha - \delta| \geq 1$ and so

$$\Big(D^\gamma(p\psi)\Big)(0) = \sum_{|\alpha|=k} a_\alpha \sum_{0\le\delta\le\gamma} b_\delta V_{\alpha-\delta}(0)(D^{(\gamma-\delta)}\psi)(0) = 0.$$

Another application of the Leibniz theorem shows that there exist constants μ_δ such that

$$D^\gamma(p\overline{p}\,\psi\overline{\psi}) = \sum_{0\le\delta\le\gamma} \mu_\delta D^\delta(p\psi)\overline{\big(D^{\gamma-\delta}(p\psi)\big)}.$$

It follows that

$$\begin{aligned}\big(D^\gamma(p\overline{p}\,\psi\overline{\psi})\big)(0) &= \sum_{0\le\delta\le\gamma} \mu_\delta \big(D^\delta(p\psi)\big)(0)\overline{\big(D^{\gamma-\delta}(p\psi)\big)(0)} \\ &= \sum_{\substack{0\le\delta\le\gamma\\ |\delta|\ge k}} \mu_\delta \big(D^\delta(p\psi)\big)(0)\overline{\big(D^{\gamma-\delta}(p\psi)\big)(0)}.\end{aligned}$$

Now since $|\delta| \geq k$ and $|\gamma| < 2k$, then $|\gamma - \delta| = |\gamma| - |\delta| < 2k - k = k$, and so $\big(D^{(\gamma-\delta)}(p\psi)\big)(0) = 0$. Thus we conclude that $\big(D^\gamma(p\overline{p}\,\psi\overline{\psi})\big)(0) = 0$, for all $\gamma \in \mathbb{Z}^n_+$ with $|\gamma| < 2k$. This has the effect that $\big|(p\overline{p}\,\psi\overline{\psi})(x)\big| = \mathcal{O}(|x|^{2k})$ as $|x| \to 0$. This allows us to determine the action of the distribution $\widehat{\phi}$ on the test function $p\overline{p}\psi\overline{\psi}$ by an integral:

$$\Big[p\overline{p}\,\widehat{\phi}, \psi\overline{\psi}\Big] = \Big[\widehat{\phi}, p\overline{p}\,\psi\overline{\psi}\Big] = \int_{\mathbb{R}^n} \frac{|p|^2|\psi|^2}{|\cdot|^{2k}w}.$$

This integral exists in the classical sense, and is non-negative, which concludes the proof. □

Space does not permit us to show how the common examples from radial basis function theory fit into the above scenario, but the reader should be assured of the fact that multiquadrics, thin-plate splines, the Gaussian, inverse multiquadrics and powers of the Eucildean norm all fit into our theory. We refer the reader to the thesis of Wayne for the details.

Acknowledgements. It is a pleasure to acknowledge that this paper is a precis of joint work with H.S.J. Wayne.

References

1. Duchon J., Sur l'erreur d'interpolation des fonctions de plusieurs variables par les D^m–splines, RAIRO Analyse Numerique **12** (4), 1978, 325–334.
2. Duchon J., Splines minimising rotation-invariant seminorms in Sobolev spaces, in *Constructive theory of functions of several variables*, Lecture Notes in Mathematics, **571**, Schempp W. and Zeller K. (eds.), Springer-Verlag (Berlin), 1977, 85–100.
3. Gelfand I. M. and Vilenkin N. Ya., *Generalised functions*, Vol. 4, Academic Press (New York), 1964.
4. Golomb M. and Weinberger H. F., Optimal approximation and error bounds, in *On numerical approximation*, Langer R.E. (ed.), University of Wisconsin Press (Madison), 1959, 117–190.
5. Hadamard J., *Lectures on Cauchy's problem in linear partial differential equations*, Yale University Press (New Haven, Conn.), 1923.
6. Hörmaoder L., On the division of distributions by polynomials, Arkiv för Mat. **53** (3), 1958, 555–568.
7. Hörmander L., *The analysis of linear partial differential operators I*, Springer Verlag (Berlin), 1983.
8. Light W. A. and Wayne H. S. J., Error estimates for approximation by radial basis functions, in *Approximation theory, wavelets and applications*, S.P. Singh, (ed.), Kluwer Academic (Dordrecht), 1995, 215-246.
9. Light W. and Wayne H. S. J., Spaces of distributions and interpolation by radial basis functions, University of Leicester Technical Report, 1996/32.
10. Madych W. and Nelson S., Multivariate interpolation and conditionally positive definite functions, Approx. Theory Appl. **4**, 1988, 77–89.
11. Madych W. R. and Nelson S. A., Multivariate interpolation and conditionally positive definite functions II, Math. Comp. **54**, 1990, 211–230.
12. Rudin W., *Functional analysis*, McGraw-Hill (New York), 1973.
13. Schaback R., Multivariate interpolation and approximation by translates of a basis function, in *Approximation Theory VIII, Vol I: Approximation and interpolation*, Chui, C.K. and Schumaker L.L. (eds.), World Scientific Publishing Co. (Singapore), 1995, 491–514.
14. Schwartz L., *Théorie des distributions*, Hermann (Paris), 1966.
15. Wayne H. S. J., *Towards a Theory of Multivariate Interpolation using Spaces of Distributions*, Ph.D. Thesis, University of Leicester, 1996.

Department of Mathematics and Computer Science
University of Leicester
Leicester LE1 7RH
England
`pwl@mcs.le.ac.uk`

On the Sup-norm Condition Number of the Multivariate Triangular Bernstein Basis.

Tom Lyche and Karl Scherer

Abstract. We give an upper bound for the L^∞ condition number of the triangular Bernstein basis for polynomials of total degree at most n in s variables. The upper bound grows like $(s+1)^n$ when n tends to infinity. Moreover the upper bound is independent of s for $s \geq n-1$.

1. Introduction

In this paper we estimate the size of the coefficients of a polynomial f of total degree n in s variables when it is represented using the triangular Bernstein basis. This basis has gained increasing popularity mainly through work in Computer Aided Geometric Design [2]. For similar estimates for univariate B-splines see [1,3,4]. We consider only estimates in the L^∞ norm in this paper. The general L^p case together with sharpness of the estimates will be published elsewhere.

The condition number of a basis can be defined quite generally.

Definition 1.1. *A basis (ϕ_j) of a normed linear space is said to be stable with respect to a vector norm if there are constants K_1 and K_2 such that for all coefficients (c_j)*

$$K_1^{-1}||(c_j)|| \leq ||\sum_j c_j\phi_j|| \leq K_2||(c_j)||. \tag{1}$$

(For simplicity we use the same symbol $||\ ||$ for the norm in the vector space and the vector norm.) The number

$$\kappa = K_1 K_2$$

with K_1 and K_2 as small as possible is called the condition number of (ϕ_j) with respect to $||\ ||$.

Multivariate Approximation and Splines
G. Nürnberger, J. W. Schmidt, and G. Walz (eds.), pp. 141–151.

ISBN 3-7643-5654-5.

Such condition numbers give an upper bound for how much an error in coefficients can be magnified in function values. Indeed, if $f = \sum_j c_j\phi_j \neq 0$ and $g = \sum_j d_j\phi_j$ then it follows immediately from (1) that

$$\frac{||f-g||}{||f||} \leq \kappa \frac{||c-d||}{||c||}.$$

Many other applications are given in [1] and it is interesting to have estimates for the size of κ.

The contents of this paper is as follows. We recall the definition of the Bernstein basis in Section 2. There we also transform the problem from a simplex to a cube. This makes it possible to analyze the problem in a tensor product fashion. The univariate case was considered in [3]. We extend these results in Section 3. In Section 4 and 5 we consider the multivariate case.

We use standard multi-index notation. Thus for tuples $\alpha = (\alpha_1, \ldots, \alpha_s)$ and $\mathbf{i} = (i_1, \ldots, i_s)$ we let $|\mathbf{i}| = i_1 + \ldots + i_s$, $\mathbf{i}! = i_1!i_2!\cdots i_s!$, and $\mathbf{i}^\alpha = (i_1^{\alpha_1}, i_2^{\alpha_2}, \ldots, i_s^{\alpha_s})$. Unless otherwise stated the indices in a sum will be nonnegative. Thus if we sum in the order $\alpha_s, \alpha_{s-1}, \ldots, \alpha_1$ then

$$\sum_{|(\alpha_1,\ldots,\alpha_s)|\leq n} = \sum_{\alpha_1=0}^{n} \sum_{\alpha_2=0}^{n-\alpha_1} \sum_{\alpha_3=0}^{n-\alpha_1-\alpha_2} \cdots \sum_{\alpha_s=0}^{n-\alpha_1-\cdots-\alpha_{s-1}}.$$

The sums $\sum_{|(i_1,\ldots,i_s)|\leq n}$ and $\sum_{|(\alpha_1,\ldots,\alpha_{s+1})|=n}$ will both contain $\binom{n+s}{s}$ terms. We denote by $||\mathbf{c}||_\infty$ and $||f||_{L^\infty(\Omega)}$ the usual sup-norms of vectors and functions defined on a set Ω, respectively. The convex hull of m points $v_1, \ldots, v_m$ is denoted $< v_1, \ldots, v_m >$. For any $x \in \mathbb{R}$ the "floor" function $[x]$ is the unique integer n so that $n \leq x < n+1$.

2. The Bernstein Basis

For the vector space

$$P_n(\mathbb{R}^s) = \{p(\mathbf{x}) = \sum_{|\mathbf{i}|\leq n} c_{\mathbf{i}}\mathbf{x}^{\mathbf{i}} : c_{\mathbf{i}} \in \mathbb{R}\}$$

of polynomials of total degree at most n in s variables $\mathbf{x} = (x_1, \ldots, x_s)$ we consider the Bernstein basis

$$\left\{\frac{n!}{\alpha!}\lambda^\alpha\right\}_{|\alpha|=n}.$$

Here $\lambda = (\lambda_1, \ldots, \lambda_{s+1})$ denotes the barycentric coordinate with respect to a nondegenerate simplex $\Sigma =< \mathbf{v}_1, \ldots \mathbf{v}_{s+1} >$ in $\mathbb{R}^s$ *i.e.*,the tuple λ corresponding to a point $\mathbf{x} \in \mathbb{R}^s$ is uniquely given by

$$\sum_{i=1}^{s+1} \lambda_i \mathbf{v}_i = \mathbf{x}, \qquad \sum_{i=1}^{s+1} \lambda_i = 1.$$

Since $\lambda \geq 0$ for each $\mathbf{x} \in \Sigma$ and $\sum_{|\alpha|=n} \frac{n!}{\alpha!}\lambda^\alpha = 1$ for any $\mathbf{x} \in \mathbb{R}^s$ we have $K_2 = 1$ in (1) so that

$$\kappa_{n,\infty}(\mathbb{R}^s) = \sup_{\mathbf{c}\neq 0} \|\mathbf{c}\|_\infty \Big/ \Big\| \sum_{|\alpha|=n} c_\alpha \frac{n!}{\alpha!}\lambda^\alpha \Big\|_{L^\infty(\Sigma)}. \tag{2}$$

For our purpose it is convenient to introduce the change of variables (see [6], p.29)

$$\mathbf{x} \to \mathbf{y} = (y_1, \ldots, y_s)$$

given by

$$\begin{aligned}
\lambda_1 &= y_1 && = y_1 \\
\lambda_2 &= y_2(1-y_1) && = y_2(1-\lambda_1) \\
\lambda_3 &= y_3(1-y_2)(1-y_1) && = y_2(1-\lambda_1-\lambda_2) \\
&\vdots && \vdots \\
\lambda_s &= y_s(1-y_{s-1})\cdots(1-y_1) && = y_s(1-\lambda_1-\lambda_2-\cdots-\lambda_{s-1}).
\end{aligned}$$

This transformation maps Σ onto the s-dimensional unit cube $[0,1]^s$. Since

$$\lambda_{s+1} = 1-\lambda_1-\lambda_2-\cdots-\lambda_s = (1-y_s)(1-y_{s-1})\cdots(1-y_1)$$

we obtain

$$\lambda^\alpha = y_1^{\alpha_1} y_2^{\alpha_2} \cdots y_s^{\alpha_s}(1-y_1)^{n-\alpha_1}(1-y_2)^{n-\alpha_1-\alpha_2}(1-y_s)^{n-\alpha_1-\cdots-\alpha_s}.$$

Combining this equation with the relation

$$\frac{n!}{\alpha!} = \binom{n}{\alpha_1}\binom{n-\alpha_1}{\alpha_2}\binom{n-\alpha_1-\alpha_2}{\alpha_3}\cdots\binom{n-\alpha_1-\cdots-\alpha_{s-1}}{\alpha_s}$$

we see that

$$\frac{n!}{\alpha!}\lambda^\alpha = B^n_{\alpha_1}(y_1) B^{n-\alpha_1}_{\alpha_2}(y_2)\cdots B^{n-\alpha_1-\cdots-\alpha_{s-1}}_{\alpha_s}(y_s), \tag{3}$$

where

$$B^n_k(x) = \binom{n}{k} x^k (1-x)^{n-k}, \quad k = 0,1,\ldots,n \quad x \in \mathbb{R} \tag{4}$$

are the usual univariate Bernstein basis polynomials. Thus every polynomial in $P_n(\mathbb{R}^s)$ can be written in tensor product manner as follows

$$\begin{aligned} f = \sum_{|\alpha|=n} c_\alpha \frac{n!}{\alpha!}\lambda^\alpha = \sum_{\alpha_1=0} B^n_{\alpha_1}(y_1)\Bigg[\sum_{\alpha_2=0}^{n-\alpha_1} B^{n-\alpha_1}_{\alpha_2}(y_2)\cdots \\ \sum_{\alpha_s=0}^{n-\alpha_1-\cdots-\alpha_{s-1}} c_\alpha B^{n-\alpha_1-\cdots-\alpha_{s-1}}_{\alpha_s}(y_s)\Bigg]. \end{aligned} \tag{5}$$

3. The Univariate Case

When $s = 1$ then (5) takes the form $f = \sum_{j=0}^{n} c_j B_j^n$ where B_j^n is the univariate Bernstein basis polynomial given by (4) and $\mathbf{c} = (c_0, \ldots, c_n)$ is called the BB-coefficient vector of f. Thus (1) takes the form

$$\kappa_{n,\infty}(\mathbb{R}^1) = \sup_{\mathbf{c}\neq 0} ||\mathbf{c}||_\infty / || \sum_{j=0}^{n} c_j B_j^n ||_{L^\infty[0,1]}.$$

The following Lemma shows that

$$\kappa_{n,\infty}(\mathbb{R}^1) = \max_{0\leq i\leq n} \gamma_{i,n}. \tag{6}$$

where $(-1)^{n-i}\gamma_{i,n}$ for $i = 0, \ldots, n$ are the BB-coefficients of the shifted Chebyshev polynomial

$$\hat{T}_n(x) := T_n(2x-1) = \sum_{j=0}^{n} (-1)^{n-j} \gamma_{j,n} B_j^n(x).$$

It is well known that the $\gamma_{i,n}$ are given by $\gamma_{0,n} = \gamma_{n,n} = 1$ for $n \geq 0$ and

$$\gamma_{i,n} = \frac{(2n-1)(2n-3)\cdots(2n-2i+1)}{1\cdot 3\cdots(2i-1)}, \quad i = 1, \ldots, n-1, \ n \geq 2. \tag{7}$$

Lemma 3.1. *For any (c_j) we have*

$$|c_i| \leq \gamma_{i,n} || \sum_{j=0}^{n} c_j B_j^n ||_{L^\infty[0,1]}, \quad i = 0, 1, \ldots, n, \tag{8}$$

where the $\gamma_{i,n}$ are given by (7).

Proof: Fix $0 \leq i \leq n$. We first show that

$$\inf_{(c_j)} \left\{ ||B_i^n - \sum_{j\neq i} c_j B_j^n ||_{L^\infty[0,1]} \right\} = ||\hat{T}_n / \gamma_{i,n}||_{L^\infty[0,1]} \tag{9}$$

Suppose $||V||_{L^\infty[0,1]} < ||\hat{T}_n/\gamma_{i,n}||_{L^\infty[0,1]}$ for some $V = B_i^n - \sum_{j\neq i} d_j B_j^n$. Then $W = V - \hat{T}_n/\gamma_{i,n}$ would change sign at the n extrema in $[0,1]$ of $\hat{T}_n$. But this would imply that $W = 0$, since $W \in \text{span}(B_j^n)_{j\neq i}$ and this set forms an order complete weak Chebyshev system on $[0,1]$. (See Theorem 4.65, the remark on p.170 and Theorem 2.42 in [5]). This contradiction establishes (9). From (9) for any nonzero c_i

$$|| \sum_{j=0}^{n} c_j B_j^n ||_{L^\infty[0,1]} = |c_i| \; ||B_i^n - \sum_{j\neq i} \frac{c_j}{c_i} B_j^n ||_{L^\infty[0,1]}$$

$$\geq |c_i| \; ||\hat{T}_n/\gamma_{i,n}||_{L^\infty[0,1]} = \frac{|c_i|}{\gamma_{i,n}}$$

from which the desired estimate follows. □

To compute $\max_{0\le i\le n}\gamma_{i,n}$ we observe that the γ's in (7) satisfy the recurrence relation

$$\gamma_{i,n} = \frac{2(n-i)+1}{2i-1}\gamma_{i-1,n}, \quad i = 1,\dots,n. \tag{10}$$

It follows that

$$\gamma_{0,n} < \cdots < \gamma_{m,n} \ge \gamma_{m+1,n} > \cdots > \gamma_{n,n}, \quad \text{with} \quad m = \left[\frac{n}{2}\right],$$

and with $n = 2m + k$ for $k \in \{0, 1\}$ we find

$$\kappa_{n,\infty}(\mathbb{R}^1) = \gamma_{m,n} = \frac{(2n-1)(2n-3)\cdots(2m+3)(2m+1)^{1-k}}{1\cdot 3\cdots(2m-1)}. \tag{11}$$

In [3] the following asymptotic bound was found

$$(1-\frac{1}{n})2^{n-1/2} \le \kappa_{n,\infty}(\mathbb{R}^1) \le (1+\frac{1}{n})2^{n-1/2}, \quad n \ge 1. \tag{12}$$

In the remaining part of the paper we extend (11) and (12) to an upper bound for $\kappa_{n,\infty}(\mathbb{R}^s)$ for $s > 1$.

4. An Upper Bound for $\kappa_{n,\infty}(\mathbb{R}^s)$

We consider now the case $s \ge 2$. We first prove a Lemma

Lemma 4.1. *For $n, s \ge 1$ we have*

$$\kappa_{n,\infty}(\mathbb{R}^s) \le K_n(\mathbb{R}^s), \tag{13}$$

where $K_n(\mathbb{R}^1) := \kappa_{n,\infty}(\mathbb{R}^1)$ and for $s \ge 2$

$$K_n(\mathbb{R}^s) := \max_{0\le i\le n} \gamma_{i,n} K_{n-i}(\mathbb{R}^{s-1}). \tag{14}$$

Here $\gamma_{i,n}$ is given by (7).

Proof: This follows from (5) by repeated application of Lemma 3.1. In order to explain the main idea we consider first the case $s = 2$. In this case (5) takes the form

$$\begin{aligned} f &= \sum_{i=0}^{n}\sum_{k=0}^{n-i} c_{i,k,n-i-k}\frac{n!}{i!k!(n-i-k)!}\lambda_1^i\lambda_2^k\lambda_3^{n-i-k} \\ &= \sum_{i=0}^{n}\left[\sum_{k=0}^{n-i} c_{i,k,n-i-k}B_k^{n-i}(y_2)\right]B_i^n(y_1). \end{aligned}$$

Using Lemma 3.1 first on the inner sum and then on the outer sum we obtain

$$\begin{aligned}|c_{i,j,n-i-j}| &\le \gamma_{j,n-i} \| \sum_{k=0}^{n-i} c_{i,k,n-i-k} B_k^{n-i} \|_{L^\infty[0,1]} \\ &\le \gamma_{j,n-i}\gamma_{i,n} \|f\|_{L^\infty[\Sigma]}.\end{aligned}$$

Thus

$$\kappa_{n,\infty}(\mathbb{R}^2) \le \max_{0\le i\le n} \{\gamma_{i,n} \max_{0\le j\le n-i} \gamma_{j,n-i}\} = \max_{0\le i\le n} \gamma_{i,n} K_{n-i}(\mathbb{R}^1).$$

This proves (14) for $s = 2$. For arbitrary s a similar argument shows that

$$|c_\alpha| \le \max_{\mathbf{i}\in J_s^n} \prod_{k=1}^{s} \gamma_{i_k, n-i_1-\cdots-i_{k-1}} \quad \text{for} \quad |\alpha| = n,$$

where $J_s^n = \{(i_1, \ldots, i_s) : 0 \le i_k \le n - i_1 - \ldots - i_{k-1}, \quad k = 1, \ldots, s\}$. Since

$$\max_{\mathbf{i}\in J_s} \prod_{k=1}^{s} \gamma_{i_k, n-i_1-\cdots-i_{k-1}} = \max_{0\le i_1\le n} \gamma_{i_1,n} \left[\max_{(i_2,\ldots,i_s)\in J_{s-1}^{n-i_1}} \prod_{k=2}^{s} \gamma_{i_k, n-i_1-\cdots-i_{k-1}} \right]$$

(14) follows by induction. □

The constant $K_n(\mathbb{R}^s)$ can be computed exactly.

Theorem 4.2. *For positive integers n and s*

$$\kappa_{n,\infty}(\mathbb{R}^s) \le K_n(\mathbb{R}^s) = \frac{(2n-1)(2n-3)\cdots(2m+3)(2m+1)^{1-k}}{1^s \cdot 3^s \cdots (2m-1)^s}, \tag{15}$$

where

$$m = \left[\frac{n}{s+1}\right], \quad \text{and} \quad k = n - (s+1)m. \tag{16}$$

Moreover, we have the alternative representations

$$K_n(\mathbb{R}^s) = \frac{(2n)!}{n!} \left(\frac{m!}{(2m)!}\right)^{s+1} (4m+2)^{-k} = \frac{\pi^{s/2}\Gamma(n+1/2)}{(m+1/2)^k \Gamma(m+1/2)^{s+1}}, \tag{17}$$

where

$$\Gamma(z) = \int_0^\infty t^{z-1} e^{-t} dt, \quad z > 0,$$

is the usual Gamma function.

Proof: By an elementary calculation it is easy to see that (15) and the leftmost formula in (17) define the same number for all $n, s \geq 1$. Note that empty products are defined to be one so that the denominator in (15) is equal to one for $m = 0$. For the rightmost formula it suffices to recall the relation (Cf. [7])

$$1 \cdot 3 \cdots (2n-1) = 2^n \pi^{-1/2} \Gamma(n + 1/2), \tag{18}$$

valid for any nonnegative integer n.

We shall prove (15) using induction on s. By (11) we see that (15) holds for $s = 1$ and all $n \geq 1$. Suppose now $s \geq 2$ and $n \geq 1$. By Lemma 4.1

$$K_n(\mathbb{R}^s) = \max_{0 \leq i \leq n} K_{i,n}(\mathbb{R}^s), \quad \text{where} \quad K_{i,n}(\mathbb{R}^s) = \gamma_{i,n} K_{n-i}(\mathbb{R}^{s-1}). \tag{19}$$

Inserting (15) for $s-1$ and (7) into (19) we find the explicit formula

$$K_{i,n}(\mathbb{R}^s) = \frac{(2n-1)(2n-3)\cdots(2l_i+3)(2l_i+1)^{1-j_i}}{1 \cdot 3 \cdots (2i-1) \cdot 1^{s-1} \cdot 3^{s-1} \cdots (2l_i-1)^{s-1}}, \tag{20}$$

where $l_i = \left[\frac{n-i}{s}\right]$ and $j_i = n - i - sl_i$. To determine the i which gives the max of this expression we show that $K_{i,n}(\mathbb{R}^s)$ satisfies the recurrence relation

$$K_{i,n}(\mathbb{R}^s) = \frac{2l_i + 1}{2i - 1} K_{i-1,n}(\mathbb{R}^s), \quad i = 1, \ldots, n. \tag{21}$$

This is clear from (20) if $0 \leq j_i \leq s-2$. For then $l_{i-1} = l_i$ and $j_{i-1} = j_i + 1$. But it also holds for the remaining case $j_i = s - 1$. In this case

$$l_{i-1} = \left[\frac{n-i+1}{s}\right] = \left[\frac{j_i + sl_i + 1}{s}\right] = l_i + 1,$$

and since $j_{i-1} = 0$ we find

$$K_{i-1,n}(\mathbb{R}^s) = \frac{(2n-1)(2n-3)\cdots(2l_i+3)}{1 \cdot 3 \cdots (2i-3) \cdot 1^{s-1} \cdot 3^{s-1} \cdots (2l_i+1)^{s-1}}.$$

Thus comparing this with (20) for $j_i = s-1$ we see that (21) also holds in this case.

From (21) it follows that

$$K_{0,n} < \cdots < K_{m,n} \geq K_{m+1,n} > \cdots > K_{n,n}.$$

Thus $K_n(\mathbb{R}^s) = K_{m,n}(\mathbb{R}^s)$ and computing this value from (20) we see that (15) also holds for s. □

Remark. The above proof also holds for the case $l_i = 0$ if we interpret the product $1 \cdot 3 \cdots (2i-1)$ as 1 if $i = 0$. In particular we obtain from (17)

$$K_n(\mathbb{R}^s) = 1 \cdot 3 \cdot 5 \cdots (2n-1), \quad \text{for} \quad s \geq n-1. \tag{22}$$

Thus $K_n(\mathbb{R}^s)$ becomes independent of the space dimension ,e.g. in the cubic case it is the same for all $s \geq 2$. This is quite remarkable and recommends the Bernstein-Bézier basis with low degree for work in highly multidimensional problems.

5. Asymptotic formulae

To derive asymptotic formulae for the constant $K_n(\mathbb{R}^s)$ we find it convenient to use the Gamma function representation of $K_n(\mathbb{R}^s)$. There is a wealth of formulas for this function see *i.e.*, the classical book [7].

The following theorem generalizes (12) and shows that the number $2^{-s/2}(s+1)^n$ is a good estimate for $K_n(\mathbb{R}^s)$ when n is large compared to s.

Theorem 5.1. *For* $n, s \geq 1$

$$K_n(\mathbb{R}^s) = 2^{-s/2}(s+1)^n(1+r_{n,s}), \quad \text{where} \quad r_{n,s} \sim \frac{s^2}{n}, \quad n \to \infty. \tag{23}$$

More precisely, for $s=2$ *and* $s=3$ *we have*

$$\frac{1}{2}3^n(1-\frac{1}{n}) \leq K_n(\mathbb{R}^2) \leq \frac{1}{2}3^n(1+\frac{1}{n}), \quad n \geq 1, \tag{24}$$

and

$$2^{-3/2}4^n(1-\frac{2}{n}) \leq K_n(\mathbb{R}^3) \leq 2^{-3/2}4^n(1+\frac{2}{n}), \quad n \geq 1. \tag{25}$$

Proof: Taking logarithms in (17) we have

$$\log K_n(\mathbb{R}^s) = \frac{s}{2}\log\pi + \log\Gamma(n+\frac{1}{2}) - (s+1)\log\Gamma(m+\frac{1}{2}) - k\log(m+\frac{1}{2}), \tag{26}$$

where m and k are such that

$$n = (s+1)m + k, \quad \text{with} \quad 0 \leq k \leq s. \tag{27}$$

From [7, p 252-253] we recall Stirlings asymptotic formula for the logarithm of the Gamma function

$$\log\Gamma(z) = (z-1/2)\log z - z + \frac{1}{2}\log 2\pi + \phi(z), \tag{28}$$

where

$$\phi(z) = \frac{1}{12z} - \frac{1}{360z^3} + \frac{1}{1260z^5} - \cdots.$$

Since the series is alternating it is shown in [7 p 253] that for $z > 0$ the upper bound $1/(12z)$ holds for $\phi(z)$. Similarly we also have a lower bound. For $z \geq 1/2$ the bounds take the form

$$\frac{1}{15z} \leq \frac{1}{12z} - \frac{1}{360z^3} < \phi(z) < \frac{1}{12z}. \tag{29}$$

Inserting (28) in (26) and using elementary properties of logarithms it follows after some calculation that

$$K_n(\mathbb{R}^s) = 2^{-s/2}(s+1)^n E_{n,s}, \tag{30}$$

where

$$\log E_{n,s} = \psi(\frac{s}{2} - k, n + \frac{1}{2}) + \phi(n + \frac{1}{2}) - (s+1)\phi(m + \frac{1}{2}), \tag{31}$$

and for any x, z such that $1 + x/z > 0$

$$\psi(x, z) = x - (z - 1/2)\log(1 + \frac{x}{z}). \tag{32}$$

For $\psi(x, z)$ we have for $-1 < x/z < 1$ the series expansion

$$\psi(x, z) = \sum_{k=1}^{\infty}(-1)^{k-1}\frac{x^k(x + \frac{k+1}{2k})}{(k+1)z^k}. \tag{33}$$

To give upper and lower bounds for $E_{n,s}$ we first show that for fixed $m \geq 0$

$$E_{m(s+1)+[\frac{s+1}{2}],s} \leq E_{m(s+1)+k,s} \leq E_{m(s+1),s}, \quad k = 0, \dots, s. \tag{34}$$

Here we define $K_0(\mathbb{R}^s) = E_{0,s} = 1$ for all $s \geq 1$. To show this we combine (30) and (15) to obtain for $n = m(s+1), \dots, m(s+1) + s$

$$\frac{E_{n+1,s}}{E_{n,s}} = \frac{K_{n+1}(\mathbb{R}^s)}{(s+1)K_n(\mathbb{R}^s)} = \frac{n + 1/2}{(s+1)(m+1/2)}. \tag{35}$$

Thus $E_{n+1,s} \leq E_{n,s}$ for $m(s+1) \leq n \leq m(s+1) + s/2$ and $E_{n+1,s} \geq E_{n,s}$ for $m(s+1) + s/2 \leq n \leq m(s+1) + s$. This shows the lower bound in (34) and also we see that the maximum of $E_{n,s}$ with n in the range $n = m(s+1), \dots, m(s+1) + s$ must occur at either $n = m(s+1)$ or at $n = m(s+1) + s$. From (35) it can be seen that

$$\frac{E_{m(s+1)+s,s}}{E_{m(s+1),s}} = \frac{(x - (s - 1/2))(x - (s - 3/2)) \cdots (x + (s - 3/2))(x + (s - 1/2))}{(x + 1/2)(x + 1/2) \cdots (x + 1/2)(x + 1/2)}$$

where $x = (s+1)m + 1/2$. It follows that $E_{m(s+1)+s,s}/E_{m(s+1),s} < 1$ for all $s \geq 1$ and all $m \geq 0$ so the upper bound in (34) follows.

In the following we do not estimate $E_{n,s}$ in (34) for any k but only in the interesting cases for the upper and lower bounds. So suppose now for the upper bound of $E_{n,s}$ that $n = m(s+1)$ for some $m \geq 1$.

For $x > 0$ the series (33) is alternating and taking only the first term in the series we obtain

$$\psi(x, z) < \frac{x(x+1)}{2z}, \tag{36}$$

valid for $x, z > 0$. Using (36), (29), and (31) we obtain with $z = n + 1/2$

$$\begin{aligned}
\log E_{n,s} &= \psi(\frac{s}{2}, z) + \phi(z) - (s+1)\phi(m + \frac{1}{2}) \\
&\leq \frac{s^2 + 2s}{8z} + \frac{1}{12z} - \frac{(s+1)^2}{15(z + s/2)} \\
&< \frac{(s+1)^2}{8z} - \frac{(s+1)^2}{16(z + s/2)} = \frac{(s+1)^2}{16z}\frac{z + s}{(z + s/2)}.
\end{aligned}$$

This proves the upper bound

$$K_n(\mathbb{R}^s) \le 2^{-s/2}(s+1)^n \exp\left(\frac{1}{16}\frac{(s+1)^2}{n+1/2}\cdot\frac{n+1/2+s}{n+1/2+s/2}\right), \quad n \ge 1. \tag{37}$$

We now show the lower bound

$$K_n(\mathbb{R}^s) \ge 2^{-s/2}(s+1)^n \exp\left(-\frac{(s+1)^2+1}{12n}\right), \quad n \ge 1. \qquad (38))$$

To show (38) there are two cases. First, if s is even then we need to consider n of the form $n = m(s+1) + s/2$ for some $m \ge 0$. From (31) and (29) we obtain with $z = n + 1/2$

$$\begin{aligned}\log E_{n,s} &= \psi(0,z) + \phi(n+\frac{1}{2}) - (s+1)\phi(m+\frac{1}{2})\\ &> 0 + \frac{1}{15z} - \frac{(s+1)^2}{12z} > -\frac{(s+1)^2}{12z}.\end{aligned} \tag{39}$$

Next if s is odd then the n which gives the lower bound is of the form $n = m(s+1) + s/2 + 1/2$ for some $m \ge 0$. From (33) we obtain

$$\psi(-\frac{1}{2},z) = -\sum_{k=1}^{\infty}\frac{1}{2^{k+1}k(k+1)z^k} > -\frac{1}{4}\sum_{k=1}^{\infty}(2z)^{-k} = -\frac{1}{8(z-1/2)}.$$

Therefore, from (31) and (29) we now obtain with $z = n + 1/2$ and $z > 4/3$

$$\begin{aligned}\log E_{n,s} &= \psi(-1/2,z) + \phi(n+\frac{1}{2}) - (s+1)\phi(m+\frac{1}{2})\\ &> -\frac{1}{8(z-1/2)} + \frac{1}{15z} - \frac{(s+1)^2}{12(z-1/2)}\end{aligned}$$

For $z > 4/3$ we have $1/(15z) > 1/(24(z-1/2))$ which gives

$$\log E_{n,s} > -\frac{1}{12(z-1/2)}(\frac{12}{8} - \frac{1}{2} + (s+1)^2) = -\frac{(s+1)^2+1}{12(z-1/2)}.$$

This bound is smaller than (39) and since $z - 1/2 = n$ we obtain (38).

Consider next the specific cases $s = 2$. Setting $s = 2$ in the upper bound (37) we find

$$\log E_{n,2} \le \frac{9}{16(n+1/2)}\frac{n+5/2}{n+3/2} =: x.$$

Since $0 < x \le 1/2$ for $n \ge 2$ we have

$$E_{n,2} \le e^x \le 1 + \frac{4}{3}x \le 1 + \frac{1}{n}$$

and the upper bound in (24) follows for $n \geq 2$. But since $K_1(\mathbb{R}^2) = 1$ (24) also holds for $n = 1$. The lower bound follows immediately from (38) and the inequality $e^x \geq 1 - x$ valid for all x.

Consider finally the case $s = 3$. With $x = (n+7/2)/(n+1/2)(n+2))$ we have $0 < x \leq 1/2$ for $n \geq 3$ and for these n we obtain

$$E_{n,3} \leq e^x \leq 1 + \frac{4}{3}x \leq 1 + \frac{2}{n}.$$

It is shown directly that the same bound is valid for $n \leq 2$. Thus the upper bound follows. For the lower bound we argue as for $s = 2$. This completes the proof. □

References

1. de Boor C., On local linear functionals which vanish at all B-splines but one, in *Theory of Approximation with Applications*, Law, A. G. and Sahney, N. B. (eds.), Academic Press, New York, 1976, 120–145.
2. Hoschek J. and Lasser D., *Fundamentals of Computer Aided Geometric Design*, AKPeters, Boston, 1993.
3. Lyche T., A note on the condition numbers of the B-spline bases, J. Approx. Theory **22** (1978), 202–205;
4. Scherer K. and Shadrin A. Yu., New upper bound for the B-Spline basis condition number. East J. Approx. **2** (1996), 331–342.
5. Schumaker L. L., *Spline Functions: Basic Theory*, Wiley, New York; 1981.
6. Stroud A. H., *Approximate Calculation of Multiple Integrals*, Prentice-Hall, Englewood Cliffs, New Jersey, 1971.
7. Whittaker E. T. and Watson G. N., *A Course of Modern Analysis*, Cambridge University Press, London, Fourth Edition 1927.

T. Lyche
University of Oslo
Institutt for informatikk
P. O. Box 1080, Blindern
0316 Oslo, Norway
tom@ifi.uio.no

K. Scherer
Rheinische Friedrich-Wilhems-Universität Bonn
Institut für angewandte Mathematik
Wegelstr. 6, 53115, Bonn
Germany
unm11c@ibm.rhrz.uni-bonn.de

Integration Methods of Clenshaw-Curtis Type, Based on Four Kinds of Chebyshev Polynomials

J. C. Mason and E. Venturino

Abstract. Previously Chebyshev polynomials of the first and second kinds have been used in interpolation methods by taking their zeros and possibly end points as abscissae, and these have then been adopted in quadrature methods of Clenshaw-Curtis and Gauss type. Here we extend these ideas to include Chebyshev polynomials of the third and fourth kinds showing that (at least) 14 discrete orthogonality formulae hold for the four kinds of Chebyshev polynomials at appropriate abscissae, the latter chosen from nine basic sets of abscissae, and each formula yields a general (double summation) Clenshaw-Curtis (C-C) integration formula for integrating $k(x)f(x)$ from -1 to 1. For each of the 9 basic sets of abscissae, there is a particular weight function $k(x)$ of Jacobi type for which the C-C formula is a Gauss formula and for which the double summation formula reduces to a single summation formula. Moreover error estimates have already been derived by Sloan and Smith, based on the equivalence at the abscissae of a Chebyshev polynomial of high degree to one of degree no more than the interpolation degree. We show that a similar equivalence can be established for all 14 choices of polynomials and abscissae. The validity of the formulae is illustrated in some numerical applications.

1. Introduction

The Clenshaw-Curtis quadrature method based on the 1st kind Chebyshev polynomials, was introduced in [1], and a number of variants have been described, such as [2] which is based on the use of 2nd kind Chebyshev polynomials. The method integrates an interpolation polynomial, and the latter is rapidly calculated by exploiting a discrete orthogonality property of Chebyshev polynomials at the chosen abscissae. The zeros of $T_{n+1}(x)$ or those of $(1-x^2)U_{n-1}(x)$ are typically adopted as abscissae. Sloan and Smith [6] give an explanation of the method, based on

Multivariate Approximation and Splines
G. Nürnberger, J. W. Schmidt, and G. Walz (eds.), pp. 153–165.

ISBN 3-7643-5654-5.

the inclusion of a general weight function, and provide some rather effective error estimation techniques.

The aims of the present paper are to provide as complete a family of Clenshaw-Curtis (C-C) methods as possible, by considering four kinds of Chebyshev polynomials and also to note all weight functions for which the C-C method coincides with a Gauss quadrature formula.

2. Chebyshev Polynomials and Properties

The Chebyshev polynomials $T_n(x), U_n(x), V_n(x), W_n(x)$, in x on $[-1,1]$ of the first, second, third and fourth kinds, respectively, are defined (Gautschi [3], Mason [4]) by the following trigonometric formulae, under the transformation $x = \cos\theta$:

$$\begin{aligned} T_n(x) &= \cos n\theta, \qquad U_n(x) = \frac{\sin(n+1)\theta}{\sin\theta}, \\ V_n(x) &= \frac{\cos(n+\frac{1}{2})\theta}{\cos\frac{1}{2}\theta}, \qquad W_n(x) = \frac{\sin(n+\frac{1}{2})\theta}{\sin\frac{1}{2}\theta}. \end{aligned} \tag{2.1}$$

They all possess minimal L_1- and L_2-properties and minimax properties with respect to appropriate weight functions (see Mason [5] for details). Also they are orthogonal on $[-1,1]$ with respect to a Jacobi weight function:

$$\int_{-1}^{1} (1-x)^\alpha (1+x)^\beta P_i(x) P_j(x) = 0, \qquad i \neq j, \tag{2.2}$$

where $\{P_i(x)\}$ is the respective system of polynomials $\{T_i(x)\}, \{U_i(x)\}, \{V_i(x)\}$ and $\{W_i(x)\}$ and α, β are, respectively,

$$\alpha = \beta = -\frac{1}{2}, \qquad \alpha = \beta = \frac{1}{2}, \qquad -\alpha = \beta = \frac{1}{2}, \qquad \alpha = -\beta = \frac{1}{2}.$$

Finally, the polynomials may be interrelated, in terms of a variable

$$y = \sqrt{\frac{1}{2}(1+x)} = \cos\frac{\theta}{2}, \tag{2.3}$$

in the form

$$\begin{aligned} V_n(x) &= y^{-1} T_{2n+1}(y), \\ W_n(x) &= U_{2n}(y). \end{aligned} \tag{2.4}$$

2.1. Chebyshev Abscissae and Discrete Orthogonality Formulae

Chebyshev polynomials also have orthogonality properties over discrete sets of abscissae. Nine sets of abscissae are adopted below, forming three groups. Each set has $(n+1)$ abscissae (for degree n interpolation) of the form,

$$x_{nk} = \cos\theta_{nk} \qquad \text{for } k = 0, 1, \ldots, n,$$

where θ_{nk} are equally spaced. The x_{nk} we rename as t_{nk}, u_{nk}, etc.

The first group include the classical Chebyshev zeros (A1) and the "natural points" (A2) that are traditionally adopted in the C-C rule, and we have also included two other natural choices with only one end point.

$$\begin{array}{ll}
\text{(A1)} & t_{nk} \text{ zeros of } T_{n+1}(x), \quad \theta_{nk} = (k+\frac{1}{2})\pi/(n+1), \\
\text{(A2)} & u^B_{nk} \text{ zeros of } (1-x^2)U_{n-1}(x), \quad \theta_{nk} = k\pi/n, \\
\text{(A3)} & v^L_{nk} \text{ zeros of } (1+x)V_n(x), \quad \theta_{nk} = (k+\frac{1}{2})\pi/(n+\frac{1}{2}), \\
\text{(A4)} & w^R_{nk} \text{ zeros of } (1-x)W_n(x), \quad \theta_{nk} = k\pi/(n+\frac{1}{2}).
\end{array}$$

The superscripts B, L, R are used to indicate, respectively, that both, left and right end points are included as abscissae.

The second group are zeros of the polynomials themselves ($T_{n+1}(x)$ has already been included as (A1)):

$$\begin{array}{ll}
\text{(A5)} & u_{nk} \text{ zeros of } U_{n+1}(x), \quad \theta_{nk} = (k+1)\pi/(n+2), \\
\text{(A6)} & v_{nk} \text{ zeros of } V_{n+1}(x), \quad \theta_{nk} = (k+\frac{1}{2})\pi/(n+\frac{3}{2}), \\
\text{(A7)} & w_{nk} \text{ zeros of } W_{n+1}(x), \quad \theta_{nk} = (k+1)\pi/(n+\frac{3}{2}).
\end{array}$$

The third group also include an end point, but with a wider gap at the other (missing) end point:

$$\begin{array}{ll}
\text{(A8)} & u^L_{nk} \text{ zeros of } (1+x)U_n(x), \quad \theta_{nk} = (k+1)\pi/(n+1), \\
\text{(A9)} & u^R_{nk} \text{ zeros of } (1-x)U_n(x), \quad \theta_{nk} = k\pi/(n+1).
\end{array}$$

There are 4 types of discrete orthogonality formulae, each of which is valid for some of the above sets, namely:

$$S = \sum_{k=0}^{n} W^2(x_{nk})P_i(x_{nk})P_j(x_{nk}) = e_{ij}, \tag{2.5}$$

for $i, j = 0, 1, \ldots, n$ where

$$e_{ij} = 0 \qquad \text{for } i \neq j$$

and

$$\begin{aligned} W(x) &= 1 && \text{for } P_i(x) = T_i(x) \\ W(x) &= (1-x^2)^{\frac{1}{2}} && \text{for } P_i(x) = U_i(x) \\ W(x) &= (1+x)^{\frac{1}{2}} && \text{for } P_i(x) = V_i(x) \\ W(x) &= (1-x)^{\frac{1}{2}} && \text{for } P_i(x) = W_i(x). \end{aligned} \tag{2.6}$$

The summation $\sum$ in (2.5) must be replaced by $\sum'$ (first term halved), $\sum''$ (first and last terms halved), and $\sum^*$ (last term halved) in the respective cases of abscissae with superscripts R, B, and L.

Theorem 2.1. *The 14 choices of polynomials and abscissae for discrete orthogonality (2.5), based on the choices of $W(x)$ in (2.6), are as follows for $i, k = 0, \ldots, n$:*

For $P_i(x) = T_i(x)$:

(D1) t_{nk}; $e_{ii} = \frac{1}{2}(n+1) \quad (i \neq 0), \quad e_{00} = n+1,$

(D2) u^B_{nk}; $e_{ii} = \frac{1}{2}n \quad (i \neq 0, n), \quad e_{00} = n, \quad e_{nn} = n,$

(D3) v^L_{nk}; $e_{ii} = \frac{1}{2}(n+1/2) \quad (i \neq 0), \quad e_{00} = n + \frac{1}{2},$

(D4) w^R_{nk}; $e_{ii} = \frac{1}{2}(n+1/2) \quad (i \neq 0), \quad e_{00} = n + \frac{1}{2}.$

For $P_i(x) = U_i(x)$:

(D5) t_{nk}; $e_{ii} = \frac{1}{2}(n+1) \quad (i \neq n), \quad e_{nn} = n+1,$

(D6) u_{nk}; $e_{ii} = \frac{1}{2}(n+2).$

For $P_i(x) = V_i(x)$:

(D7) t_{nk}; $e_{ii} = n+1,$

(D8) w^R_{nk}; $e_{ii} = n+1/2 \quad (i \neq n), \quad e_{nn} = 2n+1.$

(D9) v_{nk}; $e_{ii} = n+3/2,$

(D10) u^R_{nk}; $e_{ii} = n+1.$

For $P_i(x) = W_i(x)$:

(D11) t_{nk}; $e_{ii} = (n+1),$

(D12) v^L_{nk}; $e_{ii} = n+1/2 \quad (i \neq n), \quad e_{nn} = 2n+1,$

(D12) w_{nk}; $e_{ii} = n+3/2,$

(D14) u^L_{nk}; $e_{ii} = n+1.$

Proof: It suffices to consider only cases (D1),…,(D10). From (2.5), for the three respective polynomials T, U, V:

$$
\begin{aligned}
(i)\quad S = S_1 &= \sum \cos(i\theta_{nk})\cos(j\theta_{nk}) \\
&= \frac{1}{2}\sum \left[\cos(i+j)\theta_{nk} + \cos(i-j)\theta_{nk})\right]. \\
(ii)\quad S = S_2 &= \sum \sin(i+1)\theta_{nk}\sin(j+1)\theta_{nk} \\
&= \frac{1}{2}\sum \left[\cos(i-j)\theta_{nk} - \cos(i+j+2)\theta_{nk}\right], \\
(iii)\quad S = S_3 &= 2\sum \cos(i+1/2)\theta_{nk}\cos(j+1/2)\theta_{nk} \\
&= \sum \left[\cos(i+j+1)\theta_{nk} + \cos(i-j)\theta_{nk})\right].
\end{aligned}
$$

For $i \neq j$, each S is a sum or difference of two summations of the form

$$
\sum_{k=0}^{n} \cos r\theta_{nk} \tag{2.7}
$$

where r is an integer between 1 and $2n, 2n+2, 2n+1$ (respectively). To demonstrate orthogonality, it therefore suffices to show that the sum (2.7) is zero in each case under consideration (with $\sum$ replaced by $\sum', \sum'', \sum^*$ as appropriate).

First consider the formulae (D1)-(D5), (D7)-(D8), (D11)-D(12) which correspond to the four sets of abscissae (A1), (A2), (A3), (A4). For (A1), (A2), (A3), the sum (2.7) gives respectively

$$
\begin{aligned}
&\sum_{k=0}^{n} \cos\left[r(k+1/2)\pi/(n+1)\right], \\
&\sum_{k=0}^{n}{}'' \cos\left[rk\pi/n\right], \\
&\sum_{k=0}^{n}{}^{*} \cos\left[r(k+1/2)\pi/(n+1/2)\right].
\end{aligned} \tag{2.8}
$$

Now consider the well known formulae:

$$
\begin{aligned}
S_n^{(1)}(\theta) &= \sum_{k=0}^{n} \cos(k+1/2)\theta = \frac{1}{2}\sin(n+1)\theta/\sin\frac{1}{2}\theta, \\
S_n^{(2)}(\theta) &= \sum_{k=0}^{n}{}'' \cos k\theta = \frac{1}{2}\sin n\theta \cot\frac{1}{2}\theta.
\end{aligned}
$$

Clearly $S_n^{(1)}(\theta)$ vanishes at $r\pi/(n+1)$ and $S_n^{(2)}(\theta)$ vanishes at $r\pi/n$ and hence the first two sums of (2.8) are zero provided $\sin(\theta/2)$ does not vanish (i.e. $r \neq 2n+2, 2n$ respectively). This condition is satisfied for the cases covered here. (However, it is not satisfied, for example, for the abscissae (A2) and the polynomials U, V, although orthogonality does then hold for polynomials of degree up to $n-2, n-1$ respectively.)

Now,

$$S_{2n}^{(1)}(\theta) = 2\sum_{k=0}^{n}{}^{*} \cos(k+1/2)\theta = \sin(n+1/2)\theta\cos(n+1/2)\theta/\sin\frac{1}{2}\theta.$$

The latter follows on observing that terms for $k = n+1, \dots, 2n$ coincide with terms for $k = 0, \dots, n$ in reverse order. This sum vanishes at $r\pi/(n+1/2)$, and hence the last sum of (2.8) is zero provided $\sin(\theta/2)$ does vanish (ie. $r \neq 2n+1$). The latter condition is satisfied for the cases considered here. (It is not satisfied, for example, for the abscissae (A3) and polynomials U, V or for the abscissae (A4) and polynomials U, W, although orthogonality is valid in these cases for polynomials up to degree $n-1$.)

Now consider the cases where $i = j$. Clearly $S_1 = \sum(1), \sum{}''(1), \sum{}^{*}(1)$ for $i = j = 0$, and this gives $n+1, n, n+1/2$ for e_{00} for (D1), (D2), (D3). These values are halved for $i = j \neq 0$. Similarly, for all $i = j$, $S_2 = \frac{1}{2}\sum(1), \frac{1}{2}\sum{}''(1), \frac{1}{2}\sum{}^{*}(1)$, and $S_3 = \sum(1), \sum{}''(1), \sum{}^{*}(1)$, and this readily gives e_{ii} for the remaining cases. It now suffices to prove (D6), (D9), (D10).

Now for the abscissae (A2) and polynomials U, we have (see above)

$$\begin{aligned}\sum_{k=0}^{n}{}'' (1-x_{nk}^2)U_i(x_{nk})U_j(x_{nk}) &= 0 \qquad i \neq j \qquad (i, j = 0, \dots, n-2) \\ &= \frac{1}{2}n \qquad (i = j).\end{aligned}$$

since orthogonality is valid for polynomials of degree up to $n-2$. Here x_{nk} includes ± 1 together with the zeros of $U_{n-1}(x)$. The first and last terms of the sum vanish at ± 1. The remainder of the sum provides (D6), on replacing n by $n+2$.

Now formula (D1) with n replaced by $N = 2n+2$, gives

$$\begin{aligned}\sum_{k=0}^{N} T_{2i+1}(y_k)T_{2j+1}(y_k) &= 0 \qquad i \neq j \\ &= \frac{1}{2}(2n+3) \qquad i = j.\end{aligned}$$

Thus,

$$\begin{aligned}\sum_{k=0}^{2n+2} 2(1+x_k)V_i(x_k)V_j(x_k) &= 0 \qquad i \neq j \\ &= (2n+3) \qquad i = j\end{aligned} \tag{2.9}$$

where

$$x_k = v_{nk} = \cos\left[(k+1/2)\pi/(n+3/2)\right]$$

and

$$y_k = t_{Nk} = \cos\left[(k+1/2)\pi/(N+1)\right], \qquad 2y_k^2 = 1 + x_k.$$

This follows from (2.3)-(2.4) above.

Now the $k = n+1$ term in (2.9) is zero, and the $k = n+2, \ldots, 2n+2$ terms repeat in reverse order the $k = 0, \ldots, n$ terms. Hence (2.9) is also valid for the sum from 0 to n, and this gives (D9).

Formula (D10) may be obtained in a similar way, by replacing n by $N = 2n+2$ in (D2). In this case

$$x_k = u_{nk}^R = \cos\left[k\pi/(n+1)\right], \qquad y_k = u_{Nk}^B = \cos(k\pi/N).$$

$\sum$ is replaced by $\sum''$ in (2.9), and hence (2.9) is valid for the sum from $k = 0$ to n with $\sum$ replaced by $\sum'$. □

3. Generalised Clenshaw-Curtis Quadrature

The Clenshaw-Curtis method, generalised to include an arbitrary weight function $k(x)$, forms the integral

$$I(f) = \int_{-1}^{1} k(x) f(x) dx, \tag{3.1}$$

by replacing $f(x)$ by a polynomial of degree n, which interpolates at a selected set of abscissae, and then integrating it (see Sloan and Smith [6]). Traditionally, Chebyshev zeros or natural points, i.e. (A1) or (A2), are chosen, and the polynomial is expressed in terms of $\{T_k(x)\}$ or $\{U_k(x)\}$, by exploiting discrete orthogonality — this gives some 4 formulae. However, by using the results above, we may form 14 formulae.

We replace $f(x)$ in (3.1) by a weighted interpolating Chebyshev polynomial sum

$$f_n(x) = W(x) \sum_{i=0}^{n} b_i P_i(x), \tag{3.2}$$

where $W(x)$ is given by (2.6) for each of the choices $P_i = T_i, U_i, V_i, W_i$.

From (3.2), using any of the 14 choices of $\{x_{nk}\}$ in Theorem 2.1

$$\sum_{k=0}^{n} W(x_{nk}) f_n(x_{nk}) P_j(x_{nk}) = \sum_{i=0}^{n} b_i \sum_{k=0}^{n} W^2(x_{nk}) P_i(x_{nk}) P_j(x_{nk}) = \sum b_i e_{ii}.$$

Hence

$$b_i = e_{ii}^{-1} \sum_{k=0}^{n} W(x_{nk}) f(x_{nk}) P_i(x_{nk}), \tag{3.3}$$

since f_n interpolates f in $\{x_{nk}\}$. We note that the summation $\sum$ in (3.3) needs to be replaced by $\sum', \sum'', \sum^*$ respectively, in the case of abscissae (in Theorem 2.1) with superscripts R, B, L.

From (3.2)

$$I(f) \simeq I_n(f) = \int_{-1}^{1} k(x) f_n(x) dx = \sum_{i=0}^{n} b_i \int_{-1}^{1} k(x) W(x) P_i(x) dx.$$

So

$$I_n(f) = \sum_{i=0}^{n} a_i b_i, \tag{3.4}$$

where

$$a_i = \int_{-1}^{1} k(x) W(x) P_i(x) dx \qquad (i = 0, \ldots, n). \tag{3.5}$$

The formula (3.4) is exact if f is a polynomial of exact degree n weighted by $W(x)$. The method is intended primarily for weights $k(x), W(x)$ for which the integrals (3.5) may be calculated conveniently or indeed exactly.

3.1. Special choice of $k(x)$

Lemma 3.1. *The coefficients $a_i, (i > 0)$ given by (3.5) vanish for the choices $k(x) = (1-x^2)^{-\frac{1}{2}}, 1, (1-x)^{-\frac{1}{2}}, (1+x)^{-\frac{1}{2}}$ for $P_i(x) = T_i(x), U_i(x), V_i(x), W_i(x)$ respectively with $W(x)$ defined as in (2.6).*

Proof:

The result follows immediately from the fact that $P_i(x)$ is orthogonal to $P_0(x) = 1$ with respect to $k(x)W(x)$. □

Note from (3.4) that in the 4 cases of Lemma 3.1 only

$$I_n(f) = a_0 b_0 = a_0 e_{00}^{-1} \sum_{k=0}^{n} W(x_{nk}) f(x_{nk}) P_0(x_{nk}). \tag{3.7}$$

Thus the "double summation" formula (3.4) with (3.3) has reduced to the single summation formula (3.7) in these cases. There are of course 14 formulae that can be used in all, but some simple examples, corresponding to (i)(ii)(iii) above, are:

$$(i) \qquad I_n(f) = \frac{\pi}{n+1} \sum_{k=0}^{n} f(x_{nk}) \qquad \text{for } x_{nk} = t_{nk},$$

$$I_n(f) = \frac{\pi}{n} \sum_{k=0}^{n}{}'' f(x_{nk}) \qquad \text{for } x_{nk} = u_{nk}^{B},$$

$$(ii) \qquad I_n(f) = \frac{\pi}{n+2} \sum_{k=0}^{n} (1 - x_{nk}^2)^{\frac{1}{2}} f(x_{nk}) \qquad \text{for } x_{nk} = u_{nk},$$

$$(iii) \qquad I_n(f) = \frac{\pi}{n+3/2} \sum_{k=0}^{n} (1 + x_{nk})^{\frac{1}{2}} f(x_{nk}) \qquad \text{for } x_{nk} = v_{nk}.$$

4. Gauss Quadrature Formulae

There is one Gauss quadrature formula for each of the 9 sets of abscissae above, (A1),...,(A9), each corresponding to one of the cases (i)-(iv) of Lemma 3.1. This means that both $k(x)$ and $W(x)$ must have the fixed form prescribed in Lemma 3.1 for the relevant Chebyshev polynomial.

Theorem 4.1. *The following 9 choices provide Gauss quadrature formulae for the four respective cases (i), (ii), (iii), (iv) of Lemma 3.1:*

(i) *(D1),(D2),(D3),(D4) corresponding to (A1),(A2),(A3),(A4), respectively,*

(ii) *(D6) corresponding to (A5),*

(iii) *(D9),(D10) corresponding to (A6),(A9), respectively,*

(iv) *(D13),(D14) corresponding to (A7),(A8), respectively.*

Proof: The Clenshaw-Curtis rule is exact for weighted polynomials of degree n. It therefore becomes a Gauss quadrature formula if the polynomial, whose zeros are the abscissae, is from a system orthogonal with respect to $k(x)W(x)$.

It follows immediately that (D1),(D6),(D9),(D13) provide Gauss quadrature formulae for the respective cases (i),(ii),(iii),(iv) of Lemma 3.1. For in each case this simply corresponds to interpolation at the zeros of P_{n+1} while using $\{P_i\}$. These formulae are thus exact for polynomials of degree $2n+1$. (The results are well known for cases (i),(ii).)

Other cases follow by considering similar results for Radau/Lobatto formulae (ie including one or more end points). Consider (i), where $kW = (1-x^2)^{-\frac{1}{2}}$ is the weight function for orthogonality of $\{T_i\}$. The zeros of $(1-x^2)U_{n-1}(x)$ yield a Gauss-Lobatto formula, exact for polynomials of degree $2n-1$ since

$$(1-x^2)k(x)W(x) \equiv (1-x^2)^{\frac{1}{2}}$$

is the weight function for the orthogonality of $\{U_i\}$. More precisely, $\int kWf dx$ is already exact for polynomials of degree n. Now, for P_{2n-1} of degree $2n-1$,

$$\int kWP_{2n-1}dx = \int kW\left[(1-x^2)U_{n-1}q_{n-2} + r_n\right]dx$$

for some polynomials q_{n-2}, r_n of degrees $n-2$, n. By orthogonality,

$$\int kW(1-x^2)U_{n-1}q_{n-2}dx = 0,$$

and hence the integration is exact, since P_{2n-1} interpolates r_n in the zeros of $(1-x^2)U_{n-1}$. (This result is already known.)

Similarly for (i), the zeros of $(1+x)V_n(x)$ yield a Gauss-Radau formula, exact for polynomials of degree $2n$, since

$$(1+x)k(x)W(x) \equiv (1-x)^{-\frac{1}{2}}(1+x)^{\frac{1}{2}}$$

and is the weight function for orthogonality of $\{V_i(x)\}$ This covers (D2),(D3) for (i), and (D4) follows similarly.

Consider (iii), where $kW = (1-x)^{-\frac{1}{2}}(1+x)^{\frac{1}{2}}$ is the weight function for orthogonality of $\{V_i\}$. Then the zeros of $(1-x)U_n(x)$ yield a Gauss-Radau formula, since

$$(1-x)k(x)W(x) \equiv (1-x^2)^{\frac{1}{2}}$$

is a weight function for the orthogonality of $\{U_i\}$. This covers (D10) for (iii), and (D14) for (iv) follows similarly. □

5. Interpolation to Higher Degree Chebyshev Polynomials

Sloan and Smith [6] give an effective error estimation procedure for Clenshaw-Curtis integration, primarily for the case $(D2)$. Their error formula exploits the result, for $j > n$:

$$I_n(T_j(x)) = I_n(T_{j'}(x)), \tag{5.1}$$

where $j'(n,j)$ is at most n. For $(D2)$ they show that

$$\begin{aligned} j'(n,j) &= j \qquad (0 \le j \le n), \\ j'(n,j) &= 2n - j \qquad (n+1 \le j \le 2n), \\ j'(n,j+2n) &= j'(n,j). \end{aligned} \tag{5.2}$$

The equation (5.1) is equivalent to the statement that:

$$T_{j'} \text{ interpolates } T_j \text{ in } \{x_{nk}\}. \tag{5.3}$$

In fact a similar error estimation procedure can be developed for all 14 methods of C-C type, and in each case analogous formulae to those above occur. Specifically at the abscissae x_{nk}

$$P_{j'} = P_j \text{ with } j' = j, \qquad (j = 0, \ldots, n), \tag{5.4}$$

$$P_j = 0 \text{ for certain values of } j, \tag{5.5}$$

$$P_{j'} = \pm P_j \text{ for } j' = j + p \quad (\text{all } j), \tag{5.6}$$

$$P_{j'} = \pm P_j \text{ for } j' = j'(p,j) \quad (j = n_1, \ldots, n_2), \tag{5.7}$$

where p and $j'(p,j)$ take appropriate sets of values, as per the following lemma.

Theorem 5.1. *(5.4),(5.5),(5.6), and (5.7) hold for each of the 14 cases $(D1)-(D14)$ at $\{x_{nk}\}$, the details of the formulae being as follows:*

	(5.5)	(5.6)		(5.7)		
Choice	$P_j = 0$	p	Sign	n_1, n_2	$j'(p, j)$	Sign
$(D1)$	$j = n + 1$	$2n + 2$	$-$	$n + 2, 2n + 1$	$p - j$	$-$
$(D2)$		$2n$	$+$	$n + 1, 2n$	$p - j$	$+$
$(D3)$		$2n + 1$	$-$	$n + 1, 2n + 1$	$p - j$	$-$
$(D4)$		$2n + 1$	$+$	$n + 1, 2n + 1$	$p - j$	$+$
$(D5)$	$j = 2n + 1$	$2n + 2$	$-$	$n + 1, 2n$	$p - j - 2$	$+$
$(D6)$	$j = n + 1, 2n + 3$	$2n + 4$	$+$	$n + 2, 2n + 2$	$p - j - 2$	$-$
$(D7)$		$2n + 2$	$-$	$n + 1, 2n + 1$	$p - j - 1$	$-$
$(D8)$		$2n + 1$	$+$	$n + 1, 2n$	$p - j - 1$	$+$
$(D9)$	$j = n + 1$	$2n + 3$	$-$	$n + 2, 2n + 2$	$p - j - 1$	$-$
$(D10)$		$2n + 2$	$+$	$n + 1, 2n + 1$	$p - j - 1$	$+$
$(D11)$		$2n + 2$	$-$	$n + 1, 2n + 1$	$p - j - 1$	$+$
$(D12)$		$2n + 1$	$-$	$n + 1, 2n$	$p - j - 1$	$+$
$(D13)$	$j = n + 1$	$2n + 3$	$+$	$n + 2, 2n + 2$	$p - j - 1$	$-$
$(D14)$		$2n + 2$	$+$	$n + 1, 2n + 1$	$p - j - 1$	$-$

Proof: The proofs are elementary and are left to the reader. For example, for $(D1)$, we have

$$\begin{aligned}\cos j(k + 1/2)\pi/(n + 1) &= -\cos(2n + 2 + j)(k + 1/2)\pi/(n + 1) \\ &= -\cos(2n + 2 - j)(k + 1/2)\pi/(n + 1)\end{aligned}$$

Similarly, for $(D6)$, we have

$$\begin{aligned}\sin(j + 1)(k + 1)\pi/(n + 2) &= \sin(2n + 4 + j + 1)(k + 1)\pi/(n + 2) \\ &= -\sin(2n + 4 - j - 2 + 1)(k + 1)\pi/(n + 2)\end{aligned}$$

and the left hand side vanishes for $j = n + 1, 2n + 3$. □

6. Numerical Tests

Finally we carry out some model integrations to confirm the validity of the formulae based on choices (D1)-(D10) of Theorem 2.1, and to note the particular behaviour of Gauss formulae. The tables all show the actual errors in the results for integrals

$$\int_{-1}^{1} k(x)f(x)dx$$

where $f(x) = W(x)F(x)$.

Example 1

Here $k(x) = (1-x^2)^{-\frac{1}{2}}$, $W(x) = 1$, $P_i(x) = T_i(x)$, $F(x) = \exp(x)$, and $a_0 = \pi$ ($a_i = 0, i > 0$). The (D1), (D2), (D3), (D4) errors are entirely consistent with Gauss formulae exact for polynomials of degrees $2n+1$, $2n-1$, $2n$, $2n$ respectively.

n	(D1)	(D2)	(D3)	(D4)
2	$.141 \times 10^{-03}$	$-.171 \times 10^{-02}$	$.170 \times 10^{-02}$	$-.170 \times 10^{-02}$
3	$.625 \times 10^{-06}$	$-.141 \times 10^{-03}$	$.100 \times 10^{-04}$	$-.100 \times 10^{-04}$
4	$.172 \times 10^{-08}$	$-.625 \times 10^{-06}$	$.346 \times 10^{-07}$	$-.346 \times 10^{-07}$
5	$.326 \times 10^{-11}$	$-.172 \times 10^{-08}$	$.784 \times 10^{-10}$	$-.784 \times 10^{-10}$
6	$.399 \times 10^{-14}$	$-.326 \times 10^{-11}$	$.126 \times 10^{-12}$	$-.124 \times 10^{-12}$

Example 2

Here $k(x) = 1$, $W(x) = (1-x^2)^{\frac{1}{2}}$, $P_i(x) = U_i(x)$, $F(x) = \exp(x)$, and $a_0 = \pi/2$ ($a_i = 0, i > 0$). The (D5) and (D6) errors are consistent with Clenshaw-Curtis and Gauss formulae, respectively, though the (D5) errors are smaller than expected.

n	(D5)	(D6)
2	$-.422 \times 10^{-02}$	$.350 \times 10^{-04}$
3	$-.350 \times 10^{-04}$	$.155 \times 10^{-06}$
4	$-.155 \times 10^{-06}$	$.430 \times 10^{-09}$
5	$-.430 \times 10^{-09}$	$.814 \times 10^{-12}$
6	$-.813 \times 10^{-12}$	$.888 \times 10^{-15}$

Example 3

Here $k(x) = (1+x)^{-\frac{1}{2}}$, $W(x) = (1+x)^{\frac{1}{2}}$, $P_i(x) = V_i(x)$, and $F(x) = (x+3)^{-1}$. The coefficients are given by $a_{2m+1} = -2/(2m+1)$, $a_{2m} = 2/(2m+1)$, $m = 0, 1, \ldots$. Clearly (D7), (D8), (D9), (D10) have comparable errors, consistent with Clenshaw-Curtis rules.

n	(D7)	(D8)	(D9)	(D10)
3	$-.937 \times 10^{-04}$	$-.146 \times 10^{-03}$	$.158 \times 10^{-03}$	$.336 \times 10^{-03}$
6	$-.498 \times 10^{-07}$	$-.207 \times 10^{-06}$	$.506 \times 10^{-06}$	$.899 \times 10^{-06}$
9	$-.337 \times 10^{-09}$	$-.412 \times 10^{-09}$	$.171 \times 10^{-08}$	$.316 \times 10^{-08}$
12	$-.340 \times 10^{-12}$	$-.129 \times 10^{-11}$	$.700 \times 10^{-11}$	$.122 \times 10^{-10}$
15	$-.333 \times 10^{-14}$	$-.366 \times 10^{-14}$	$.276 \times 10^{-13}$	$.498 \times 10^{-13}$

References

1. Clenshaw C. W. and Curtis A. R., A method for numerical integration on an automatic computer, Numer. Math. **2** (1960), 197–205.
2. Filippi S., Angenäherte Tschebyscheff-Approximation einer Stammfunktion - eine Modifikation des Verfahrens von Clenshaw-Curtis, Numer. Math. **6** (1964), 320.
3. Gautschi W., On mean convergence of extended Lagrange interpolation, J. Comput. Appl. Math. **43** (1992), 19–35.
4. Mason J. C., Chebyshev polynomials of the second, third and fourth kinds in approximation, indefinite integration and integral transforms. J. Comput. Appl. Math. **49** (1993), 169–178.
5. Mason J. C., Minimality properties and applications of four kinds of Chebyshev polynomials, in *Approximation Theory*, Müller W. M., Felten M. and Mache D. H. (eds.), Mathematical Research Vol 86, Akademic Verlag, Berlin, 1995, 231–250.
6. Sloan I. H. and Smith W. E., Product integration with the Clenshaw-Curtis points: implementation and error estimates. Numer. Math. **34** (1980), 387–401.

Professor J. C. Mason.
Head of Mathematics and Statistics Department,
School of Computing and Mathematics,
University of Huddersfield,
Queensgate,
Huddersfield,
West Yorkshire,
HD1 3DH
U.K.
`j.c.mason@hud.ac.uk`

Dr. E. Venturino.
School of Mathematics,
University of Leeds,
Leeds,
LS2 9JT
U.K.
`egv@amsta.leeds.ac.uk`

References

Tensor Products of Convex Cones

Bernd Mulansky

Abstract. Motivated by problems of shape preserving tensor product interpolation, tensor products of convex cones in finite dimensional linear spaces are studied in this paper. We recall the notions of projective and injective tensor product cones and derive some of their properties. It is shown that the cones usually considered in shape preserving tensor product interpolation can be represented as intersections of injective tensor product cones. Consequently, sufficient conditions for the fulfillment of the shape constraints are easily derived from corresponding conditions in the univariate case.

1. Introduction

To motivate the following considerations of tensor products of convex cones, we start with a brief discussion of shape preserving tensor product interpolation.

Given a finite dimensional linear space $S^x \subset C[0,1]$, a linear map $A^x \in L(S^x, \mathbb{R}^n)$, a nonempty convex set $C^x \subset S^x$, and data $d \in \mathbb{R}^n$, a problem of (univariate) shape preserving interpolation is to find a function $s^x \in S^x$ fulfilling the interpolation conditions $A^x s^x = d$ and the shape constraints $s^x \in C^x$. Frequently, the set C^x is a closed cone (we concentrate on this case here), *e.g.*, the cone of nonnegative, monotone, or convex functions, or the intersection of a number of shifts of such cones.

As a specific example we mention nonnegativity preserving Lagrange interpolation. In this problem, the map A^x is given by $A^x s^x = (s^x(x_i))_{i=1}^n$, where $0 = x_1 < x_2 < \cdots < x_n = 1$ is a grid of interpolation points. The constraints are described by $C^x = C^x_{pos} = \{s^x \in S^x : \ s^x(u) \geq 0, \ u \in [0,1]\}$, the cone of nonnegative functions in S^x. Usually, the construction of nonnegative interpolants is accomplished by choosing a suitable space S^x of, *e.g.*, polynomial or rational spline functions, and a subcone $\tilde{C}^x \subset C^x$ defined by simple sufficient nonnegativity conditions, such that $A^x[\tilde{C}^x] = \mathbb{R}^n_+$. This just means the existence of nonnegative

Multivariate Approximation and Splines
G. Nürnberger, J. W. Schmidt, and G. Walz (eds.), pp. 167–176.

ISBN 3-7643-5654-5.

interpolants (even in $\tilde{C}^x$) for all nonnegative data d. The selection of a particular solution is often based on the minimization of a smoothness functional.

A possible way to extend univariate interpolation methods to bi- and multivariate problems is the tensor product construction. For that, let additionally be given a space $S^y \subset C[0,1]$ in the y-direction and a map $A^y \in L(S^y, \mathbb{R}^m)$. Given the data values $d \in \mathbb{R}^n \otimes \mathbb{R}^m$, we are looking for a function $s \in S^x \otimes S^y \subset C([0,1]^2)$ fulfilling the tensor product interpolation conditions $(A^x \otimes A^y)s = d$ and the shape constraints $s \in C$ prescribed by a convex cone $C \subset S^x \otimes S^y$. For the tensor product notions used throughout this paper we refer to [7,8,11].

In the problem of nonnegative tensor product Lagrange interpolation, the map $A^x \otimes A^y$ on $S^x \otimes S^y$ is obtained as $(A^x \otimes A^y)\, s = (s(x_i, y_j))_{i=1,j=1}^{n,m}$, *i.e.*, the interpolation conditions are given on a rectangular grid. Obviously, we take $C = C_{pos} = \{s \in S^x \otimes S^y :\ s(u,v) \geq 0,\ (u,v) \in [0,1]^2\}$, the cone of nonnegative functions in $S^x \otimes S^y$.

Further shape constraints of interest in tensor product interpolation are described, *e.g.*, by the cone of bi-monotone or bi-convex functions, *i.e.*, the cone of functions which are monotone respective convex in the x-direction as well as in the y-direction.

The efficient utilization of the tensor product structure in the computation of the interpolants is well established in the unconstrained case, *i.e.*, in the absence of shape constraints [7]. For shape preserving tensor product interpolation, some questions arise naturally. How are the usual cones $C \subset S^x \otimes S^y$ related to corresponding cones $C^x \subset S^x$ and $C^y \subset S^y$? Can sufficient conditions for $s \in C$ be easily obtained from sufficient conditions established for the univariate case?

These questions lead us to the study of tensor products of convex cones. In section 2, the projective and the injective tensor product cone of two convex cones is introduced, and some properties of these notions are presented. In section 3 we show that the cones C usually considered in shape preserving tensor product interpolation are obtained as the intersection of injective tensor product cones. Therefore, sufficient conditions for $s \in C$ can be easily generated using sufficient conditions for the univariate case. Finally, we give some hints on possible applications and further developments.

Particular instances of shape preserving interpolation by tensor product spline functions have been considered in [1,4–6,9,17–20]. We have presented a first attempt to a more abstract approach (as proposed here) in [13], this approach is further utilized in [14,21,22].

2. Projective and Injective Tensor Product Cones

At first, we recall some notions and well-known facts concerning convex cones in finite dimensional spaces.

Let E be a finite dimensional linear space (we consider real spaces only) and let E^* denote the dual space of E, *i.e.*, the space of all linear functionals on E. The bidual E^{**} of E is isomorphic to E, and we identify E^{**} and E. For a set

$M \subset E$, the *convex conical hull* con M of M is defined by

$$\text{con } M := \left\{ \sum_{i=1}^{r} \alpha_i e_i : \, r \in \mathbb{N}, e_i \in M, \alpha_i \in \mathbb{R}_+ \right\},$$

and the *dual (polar) cone* $M^* \subset E^*$ of M by

$$M^* := \{\lambda \in E^* : \, \lambda e \geq 0, \, e \in M\},$$

which is a closed cone (see below). On several occasions we will use the famous bipolarity theorem.

Theorem 1. *For every set $M \subset E$, it holds*

$$M^{**} = \text{cl}\,(\text{con } M).$$

A set $C \subset E$ is called a (convex) *cone*, if $C = \text{con } C$. A cone $C \subset E$ is said to be *pointed*, if $C \cap (-C) = \{0\}$; *solid*, if int $C \neq \emptyset$; and *generating*, if $C - C = E$.

Proposition 2. *Let $C \subset E$ be a cone.*

(i) *The cone C is solid if and only if C is generating.*

(ii) *If* cl C *is pointed, then C^* is solid.*

(iii) *If C is solid, then C^* is pointed.*

A cone $C \subset E$ is called *proper*, if it is closed, pointed, and solid; then C^* is also proper. If $C = \text{con } M$ for some finite set M, then C is *polyhedral.* Polyhedral cones are always closed, and the dual cone C^* of a polyhedral cone C is polyhedral. In particular, a cone C is *simplicial (minihedral)*, if C is solid and $C = \text{con } M$ for some set M of cardinality equal to the dimension of E (in other words, if the number of extreme rays of C equals the dimension of E). If C is simplicial, then C^* simplicial. Since this result may be less known, we refer to Proposition 4 below for a proof outline and to [10] for historical comments and further references. It is worthwhile to mention that in a two-dimensional space E the only proper cones are the simplicial ones.

The bipolarity theorem serves to formulate the reverse implications of the statements above.

Now we turn to the consideration of tensor products of cones. Given the finite dimensional linear spaces E and F, the (algebraic) *tensor product* $E \otimes F$ of E and F consists of all linear combinations of tensor products $e \otimes f$, where $e \in E$ and $f \in F$. In case of function spaces E and F, e.g., $E, F \subset C[0,1]$, the tensor product of elements can be simply defined by

$$(e \otimes f)\,(u,v) = e(u) \cdot f(v), \quad (u,v) \in [0,1]^2.$$

The tensor product $\lambda \otimes \mu$ of linear functionals $\lambda \in E^*$ and $\mu \in F^*$ is obtained by bilinear extension from the basic rule

$$(\lambda \otimes \mu)(e \otimes f) = (\lambda e) \cdot (\mu f).$$

The spaces $E^* \otimes F^*$ and $(E \otimes F)^*$ are isomorphic and can be identified.

There are (at least) two suggested ways to introduce a tensor product cone of cones $C \subset E$ and $D \subset F$, see, e.g., [12,16,24]. The *projective* tensor product cone $C_p(C, D)$ of C and D is defined by

$$C_p(C, D) := \text{con}\,\{e \otimes f : \; e \in C, \, f \in D\},$$

and the *injective* tensor product cone $C_i(C, D)$ of C and D is defined by

$$C_i(C, D) := \{z \in E \otimes F : \; (\lambda \otimes \mu)\, z \geq 0, \quad \lambda \in C^*, \; \mu \in D^*\}.$$

As stated in the following proposition, these notions are dual to each other.

Proposition 3. *Let $C \subset E$ and $D \subset F$ be cones, then*

(i) $C_i(C, D) = C_p(C^*, D^*)^*$,

(ii) $\text{cl}\; C_p(C, D) = C_i(C^*, D^*)^*$.

Proof: Since always $M^* = (\text{con}\; M)^*$, (i) follows immediately from the definitions. By (i) and the bipolarity theorem, we get

$$C_i(C^*, D^*)^* = C_p(C^{**}, D^{**})^{**} = \text{cl}\; C_p(\text{cl}\; C, \text{cl}\; D) = \text{cl}\; C_p(C, D). \qquad \square$$

It should be mentioned that the projective cone $C_p(C, D)$ is not necessarily closed, even if C and D are proper. Further properties of the projective and injective tensor product cones are collected in the next proposition.

Proposition 4. *For cones $C \subset E$ and $D \subset F$, the following results hold:*

(i) *if C and D are proper, then* $\text{cl}\; C_p(C, D)$ *and* $C_i(C, D)$ *are proper,*

(ii) *if C and D are polyhedral, then $C_p(C, D)$ and $C_i(C, D)$ are polyhedral (in particular, $C_p(C, D)$ is closed),*

(iii) $\text{cl}\; C_p(C, D) \subset C_i(C, D)$,

(iv) *if C or D is simplicial, then* $\text{cl}\; C_p(C, D) = C_i(C, D)$,

(v) *if C and D are simplicial, then $C_p(C, D) = C_i(C, D)$ is simplicial,*

(vi) *if C is a proper cone, then* $\text{cl}\; C_p(C^*, C) = C_i(C^*, C)$ *if and only if C is simplicial.*

Proof: Apart from (vi), the proofs can be easily given using Propositions 2 and 3, the results recalled after Proposition 2, and the bipolarity theorem. We only give a brief outline. The statement (iii) is well-known [12,24], and its proof is straightforward, see [13]. It is obvious that $C_p(C,D)$ is generating if C and D are. By (iii), also $C_i(C,D)$ is generating, and (i) can be obtained. Furthermore, it is also clear that $C_p(C,D)$ is polyhedral whenever C and D are polyhedral, and this gives (ii).

To prove (iv), we have to show that $C_i(C,D) \subset \text{cl}\ C_p(C,D)$, if C is simplicial. Then there is a basis $\{e_1,\dots,e_n\}$ of E such that $C = \text{con}\ \{e_1,\dots,e_n\}$, and a dual basis $\{\lambda_1,\dots,\lambda_n\}$ of E^*, i.e., $\lambda_i e_j = \delta_{ij}$. Hence $C^* = \text{con}\ \{\lambda_1,\dots,\lambda_n\}$. Using a basis $\{f_1,\dots,f_m\}$ of F, every element $z \in E \otimes F$ has a unique representation

$$z = \sum_{i=1}^{n}\sum_{k=1}^{m} \alpha_{ik}\,(e_i \otimes f_k), \quad \alpha_{ik} \in \mathbb{R},$$

and $z \in C_i(C,D)$ means

$$(\lambda_j \otimes \mu)\,z = \mu\left(\sum_{k=1}^{m} \alpha_{jk} f_k\right) \geq 0$$

for $j = 1,\dots,n$ and all $\mu \in D^*$. By the bipolarity theorem, this implies

$$\sum_{k=1}^{m} \alpha_{jk} f_k \in \text{cl}\ D$$

for $k = 1,\dots,m$, hence

$$z = \sum_{j=1}^{n} e_j \otimes \left(\sum_{k=1}^{m} \alpha_{jk} f_k\right) \in \text{cl}\ C_p(C,D).$$

The proof of (v) requires similar ideas, see [13]. Concerning (vi) we refer to the discussion below. □

Discussion:

1. A proper cone $C \subset E$, taken as the cone of positive elements, induces an order in E. It is known that E ordered by C is a vector lattice if and only if C is simplicial [12,15,16]. Therefore, (v) can be (roughly) restated as follows: the tensor product of finite dimensional vector lattices is a vector lattice. In this case, it does not matter whether the projective or the injective tensor product cone is used to generate the order in $E \otimes F$, since they are equal. Such results also hold in the infinite dimensional case, e.g., for Banach lattices [12,16,24]. In particular, the (topolgical) tensor product of two copies of the Banach lattice $C[0,1]$, ordered by the cone of nonnegative functions, is a Banach lattice

isomorphic to $C([0,1]^2)$. However, finite dimensional subspaces of $C[0,1]$ are usually not vector lattices with respect to the induced order, see [15] for details. Hence, the (closure of the) projective and the injective tensor product cone of the cones of nonnegative functions in finite dimensional subspaces of $C[0,1]$ will not be the same, in general.

2. The space $L(E,F)$ of linear maps from E into F can be identified with $E^* \otimes F$, the isomorphism given by bilinear extension of

$$\lambda \otimes f \longleftrightarrow [e \mapsto (\lambda e) \cdot f].$$

Using this identification, we get

$$C_i(C^*, D) = \{A \in L(E,F) : \; A[C] \subset D\},$$

which is denoted by $\pi(C,D)$ in [3]. Hence, all statements on injective and projective tensor product cones can be translated into statements on generalized nonnegative matrices and vice versa. Taking the identification of $E^* \otimes F$ and $L(E,F)$ into account, all the results stated in Propositions 3 and 4 (except (iv) of Proposition 4) can be found in [3, Chapter 1] and the references given there, in particular in [2,23]. In this context, the cone $\pi(C,C)$ is of particular interest, and (vi) of Proposition 4 has been proved in [2]. This result allows the construction of polyhedral cones C and D such that $C_p(C,D) \neq C_i(C,D)$ or even $C_p(C,C) \neq C_i(C,C)$, see [3], or [23] for an earlier example.

3. The Injective Tensor Product Cone and Sufficient Conditions

We intend to show that the cones usually considered in shape preserving tensor product interpolation are intersections of injective cones of corresponding cones in the univariate case. This is based on the following lemma.

Lemma 5. *Suppose the cones $C \subset E$, $D \subset F$ and sets $L \subset E^*$, $M \subset F^*$ fulfil $C = L^*$ and $D = M^*$. Then*

$$C_i(C,D) = \{z \in E \otimes F : \; (\lambda \otimes \mu)\, z \geq 0, \quad \lambda \in L, \, \mu \in M\}.$$

Proof: Using the bipolarity theorem again, we get $C^* = L^{**} = \mathrm{cl}\,(\mathrm{con}\, L)$ and $D^* = M^{**} = \mathrm{cl}\,(\mathrm{con}\, M)$. To prove the nontrivial part of the stated equality, suppose that $z \in E \otimes F$ is contained in the set on the right hand side. Let $\lambda \in C^*$, $\mu \in D^*$, then there are sequences

$$\sum_{i=1}^{n} \alpha_i \lambda_i, \quad \alpha_i \in \mathbb{R}_+, \, \lambda_i \in L, \qquad \sum_{j=1}^{m} \beta_j \mu_j, \quad \beta_j \in \mathbb{R}_+, \, \mu_j \in M,$$

converging to λ respective μ for $n \to \infty$ respective $m \to \infty$. Hence,

$$\left(\Big(\sum_{i=1}^{n} \alpha_i \lambda_i\Big) \otimes \Big(\sum_{j=1}^{m} \beta_j \mu_j\Big) \right) z = \sum_{i=1}^{n} \sum_{j=1}^{m} \alpha_i \beta_j \, (\lambda_i \otimes \mu_j)\, z \geq 0,$$

which implies

$$(\lambda \otimes \mu)\, z \geq 0. \qquad \square$$

We want to apply this lemma to our introductory example. Taking

$$L = \{\lambda_u : u \in [0,1]\}, \qquad M = \{\mu_v : v \in [0,1]\},$$

where λ_u and μ_v are point functionals on S^x respective S^y defined by $\lambda_u s^x = s^x(u)$ for $s^x \in S^x$ respective $\mu_v s^y = s^y(v)$ for $s^y \in S^y$, we have $C^x_{pos} = M^*$ and $C^y_{pos} = L^*$. Since $(\lambda_u \otimes \mu_v)\, s = s(u,v)$ for all $s \in S^x \otimes S^y$, we obtain

$$C_{pos} = \{s \in S^x \otimes S^y : s(u,v) \geq 0, \ (u,v) \in [0,1]^2\} = C_i(C^x_{pos}, C^y_{pos}),$$

i.e., the cone of nonnegative functions in $S^x \otimes S^y$ is the injective tensor product cone of the cones of nonnegativity functions in S^x and S^y.

Let C^x_{mon} and C^x_{conv} denote the cone of monotone respective convex functions in S^x; similar notations are used for the corresponding cones in S^y. By Lemma 5, it is easily established that

$C_i(C^x_{mon}, C^y_{pos})$ is the cone of functions monotone in x-direction,

$C_i(C^x_{mon}, C^y_{pos}) \cap C_i(C^x_{pos}, C^y_{mon})$ is the cone of bi-monotone functions,

$C_i(C^x_{conv}, C^y_{pos})$ is the cone of functions convex in x-direction,

$C_i(C^x_{conv}, C^y_{pos}) \cap C_i(C^x_{pos}, C^y_{conv})$ is the cone of bi-convex functions, and

$C_i(C^x_{conv}, C^y_{pos}) \cap C_i(C^x_{pos}, C^y_{conv}) \cap C_i(C^x_{mon}, C^y_{mon})$ is the cone of S-convex functions [16] in $S^x \otimes S^y$.

Hence, all the shape constraints which have been successfully considered in tensor product interpolation so far, can be represented by intersetions of injective tensor product cones of usual univariate cones. Sufficient conditions for this type of constraints can be easily generated using corresponding sufficient conditions obtained in the univariate case.

Lemma 6. *Let $C \subset E$ and $D \subset F$ be cones, and suppose the cones $\tilde{C}$ and $\tilde{D}$ fulfil $\tilde{C} \subset C$ and $\tilde{D} \subset D$. Then*

$$C_i(\tilde{C}, \tilde{D}) \subset C_i(C, D).$$

Proof: We have $C^* \subset \tilde{C}^*$ and $D^* \subset \tilde{D}^*$, hence

$$C_p(C^*, D^*) \subset C_p(\tilde{C}^*, \tilde{D}^*)$$

and therefore

$$C_p(\tilde{C}^*, \tilde{D}^*)^* \subset C_p(C^*, D^*)^*,$$

which, by Proposition 3, proves the lemma. □

The case of a finite number of linear sufficient conditions is particularly useful, and we state the corresponding result separately.

Corollary 7. *Let $\lambda_1, \ldots, \lambda_n \in E^*$ be such that $\{\lambda_1, \ldots, \lambda_n\}^* \subset C$, i.e., the conditions $\lambda_i e \geq 0$, $i = 1, \ldots, n$, are sufficient for $e \in C$. Analogously, let $\mu_1, \ldots, \mu_m \in F^*$ be such that $\{\mu_1, \ldots, \mu_m\}^* \subset D$. Then*

$$\left(\{\lambda_i \otimes \mu_j\}_{i=1,j=1}^{n,m}\right)^* \subset C_i(C, D),$$

i.e., the conditions $(\lambda_i \otimes \mu_j)\, z \geq 0$, $\quad i = 1, \ldots, n$, $j = 1, \ldots, m$, are sufficient for $z \in C_i(C, D)$.

Obviously, the corollary also serves to obtain sufficient conditions for intersections of injective tensor product cones. Indeed, all the sufficient conditions used in the papers on shape preserving tensor product interpolation mentioned in the introduction can be easily generated in this way, since the corresponding conditions in the univariate case are available.

4. Applications and further remarks

We want to mention that the results above can also be applied to generate sufficient conditions for shape constraints given by intersections of shifts of injective tensor product cones. For explanation, let us consider the problem of range restricted interpolation. Given (sufficiently simple) lower und upper bounds $L, U \in C([0,1]^2)$, we are looking for an interpolant $s \in S^x \otimes S^y$ fulfilling the range restictions

$$L(u,v) \leq s(u,v) \leq U(u,v), \quad (u,v) \in [0,1]^2.$$

Enlarging the spaces S^x and S^y if necessary such that $L, U \in S^x \otimes S^y$, the range restrictions are seen to be equivalent to

$$s \in (L + C_{pos}) \cap (U - C_{pos}).$$

Obviously, the sufficient nonnegativity conditions derived from the corresponding conditions in the univariate case have to be applied to $s - L$ and $U - s$. We refer to [14,22] for particular examples.

In univariate shape preserving interpolation, it is often necessary to adapt the linear space of interpolants to the given data. For that, nonlinear parameters are introduced in the representation, *e.g.*, variable additional knots in polynomial splines or rationality parameters in rational splines. For historical reasons related to exponential splines, these parameters are usually called tension parameters. Now, the presented results can be used for fixed tension parameters to obtain sufficient conditions in the tensor product case, see [14,21,22]. These sufficient conditions form a nonlinear inequality system due to the nonlinear tension parameters. For given compatible data values, we have to assure the solvability of this system by choosing the tension parameters appropriately. Based on corresponding univariate methods, this has been accomplished for particular examples in [14,21,22]. Certainly, a more abstract approach subsuming these particular methods would be of interest.

The efficient computation of tensor product interpolants in the unconstrained case is based on the fact that the tensor product interpolation operator is just the tensor product operator of the *linear* univariate interpolation operators, see [7,8]. But for shape preserving interpolation, the corresponding interpolation operators are usually not linear any more. We have shown how to utilize the tensor product structure in order to derive sufficient conditions for the shape properties from known conditions for corresponding constraints in the univariate case. It remains an open question, to what extend the tensor product structure can be used in the actual computation of shape preserving tensor product interpolants.

Acknowledgements. The author thanks J. W. Schmidt and M. Walther for their stimulating cooperation.

References

1. Asaturyan S. and Unsworth G., A C^1 monotonicity preserving surface interpolation scheme, in *Mathematics of Surfaces III*, Handscomb D. C. (ed.), Oxford University Press, Oxford, 1989, 243–266.
2. Barker G. P. and Loewy R., The structure of cones of matrices, Linear Algebra Appl. **12** (1975), 87–94.
3. Berman A. and Plemmons R. J., *Nonnegative Matrices in the Mathematical Sciences*, SIAM, Philadelphia, 1994.
4. Brodlie K. W., Butt S., and Mashwama P., Visualization of surface data to preserve positivity and other simple constraints, Computers and Graphics **19** (1995), 585–594.
5. Carlson R. E. and Fritsch F. N., Monotone piecewise bicubic interpolation, SIAM J. Numer. Anal. **22** (1985), 386–400.
6. Costantini P. and Fontanella F., Shape-preserving bivariate interpolation, SIAM J. Numer. Anal. **27** (1990), 488–506.
7. de Boor C., *A Practical Guide to Splines*, Springer, New York, 1978.
8. Ewald S., Mühlig H., and Mulansky B., Bivariate interpolating and smoothing tensor product splines, in *Splines in Numerical Analysis*, Schmidt J. W. and Späth H. (eds.), Akademie-Verlag, Berlin, 1989, 55–68.
9. Heß W. and Schmidt J. W., Positive quartic, monotone quintic C^2-spline interpolation in one and two dimensions, J. Comput. Appl. Math. **55** (1994), 51–67.
10. Isac G. and Nemeth A. B., Monotonicity of metric projections onto positive cones of ordered Euclidean spaces, Arch. Math. **46** (1986), 568–576.
11. Köthe G., *Topologische Lineare Räume I*, Springer, Berlin, 1960.
12. Meyer-Nieberg P., *Banach Lattices*, Springer, Berlin, 1991.

13. Mulansky B. and Schmidt J. W., Nonnegative interpolation by biquadratic splines on refined rectangular grids, in *Wavelets, Images and Surface Fitting*, Laurent P. J., Le Méhauté A., and Schumaker L. L. (eds.), A K Peters, Wellesley, 1994, 379–386.

14. Mulansky B., Schmidt, J. W., and Walther M., Tensor product spline interpolation subject to piecewise bilinear lower and upper bounds, in *Advanced Course on FAIRSHAPE*, Hoschek J. and Kaklis P. (eds.), Teubner, Stuttgart, 1996, 201–216.

15. Polyrakis I. A., Finite-dimensional lattice-subspaces of $C(\Omega)$ and curves of $\mathbb{R}^n$, Trans. Amer. Math. Soc. **348** (1996), 2793–2810.

16. Schaefer H. H., *Banach Lattices and Positive Operators*, Springer, Berlin, 1974.

17. Schmidt J. W., Rational biquadratic C^1-splines in S-convex interpolation, Computing **47** (1991), 87–96.

18. Schmidt J. W., Positive, monotone, and S-convex C^1-interpolation on rectangular grids, Computing **48** (1992), 363–371.

19. Schmidt J. W., Positive, monotone, and S-convex C^1-histopolation on rectangular grids, Computing **50** (1993), 19–30.

20. Schmidt J. W. and Heß W., S-convex, monotone, and positive interpolation with rational bicubic splines of C^2-continuity, BIT **33** (1993), 496–511.

21. Schmidt J. W. and Walther M., Tensor product splines in S-convexity preserving interpolation, submitted.

22. Schmidt J. W. and Walther M., Gridded data interpolation with restrictions on the first order derivatives, this proceedings.

23. Schneider H. and Vidyasagar M., Cross-positive matrices, SIAM J. Numer. Anal. **7** (1970), 508–519.

24. Wittstock G., Eine Bemerkung über Tensorprodukte von Banachverbänden, Arch. Math. **25** (1974), 627–634.

Bernd Mulansky
Technische Universität Dresden
Institut für Numerische Mathematik
01062 Dresden
Germany
`mulansky@math.tu-dresden.de`

The Curse of Dimension and a Universal Method For Numerical Integration

Erich Novak and Klaus Ritter

Abstract. Many high dimensional problems are difficult to solve for any numerical method. This curse of dimension means that the computational cost must increase exponentially with the dimension of the problem. A high dimension, however, can be compensated by a high degree of smoothness. We study numerical integration and prove that such a compensation is possible by a recently invented method. The method is shown to be universal, i.e., simultaneously optimal up to logarithmic factors, on two different smoothness scales. The first scale is defined by isotropic smoothness conditions, while the second scale involves anisotropic smoothness and is related to partially separable functions.

1. Introduction

Several applications require the computation of high dimensional integrals. They are present, for example, in statistical mechanics, see [28] for an introduction. Another important example is the fast valuation of financial derivatives, see [16]. Some applications even require approximate values of path integrals, i.e., integrals over infinite dimensional spaces, see [30].

Limits of practical computability are often related to the high dimension of the problem. There is a "curse of dimension" which means that the minimal cost of computing an approximation grows exponentially in the dimension of the problem. Hence, many high dimensional numerical problems are practically impossible to solve. The curse of dimension was already observed by Bellman [3].

In addition to the number of variables, i.e., to the dimension, also the smoothness and the structure of the function is of importance. To a certain extent, a high dimension can be compensated by a high degree of smoothness and/or a favorable structure. As a rule, such a compensation cannot be achieved by means of classical methods. Hence we have to invent new methods.

Multivariate Approximation and Splines
G. Nürnberger, J. W. Schmidt, and G. Walz (eds.), pp. 177–188.

ISBN 3-7643-5654-5.

We want to expand the limits of practical computability by using, in an optimal way, the smoothness and the structure of the problem. We want to achieve this optimality even if the smoothness and the structure are unknown.

By smoothness we mean that the underlying functions have certain derivatives whose norms are not too large. By structure we mean certain geometrical properties, like convexity. Later we will discuss partially separable problems, another structural property. We will see that this property is strongly related to certain nonclassical smoothness classes of Nikolskii, Korobov, and Babenko.

It is a major goal of scientific computing to expand present limits of practical computability. To reach this goal, one should use the art and the techniques of computer science, such as parallel processing. First of all, however, it is very important to optimize the underlying algorithms with respect to their cost and accuracy. Optimal numerical methods can only be realized if the analytical model and the available information is fully used. This information-based view can exhibit the fundamental limits of practical computability. It is remarkable that these limits were already discussed by Kolmogorov, Bakhvalov, Nikolskii and Babenko, almost 40 years ago. Now these limits are systematically studied in the field of "information-based complexity", see [12,19,26,27,30].

In this paper we study numerical integration and further analyze a method that was introduced in [14]. In particular we prove that our method is almost optimal for each of the classical smoothness spaces C_d^r. The analogous result for the spaces F_d^r of functions with bounded mixed derivatives as well as many numerical examples are contained in our earlier paper. It is well known that classical methods, like product Gaussian rules, achieve the optimal order of convergence for all spaces C_d^r. However, these methods are useless if $d \geq 40$, say, while our method works with a relatively small number of knots even if the dimension d is large.

2. The Curse of Dimension

The curse of dimension occurs for many numerical problems. Assume that we want to compute a value $S(f)$ of a (linear or nonlinear) operator

$$S : X \to G$$

between two normed spaces. Often X is a space of functions f and we consider discretizations of S of the form

$$S_n(f) = \varphi(f(\mathbf{x}_1), \ldots, f(\mathbf{x}_n)). \tag{1}$$

As the error criterion we often take the worst case error

$$e(S_n, X) = \sup_{f \in F} \|S(f) - S_n(f)\|$$

over the unit ball F of X. If S and S_n are linear then $e(S_n, X)$ is the operator norm of $S - S_n$. The total cost of computing $S_n(f)$ is at least n and often proportional to n. For simplicity we use here the number n of function evaluations as a measure for the cost. It should be stressed, however, that certain adaptive methods have a total cost that increases much faster than the number of function evaluations. The optimal error bounds

$$e_n(S,X) = \inf_{S_n} e(S_n, X)$$

indicate the best possible accuracy that can be achieved by any method with cost at most n.

Consider the standard smoothness classes

$$C_d^r = \{f : [-1,1]^d \to \mathbb{R} : \max_{|\alpha|\le r} \|f^{(\alpha)}\|_\infty < \infty\},$$

where $\alpha = (\alpha_1, \ldots, \alpha_d) \in \mathbb{N}_0^d$ is used to denote a partial derivative of order $|\alpha| = \alpha_1 + \ldots + \alpha_d$. For many problems, i.e, for many operators S, the optimal error bounds on the unit ball

$$F = \{f \in C_d^r : \max_{|\alpha|\le r} \|f^{(\alpha)}\|_\infty \le 1\}$$

in C_d^r satisfy

$$e_n(S, C_d^r) \asymp n^{-r/d}, \tag{2}$$

see, e.g., [12]. Here we use $\asymp$ to denote the weak equivalence of sequences, i.e., $v_n \asymp w_n$ iff $c_1 \le v_n/w_n \le c_2$ for sufficiently large n and positive constants c_i. The bound (2) holds, in particular, for the problem of numerical integration where

$$S(f) = \int_{[-1,1]^d} f(\mathbf{x})\, d\mathbf{x}.$$

Suppose that we want to construct an approximation S_n with error at most ε for a given positive and usually small or moderate ε. The cost n should be as small as possible. Then (2) yields

$$n \asymp \varepsilon^{-d/r}.$$

The curse of dimension means that the minimal cost n increases exponentially with the dimension d. Hence, the curse of dimension is present for many problems defined for the class C_d^r.

The proof of (2) consists of two parts. On the one hand we must prove a lower bound that is true for arbitrary methods of the form (1). On the other hand we must prove that there is a sequence $(S_n)_n$ of methods such that

$$e(S_n, C_d^r) \le c_{r,d} \cdot n^{-r/d} \tag{3}$$

for positive constants $c_{r,d}$. The upper bound (3) can be achieved by methods that use the function values from a regular grid. For the problem of numerical integration, for example, product formulas with the Gaussian method lead to the optimal order $n^{-r/d}$ for every r and every d.

Using the function values from a grid is a very classical idea that leads to practicable methods if d is small, say $d < 6$. Product rules cannot be applied, however, if d is large. Consider, for example, the case $d = 40$. If r is large, say $r = 80$ or even $r = \infty$, then the optimal order of convergence is not too bad, see (2). It is not clear, however, how to get practical methods since the smallest product method already needs 2^{40} points which is too large. Therefore the optimal order of product methods is only a theoretical result and we need new methods that realize (almost) the optimal order of convergence and can be used with a relatively small number of function values.

Known methods cannot fully use the existing smoothness if d is large, at least not for reasonable small values of n. Often Monte Carlo methods and/or number theoretic methods are used in high dimensions. Such methods can be applied for relatively small n but they can not fully use the smoothness properties of f, hence they cannot be used if a high accuracy is needed.

The error estimate (2) is a "negative" result if $d \gg r$. In this case the optimal error bounds $e_n(S, C_d^r)$ converge slowly and we should look for "smaller" classes $X \subset C_d^r$ that allow smaller errors and still contain many interesting functions.

3. Partially Separable Problems and Non-classical Function Spaces

Many high dimensional problems have a particular structure; they are partially separable. This means that f may depend on a huge number of variables but can be written as a sum of functions that only depend on few variables. A simple case would be

$$f(\mathbf{x}) = \sum_{1 \le i,j \le d} f_{i,j}(x_i, x_j) \tag{4}$$

where $\mathbf{x} = (x_1, \ldots, x_d)$. Here f is a sum of functions that only depend on two variables. At first glance it is not clear how we can use property (4) for numerical integration, if only function values of f can be computed.

Assume that $f_{i,j} \in C_2^2$ for each $f_{i,j}$ in the sum (4). Then $f \in C_d^2$, but in addition many higher order mixed derivatives exist. In particular, the derivative $f^{(\alpha)}$ with

$$\alpha = (1, \ldots, 1)$$

exists. Therefore f is in a class of functions with bounded mixed derivatives. Such classes were already studied by Nikolskii, Korobov, and Babenko in the sixties.

We consider the following scale of function spaces

$$F_d^r = \{f : [-1,1]^d \to \mathbb{R} : \max_{\alpha \le r} \|f^{(\alpha)}\|_\infty < \infty\}$$

where $\alpha \le r$ means $\alpha_i \le r$ for every $i = 1, \ldots, d$. Let $f \in F_d^r$. Then $f^{(\alpha)}$ exists for

$$\alpha = (r, \ldots, r),$$

but the derivative $f^{(\alpha)}$ with

$$\alpha = (r+1, 0, \ldots, 0)$$

may not exist. Hence the class F_d^r is not isotropic. We stress that being of the form (4) with $f_{i,j} \in C_2^2$ is sufficient but not necessary for f to belong to the class F_d^1.

Because of their tight connection to partially separable functions we want to study the classes F_d^r, as well as the more traditional classes C_d^r. A given "smooth" function f usually is an element of several different classes F_d^r and/or C_d^r, the exact degree of smoothness often is not known to us. Moreover, different norms are used to define the unit balls in the scales F_d^r and C_d^r. The norms are nondecreasing in r while the error bounds are nonincreasing in r. Hence it is not obvious which choice of r in the condition $f \in F_d^r$ or $f \in C_d^r$ leads to the best error bound. This problem is sometimes called the fat F problem, see [31].

It is therefore important to have a "universal" method that is optimal (or almost optimal) for many different classes. Hence the following problem seems to be of great practical importance.

Problem: Construct optimal (or almost optimal) algorithms for C_d^r and F_d^r. In particular, find such algorithms

- simultaneously for all classes C_d^r and F_d^r;
- with good performance for a relatively small number n of knots.

Similar problems were studied in the univariate case by [4,17] and in the multivariate periodic case by [1,11,21,24,25,26]. We are interested in the general (i.e. nonperiodic) case and we have suggested a method in [14]. This method uses the full smoothness of the underlying function; it is not necessary to know the degree of smoothness in advance. Also the structural property "partially separable" is exploited optimally.

4. A Universal Method

We begin with the construction of Smolyak [23]. Special cases of this construction were invented independently for different applications and are known under different names, such as hyperbolic cross points methods, Boolean methods, discrete blending methods, and sparse grid methods. We give a partial list of references. Integration and approximation of functions is studied in [2,5,8,18]. See [6,7,9,33] for operator equations and [13] for global optimization. Problems for random functions are analyzed in [10,19,20,29]. Tractability of multidimensional problems is studied in [15,29,30,32].

We assume that for $d = 1$ a sequence of formulas

$$U^i(f) = \sum_{j=1}^{m_i} a_j^i \, f(x_j^i)$$

is given. In the case of numerical integration the a_j^i are just numbers. The method U^i uses m_i function values and we assume that U^{i+1} has smaller error than U^i and $m_{i+1} > m_i$. Define then, for $d > 1$, the tensor product formulas

$$(U^{i_1} \otimes \cdots \otimes U^{i_d})(f) = \sum_{j_1=1}^{m_{i_1}} \cdots \sum_{j_d=1}^{m_{i_d}} a_{j_1}^{i_1} \cdots a_{j_d}^{i_d} \, f(x_{j_1}^{i_1}, \ldots, x_{j_d}^{i_d}).$$

A tensor product formula clearly needs

$$m_{i_1} \cdot m_{i_2} \cdot \ldots \cdot m_{i_d}$$

function values, sampled on a regular grid. It is known that all tensor product formulas $S_n = U^{i_1} \otimes \cdots \otimes U^{i_d}$ lead to relatively large errors for the classes F_d^r, the optimal order of such methods is

$$e(S_n, F_d^r) \asymp e(S_n, C_d^r) \asymp n^{-r/d}.$$

The Smolyak formulas $A(q, d)$ are clever linear combinations of tensor product formulas such that:

- only tensor products with a relatively small number of knots are used;
- the linear combination is chosen in such a way that an interpolation property for $d = 1$ is preserved for $d > 1$.

The Smolyak formulas are defined by

$$A(q,d) = \sum_{q-d+1 \le |\mathbf{i}| \le q} (-1)^{q-|\mathbf{i}|} \cdot \binom{d-1}{q-|\mathbf{i}|} \cdot (U^{i_1} \otimes \cdots \otimes U^{i_d}),$$

where $q \ge d$.

The following method was suggested in [14]. Use, for $d > 1$, the Smolyak construction and start, for $d = 1$, with the classical Clenshaw-Curtis formula with

$$m_1 = 1 \quad \text{and} \quad m_i = 2^{i-1} + 1 \;\text{ for }\; i > 1.$$

The Clenshaw-Curtis formulas

$$U^i(f) = \sum_{j=1}^{m_i} a_j^i \, f(x_j^i)$$

use the knots

$$x_j^i = -\cos \frac{\pi(j-1)}{m_i - 1} \tag{5}$$

(and $x_1^1 = 0$). Hence we use nonequidistant knots. The weights a_j^i are defined in such a way that U^i is exact for all (univariate) polynomials of degree at most m_i. The resulting method has several remarkable properties, see [14]. In the next section we prove that it is almost optimal for the scale C_d^r; the analogous result for the scale F_d^r is already known.

5. Results

Let $A(q,d)$ denote the method from Section 4. First we cite an estimate on the number $n(q,d)$ of knots that are used by $A(q,d)$. We use $\approx$ to denote the strong equivalence of sequences, i.e., $v_n \approx w_n$ iff $\lim_{n\to\infty} v_n/w_n = 1$.

Lemma 1. *[10, Lemma 1]. For $k \to \infty$ and fixed d*

$$n(k+d,d) \approx \frac{1}{(d-1)! \cdot 2^{d-1}} \cdot 2^k\, k^{d-1}.$$

Smolyak's construction leads to cubature formulas with negative weights, even if positive weights are used in the univariate case. We show, however, that the weights are relatively small in absolute value. We use c_d and $c_{r,d}$ to denote different positive constants.

Lemma 2. *There exists a constant $c_d > 0$ such that*

$$\|A(q,d)\|_\infty \le c_d \cdot (\log(n(q,d)))^{d-1}.$$

Here $\|A(q,d)\|_\infty$ denotes the sum of the absolute values of the weights of $A(q,d)$.

Proof: Observe that

$$\#\{\mathbf{i} \in \mathbb{N}^d : \ |\mathbf{i}| = \ell\} = \binom{\ell-1}{d-1}.$$

Since the Clenshaw-Curtis formulas have positive weights, we conclude

$$\|A(q,d)\|_\infty \le c_d \cdot \sum_{\ell=q-d+1}^{q} \binom{\ell-1}{d-1} \le d \cdot \binom{q-1}{d-1} \le c_d \cdot q^{d-1}.$$

Due to Lemma 1,

$$\log n(q,d) \ge c_d \cdot q,$$

and the statement follows. □

Now we present error bounds for our method $A(q,d)$ and the integration problem, i.e.,

$$S(f) = \int_{[-1,1]^d} f(\mathbf{x})\, d\mathbf{x}.$$

Theorem. *For $d, r \in \mathbb{N}$ there exists $c_{r,d} > 0$ such that*

$$e(A(q,d), F_d^r) \le c_{r,d} \cdot n^{-r} \cdot (\log n)^{(d-1)\cdot(r+1)}$$

and

$$e(A(q,d), C_d^r) \le c_{r,d} \cdot n^{-r/d} \cdot (\log n)^{(d-1)\cdot(r/d+1)},$$

where $n = n(q,d)$.

Proof: The error bound for the classes F_d^r is proven in [14]. In the sequel we analyze the error on C_d^r. Let $\mathbb{P}(\ell)$ be the space of all polynomials in one variable of degree at most ℓ. It turns out that $A(q,d)(f) = S(f)$ for all

$$f \in \sum_{|\mathbf{i}|=q} \mathbb{P}(m_{i_1}) \otimes \cdots \otimes \mathbb{P}(m_{i_d}),$$

see [14]. We conclude that $A(kd,d)$ is exact on

$$Z_k = \mathbb{P}(m_k) \otimes \cdots \otimes \mathbb{P}(m_k),$$

the space of polynomials of degree at most m_k in each variable. Therefore

$$e(A(kd,d), C_d^r) \leq (2^d + \|A(kd,d)\|_\infty) \cdot \sup_{f\in F} \operatorname{dist}(f, Z_k), \tag{6}$$

where F denotes the unit ball in C_d^r and dist denotes the distance in the maximum norm. It is well known that

$$\sup_{f\in F} \operatorname{dist}(f, Z_k) \leq c_{r,d} \cdot m_k^{-r}.$$

Applying Lemma 2 and $m_k \approx 2^{k-1}$ we obtain

$$e(A(kd,d), C_d^r) \leq c_{r,d} \cdot (\log n(kd,d))^{d-1} \cdot 2^{-rk}.$$

Lemma 1 yields

$$\log n(kd,d) \approx kd + (d-1)\log k,$$

and therefore

$$e(A(kd,d), C_d^r) \leq c_{r,d} \cdot (\log n(kd,d))^{(d-1)\cdot(r/d+1)} \cdot n(kd,d)^{-r/d}. \qquad \square$$

Definition. *A sequence of quadrature formulas is called* universal *for a given scale of function spaces if it achieves, simultaneously for each member of the scale, the optimal order of convergence – up to logarithmic factors.*

Corollary. *The method $A(q,d)$ is universal for both the scales C_d^r and F_d^r.*

Proof: The upper bounds from the Theorem match the well known lower bounds for $e_n(S, C_d^r)$ and $e_n(S, F_d^r)$, respectively, up to logarithmic factors. This follows from the lower bound in (2) and $e_n(S, C_1^r) = e_n(S, F_1^r) \leq e_n(S, F_d^r)$. $\square$

Remark 1. An analogous estimate to (6) shows that $A(q,d)(f)$ converges to the integral $S(f)$ whenever f is Hölder continuous for some exponent $\beta > 0$. More precisely,

$$e(A(q,d), X) \leq c_{\beta,d} \cdot n^{-\beta/d} \cdot (\log n)^{(d-1)\cdot(\beta/d+1)}$$

on the respective unit ball. Nevertheless, in case of such low regularity equal weight formulas are likely to give better results.

On the other hand, consider classes X of analytic functions, where the distance

$$\sup_{f\in F} \operatorname{dist}(f, Z_k)$$

converges exponentially. Then we have exponential convergence of $e(A(q,d),X)$, again by the analogue to (6).

Remark 2. Our method is easily modified to work for weighted integrals

$$S(f) = \int_{[-1,1]^d} f(\mathbf{x}) \cdot \varrho(\mathbf{x})\, d\mathbf{x},$$

where

$$\varrho = \omega_1 \otimes \ldots \otimes \omega_d$$

is a tensor product. We only need to redefine the weights in the cubature formulas $A(q,d)$. In the univariate case we consider generalized Clenshaw-Curtis formulas

$$U^{i,\omega}(f) = \sum_{j=1}^{m_i} f(x_j^i) \cdot a_j^{i,\omega}$$

which use the knots x_j^i from (5) and are exact for all polynomials of degree less than m_i. Again we choose $m_1 = 1$ and $m_i = 2^{i-1}+1$ for $i > 1$ and apply Smolyak's construction to obtain cubature formulas $A(q,d)$. Observe that the set of knots is not changed.

It is known that

$$\lim_{i\to\infty} \sum_{j=1}^{m_i} |a_j^{i,\omega}| = \int_{-1}^{1} |\omega(x)|\, dx$$

if $\omega \in L_p([-1,1])$ for some $p > 1$, see [22]. In particular, the functionals $U^{i,\omega}$ are uniformly bounded on $C([-1,1])$. Assume that $\omega_1,\ldots,\omega_d \in L_p([-1,1])$ for some $p > 1$. As in the case $\varrho \equiv 1$ Lemma 2 and the error bound

$$e(A(q,d), C_d^r) \le c_{r,d} \cdot n^{-r/d} \cdot (\log n)^{(d-1)\cdot(r/d+1)}$$

follows. For the classes F_d^r the bound

$$e(A(q,d), F_d^r) \le c_{r,d} \cdot n^{-r} \cdot (\log n)^{(d-1)\cdot(r+1)}$$

is established in [14].

Acknowledgements. We thank Volodya Temlyakov for his help.

References

1. Babuška I., Über universal optimale Quadraturformeln, Apl. Mat. **13** (1968), 304–338 and 388–404.
2. Baszenski G. and Delvos F.-J., Multivariate Boolean trapezoidal rules, in *Approximation, Probability and Related Fields*, Anastassiou G. (ed.), Plenum, New York, 1994, 109–117.
3. Bellman R., *Adaptive Control Processes: a Guided Tour*, Princeton University, Princeton, 1961.
4. Brass H., Universal quadrature rules in the space of periodic functions, in *Numerical Integration III*, Brass H. and Hämmerlin G. (eds.), ISNM **85** Birkhäuser, Basel, 1988, 16–24.
5. Delvos F.-J. and Schempp W., *Boolean Methods in Interpolation and Approximation*, Pitman Research Notes in Mathematics Series **230**, Longman, Essex, 1989.
6. Griebel M. and Oswald P., Tensor product type subspace splitting and multilevel iterative methods for anisotropic problems, Advances in Comp. Math. **4** (1995), 171–206.
7. Griebel M., Schneider M., and Zenger Ch., A combination technique for the solution of sparse grid problems, in *Iterative Methods in Linear Algebra*, Beauwens R. and de Groen P. (eds.), Elsevier, North-Holland, 1992, 263–281.
8. Frank K. and Heinrich S., Computing discrepancies of Smolyak quadrature rules, J. Complexity **12** (1996), 287–314.
9. Frank K., Heinrich S., and Pereverzev S., Information complexity of multivariate Fredholm integral equations in Sobolev classes, J. Complexity **12** (1996), 17–34.
10. Müller-Gronbach Th., Hyperbolic cross designs for approximation of random fields, Preprint-Nr. 1811, Fachbereich Mathematik, TH Darmstadt, 1996.
11. Niederreiter H., *Random Number Generation and Quasi-Monte Carlo Methods*, SIAM, Philadelphia, 1992.
12. Novak E., *Deterministic and Stochastic Error Bounds in Numerical Analysis*, Lecture Notes in Mathematics **1349**, Springer, Berlin, 1988.
13. Novak E. and Ritter K., Global optimization using hyperbolic cross points, in *State of the Art in Global Optimization*, Floudas C. A. and Pardalos P. M. (eds.), Kluwer, Dordrecht, 1996, 19–33.
14. Novak E. and Ritter K., High dimensional integration of smooth functions over cubes, Numer. Math. **75** (1996), 79–97.
15. Novak E., Sloan I. H., and Woźniakowski H., Tractability of tensor product linear operators, preprint.
16. Paskov S. H. and Traub J. F., Faster valuation of financial derivatives, J. Portfolio Management **22** (1995), 113–120.
17. Petras K., On the universality of the Gaussian quadrature formula, East J. Approx. **2** (1996), 427–438.

18. Pöplau G. and Sprengel F., Some error estimates for periodic interpolation on full and sparse grids, in *Proceedings of Chamonix 1996*, Le Méhauté A., Rabut C., and Schumaker L. L. (eds.), Vanderbilt, Nashville, 1996, to appear.
19. Ritter K., *Average Case Analysis of Numerical Problems*, Habilitationsschrift, Erlangen, 1995.
20. Ritter K., Wasilkowski G. W. and Woźniakowski H., Multivariate integration and approximation for random fields satisfying Sacks-Ylvisaker conditions, Ann. Appl. Prob. **5** (1995), 518–540.
21. Sloan I. H. and Joe S., *Lattice Methods for Multiple Integration*, Clarendon, Oxford, 1994.
22. Sloan I. H. and Smith W. E., Product-integration with the Clenshaw-Curtis and related points, Numer. Math.**30** (1978), 415–428.
23. Smolyak S. A., Quadrature and interpolation formulas for tensor products of certain classes of functions, Soviet Math. Dokl. **4** (1963), 240–243.
24. Temlyakov V. N., On universal cubature formulas, Soviet Math. Dokl. **43** (1991), 39–42.
25. Temlyakov V. N., On a way of obtaining lower estimates for the errors of quadrature formulas, Math. USSR Sbornik **71** (1992), 247–257.
26. Temlyakov V. N., *Approximation of Periodic Functions*, Nova Science, New York, 1994.
27. Traub J. F., Wasilkowski G. W., and Woźniakowski H., *Information-Based Complexity*, Academic Press, San Diego, 1988.
28. Vesely F. J., *Computational Physics, an Introduction*, Plenum, New York, 1994.
29. Wasilkowski G. W. and Woźniakowski H., Explicit cost bounds of algorithms for multivariate tensor product problems, J. Complexity **11** (1995), 1–56.
30. Wasilkowski G. W. and Woźniakowski H., On tractability of path integration, J. Math. Phys. **37** (1996), 2071–2088.
31. Woźniakowski H., Information-based complexity, Ann. Rev. Comput. Sci. **1** (1986), 319–380.
32. Woźniakowski H., Tractability and strong tractability of linear multivariate problems, J. Complexity **10** (1994), 96–128.
33. Zenger Ch., Sparse grids, in *Parallel Algorithms for Partial Differential Equations*, Hackbusch W. (ed.), Vieweg, Braunschweig, 1991, 241–251.

Erich Novak
Mathematisches Institut
Universität Erlangen-Nürnberg
Bismarckstr. 1 1/2
D-91054 Erlangen, Germany
`novak@mi.uni-erlangen.de`

Klaus Ritter
Mathematisches Institut
Universität Erlangen-Nürnberg
Bismarckstr. 1 1/2
D-91054 Erlangen, Germany
`ritter@mi.uni-erlangen.de`

Interpolation by Bivariate Splines on Crosscut Partitions

G. Nürnberger, O. V. Davydov, G. Walz, and F. Zeilfelder

Abstract. We give a survey of recent methods to construct Lagrange interpolation points for splines of arbitrary smoothness r and degree q on general crosscut partitions in $\mathbb{R}^2$. For certain regular types of partitions, also results on Hermite interpolation sets and on the approximation order of the corresponding interpolating splines are given.

1. Introduction

We consider bivariate spline spaces of the following type. Let a convex compact region $\Omega \in \mathbb{R}^2$ be given, which is subdivided by a finite number of straight lines (crosscuts) into convex subregions $\{T\}$, called a *partition* Δ of Ω. The space of all functions $s \in C^r(\Omega)$, whose restriction to each T is a bivariate polynomial of degree q, is denoted as the spline space $S_q^r(\Delta)$.

In Section 2, a construction method for point sets $\{z_1, \ldots, z_{\dim S_q^r(\Delta)}\}$ on a general partition Ω, which admit unique Lagrange interpolation w.r.t. $S_q^r(\Delta)$, is described.

The construction of Hermite interpolation sets (which can be considered as limits of Lagrange interpolation sets) for certain rectangular types of partitions Δ, denoted as Δ^1 resp. Δ^2 partitions, is given in Section 3. Here, Δ^1 denotes the partitions where to each subrectangle the same diagonal is added, while Δ^2 means that both diagonals are added.

We also give results on the approximation order of the interpolating spline function. The approximation order for $S_q^r(\Delta^1)$ equals $q+1$ (which is optimal), if $q \geq 3.5r+1$, $r \geq 1$, and q, if $r = 1$ and $q = 4$. For $S_q^1(\Delta^2)$, we get the optimal order $q+1$, if $q \geq 4$, and the order q, if $q = 2, 3$.

Multivariate Approximation and Splines
G. Nürnberger, J. W. Schmidt, and G. Walz (eds.), pp. 189–203.

ISBN 3-7643-5654-5.

2. Construction of Lagrange Interpolation Sets for Crosscut Partitions

Let $\Omega \subset \mathbb{R}^2$ be a convex compact domain. Any finite set of straight lines (called *crosscuts*) $l_1, \ldots, l_M$ having nonempty intersections with the interior of Ω, produces a *partition* Δ of Ω into convex compact subregions (called *cells*) with pairwise disjoint interiors. We denote by $\mathcal{T}$ the set of all cells. The straight line boundaries of each cell are called *edges* and their endpoints *vertices.* Let $\{v_1, \ldots, v_L\}$ be the set of all *interior vertices* of the partition Δ. Thus, each v_i is an intersection point of two or more crosscuts which lies in the interior of Ω.

The space of polynomial splines of degree q and smoothness r, $0 \le r < q$, with respect to the partition Δ is defined by

$$S_q^r(\Delta) := \{s \in C^r(\Omega) : \; s|_T \in \tilde{\Pi}_q \, , \; T \in \mathcal{T}\} ,$$

where

$$\tilde{\Pi}_q := \operatorname{span}\{x^i y^j : \; i \ge 0, \; j \ge 0, \; i + j \le q\}$$

is the space of bivariate polynomials of total degree q.

The dimension of $S_q^r(\Delta)$ was determined by Chui & Wang [6].

We consider the following problem. Determine points $z_1, \ldots, z_N$ in Ω, where $N = \dim S_q^r(\Delta)$, such that for any $f \in C(\Omega)$, the Lagrange interpolation problem $s(z_i) = f(z_i)$, $i = 1, \ldots, N$, has a unique solution s in $S_q^r(\Delta)$. Such sets $\{z_1, \ldots, z_N\}$ are called *Lagrange interpolation sets for* $S_q^r(\Delta)$. (Sometimes we say that $z_1, \ldots, z_N$ are *Lagrange interpolation points for* $S_q^r(\Delta)$.)

First we describe bases and interpolation points for three types of subspaces of $S_q^r(\Delta)$ and then show how these points can be combined into an interpolation set for the whole space.

A. The space of bivariate polynomials $\tilde{\Pi}_q$. It is well known that $\dim \tilde{\Pi}_q = \binom{q+2}{2}$, and a basis of $\tilde{\Pi}_q$ is given by

$$\{x^i y^j : \; i \ge 0, \; j \ge 0, \; i + j \le q\} .$$

Interpolation sets for $\tilde{\Pi}_q$ can be obtained in the following way (see, for example, Nürnberger [11]). Let $\gamma_0, \ldots, \gamma_q$ be distinct parallel lines and $z_{i,0}, \ldots, z_{i,q-i}$ distinct points on γ_i, $i = 0, \ldots, q$ (see Fig. 2.1). Then

$$\{z_{0,0}, \ldots, z_{0,q}, \ldots, z_{q-1,0}, z_{q-1,1}, z_{q,0}\}$$

is a Lagrange interpolation set for $\tilde{\Pi}_q$, and the points are called *interpolation points of type A*.

B. The space of truncated power functions. Let l be any crosscut dividing Ω into two subdomains Ω_0 and Ω_1. We consider the following space:

$$T_q^r := \{s \in C^r(\Omega) : \; s|_{\Omega_0} \equiv 0 \, , \; s|_{\Omega_1} \in \tilde{\Pi}_q\} .$$

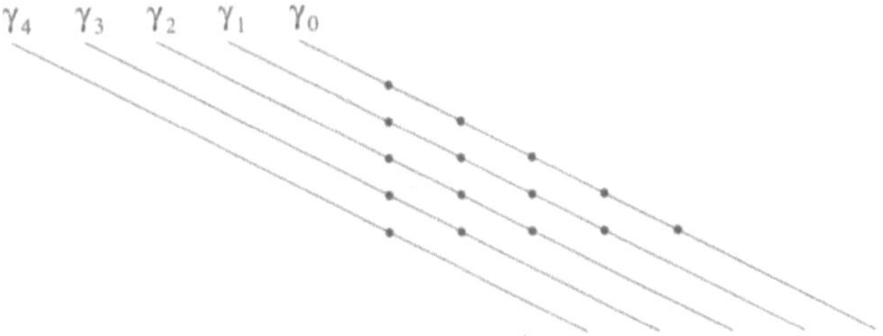

Fig. 2.1. Interpolation points of type A.

Suppose

$$l = \{(x,y) \in \mathbb{R}^2 : \; ax + by + c = 0\},$$

such that

$$ax + by + c > 0, \qquad (x,y) \in \operatorname{int} \Omega_1 .$$

A basis for T_q^r is given by truncated powers multiplied with polynomials,

$$\{(ax+by+c)_+^{r+1} x^i y^j : \; i \geq 0, \; j \geq 0, \; i+j \leq q-r-1\},$$

where

$$w_+^k := \begin{cases} w^k, & w \geq 0, \\ 0, & w < 0. \end{cases}$$

Therefore,

$$\dim T_q^r = \binom{q-r+1}{2}.$$

In order to determine interpolation sets for T_q^r, we choose $q-r$ parallel lines $\gamma_0, \ldots, \gamma_{q-r-1}$ intersecting l and put $q-r-i$ distinct points $z_{i,0}, \ldots, z_{i,q-r-1-i}$ on $\gamma_i \cap (\Omega_1 \setminus l)$, $i = 0, \ldots, q-r-1$ (see Fig. 2.2). Then

$$\{z_{0,0}, \ldots, z_{0,q-r-1}, \ldots, z_{q-r-2,0}, z_{q-r-2,1}, z_{q-r-1,0}\}$$

is a Lagrange interpolation set for T_q^r, and the points are called *interpolation points of type B.*

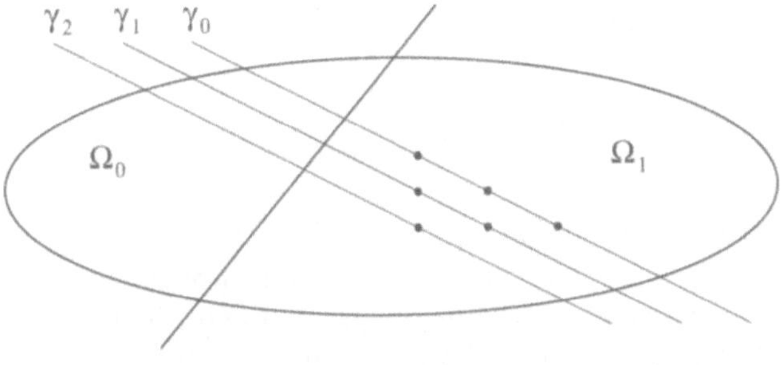

Fig. 2.2. Interpolation points of type B.

C. The space of cone splines. Let v be any interior vertex such that m crosscuts $l_1, \ldots, l_m$ intersect at v. Thus, $2m$ rays originate at v. We take m consecutive rays $r_1, \ldots, r_m$ (so that all m crosscuts are involved) and divide Ω into m subdomains $\Omega_0, \Omega_1, \ldots, \Omega_{m-1}$ as in Fig. 2.3.

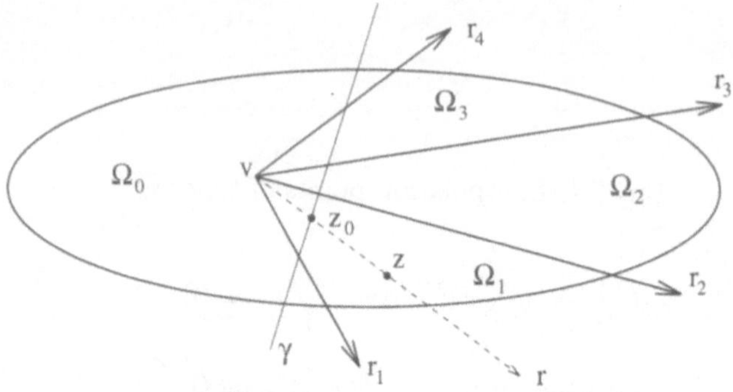

Fig. 2.3.

The space of cone splines K_q^r is defined as follows.

$$K_q^r := \{s \in C^r(\Omega) : \ s|_{\Omega_0} \equiv 0 \,, \ s|_{\Omega_i} \in \tilde{\Pi}_q \,, \ i = 1, \ldots, m-1\} \,.$$

Thus, all splines in K_q^r are zero outside the cone $\bigcup_{i=1}^{m-1} \Omega_i$. We now define a basis for K_q^r. We choose a line γ which intersects all the rays $r_1, \ldots, r_m$, with $v \notin \gamma$ (see Fig. 2.3). For $n = q, q-1, \ldots$, we consider the univariate spline spaces $K_n^r|_\gamma$ on γ such that $\dim K_n^r|_\gamma > 0$. Then we extend each univariate B-spline B in $K_n^r|_\gamma$ to a function in K_n^r as follows. Let $B(z) \equiv 0$, $z \in \Omega_0$. For each ray r in $\Omega \setminus \Omega_0$ passing through v, we define B to be the univariate truncated power function t_+^n multiplied with an appropriate constant. More precisely, let $z \in \Omega \setminus \Omega_0$ and let z_0 be the intersection point of γ and the line through v and z (see Fig. 2.3). Then $B(z) := B(z_0)|z - v|^n / |z_0 - v|^n$. All the bivariate splines in K_n^r, $n = q, q-1, \ldots$, obtained in this way (called *cone splines*), form a basis of K_q^r (see Chui & Wang [6] and Dahmen & Micchelli [7]). Therefore,

$$\dim K_q^r = \sum_{n \le q} \dim K_n^r|_\gamma \,.$$

We now describe interpolation points for K_q^r. Let $\gamma_0, \ldots, \gamma_p$ be parallel lines which intersect the rays $r_1, \ldots, r_m$ and do not pass through the vertex v, where p is determined by the conditions $\dim K_{q-p}^r|_\gamma > 0$ and $\dim K_{q-p-1}^r|_\gamma = 0$. We choose $\dim K_{q-i}^r|_\gamma$ points on $\gamma_i \cap (\Omega \setminus \Omega_0)$, $i = 0, \ldots, p$, so that these points satisfy Schoenberg-Whitney condition for the univariate spline space $K_{q-i}^r|_{\gamma_i}$ (see Fig. 2.4, a)).

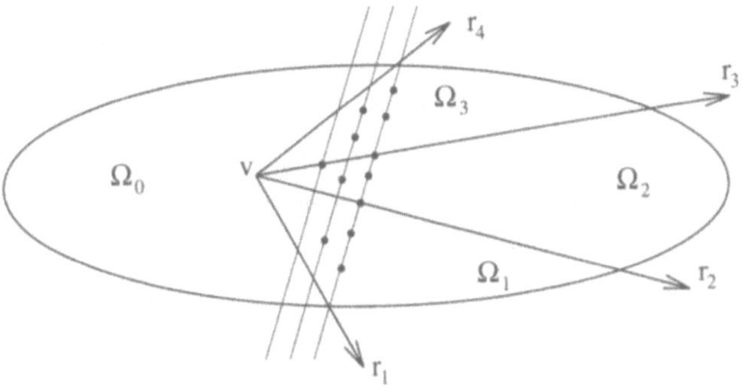

a)

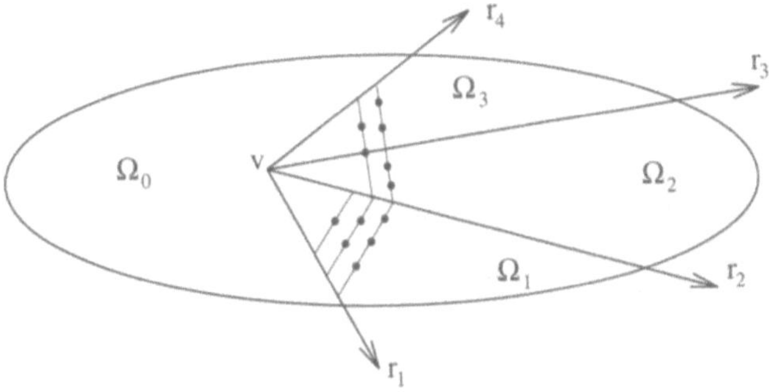

Fig. 2.4. Interpolation points of type C.

In other words, we take a point in a support of each univariate B-spline in $K^r_{q-i}|_{\gamma_i}$. It was shown by Nürnberger & Riessinger [14] that each set of points obtained in this way admits unique Lagrange interpolation from K^r_q. We call them *interpolation points of type C.*

A more general scheme of constructing interpolation points of type C (see Fig. 2.4, b)) was also proposed in [14]. The cone formed by r_1 and r_m can be divided into several subcones. Then the points are chosen on line segments inside the subcones according to some Schoenberg-Whitney type conditions for the spaces K^r_n restricted to the line segments as in Fig. 2.4, b). (For details see [14].) Another configuration of interpolation points of type C was given by Adam [1], where the points are lying on certain rays.

Now we are able to describe a basis for $S^r_q(\Delta)$ which is due to Chui & Wang [6] and Dahmen & Micchelli [7].

Theorem 2.1. *A basis of $S^r_q(\Delta)$ is given by the following functions:*

A. *Polynomial basis functions* $x^i y^j$, $i \geq 0$, $j \geq 0$, $i+j \leq q$.

B. *Truncated power functions, for each crosscut* l_i, $i = 1, \ldots, M$.

C. *Cone splines, as described above, for each interior vertex* v_i, $i = 1, \ldots, L$.

We note that there is some freedom in choosing basis functions in B. and C. of Theorem 2.1 since there are two possible spaces of truncated powers T_q^r with respect to a given crosscut, and also the rays which define cone splines can be choosen differently. On the other hand, the interpolation points of type A, B and C cannot be freely combined to obtain a Lagrange interpolation set for $S_q^r(\Delta)$. A method to assign a type A, B or C to each cell of the partition so that the combination of corresponding interpolation points on the cells is an interpolation set for the whole spline space was proposed by Nürnberger & Riessinger [13,14] for rectangular partitions with diagonals and extended to arbitrary crosscut partitions by Adam [1]. Their construction depends upon an order of the cells which is the natural ordering with respect to rows and columns in the case of rectangular partitions. We now describe this order in the general case of crosscut partitions (see Adam [1]).

The *lexicographical order* of the points in $\mathbb{R}^2$ is defined as follows. Given two points $z' = (x', y')$, $z'' = (x'', y'') \in \mathbb{R}^2$ we say that $z' \leq z''$ if

$$x' < x'' \quad \text{or} \quad (x' = x'' \text{ and } y' \leq y'') .$$

As usual, $z' < z''$ if $z' \leq z''$ and $z' \neq z''$. For any compact set $K \subset \mathbb{R}^2$, $m(K) \in \mathbb{R}^2$ denotes the minimal point of K with respect to the lexicographical order.

The total order of cells $T \in \mathcal{T}$ of the partition Δ is defined as follows. In the case when $m(T') < m(T'')$ (in lexicographical order), $T', T'' \in \mathcal{T}$, we set $T' < T''$. In the case when several cells have the same minimal point m, they are situated to the right of m and separated from each other by ray segments originating at m. Therefore, they can be ordered either clockwise or counterclockwise. We choose every time one of these two orders according to the following rule:

Case 1. $m = m(\Omega)$ or $m \in \operatorname{int} \Omega$. The cells $\{T \in \mathcal{T} : \; m(T) = m\}$ are ordered *counterclockwise* (see Fig. 2.5).

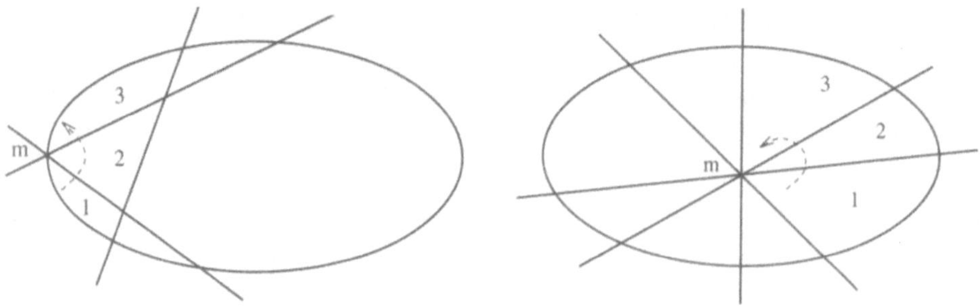

Fig. 2.5. $m = m(\Omega)$ or $m \in \operatorname{int} \Omega$.

Case 2. $m \in \partial\Omega \setminus \{m(\Omega)\}$. The cells $\{T \in \mathcal{T} : \ m(T) = m\}$ are ordered *clockwise* if $m = \min(\Omega \cap \xi_m)$ and *counterclockwise* if $m = \max(\Omega \cap \xi_m)$, where ξ_m is the vertical line through m (see Fig. 2.6).

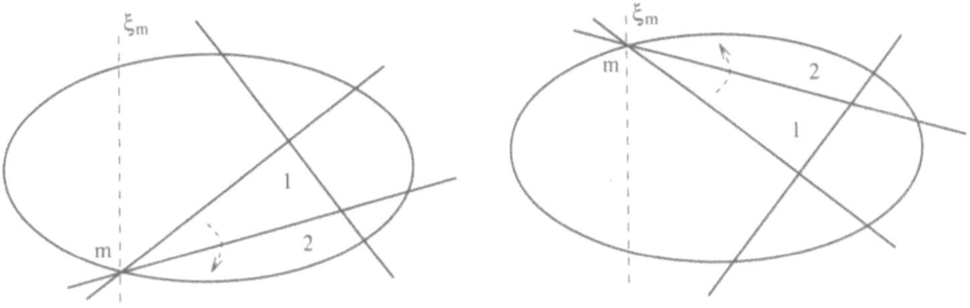

Fig. 2.6. $m \in \partial\Omega \setminus \{m(\Omega)\}$.

Thus, a total order for the elements of $\mathcal{T}$ has been defined, and we can write

$$\mathcal{T} = \{T_1, \ldots, T_n\}, \quad \text{where} \quad T_i < T_{i+1}, \ i = 1, \ldots, n-1.$$

We now assign a type A, B or C to each cell T_i and then choose interpolation points on T_i's according to their types. Namely, T_1 is the only *cell of type A* (since $m(T_1) = m(\Omega)$), T_i is a *cell of type B* if $m(T_i) \in \partial\Omega \setminus \{m(\Omega)\}$, and a *cell of type C* if $m(T_i) \in \operatorname{int}\Omega$. We put interpolation points of type A (as in Fig. 2.1) on T_1. It is easy to see that if T_i is of type B, then

$$I_i := \partial T_i \cap \partial \left(\bigcup_{j=1}^{i-1} T_j \right)$$

is a nondegenerate line segment with an endpoint at $m(T_i)$. We choose interpolation points of type B on T_i, as in Fig. 2.2, where $T_i \subset \Omega_1$, $\bigcup_{j=1}^{i-1} T_j \subset \Omega_0$, $I_i \subset l$.

Let now $T_i, T_{i+1}, \ldots, T_k$ be a family of cells of type C with the same minimal point $m \in \operatorname{int}\Omega$, so that $m(T_{i-1}) < m$, $m(T_i) = m(T_{i+1}) = \cdots = m(T_k) = m$, $m(T_{k+1}) > m$. Then $\bigcup_{j=1}^{k} T_j$ is a subset of a cone, and we choose interpolation points of type C (as in Fig. 2.4, a) or b)) on it.

Theorem 2.2. [1] *The set of all points chosen on the cells T_i, $i = 1, \ldots, n$, as described above, is a Lagrange interpolation set for $S_q^r(\Delta)$.*

We briefly describe the idea of the proof of Theorem 2.2. Let $z_1, \ldots, z_N$ be all the points chosen on the cells in accordance with the above procedure. Then it follows from Theorem 2.1 that $N = \dim S_q^r(\Delta)$. Therefore, in order to prove that $\{z_1, \ldots, z_N\}$ is a Lagrange interpolation set for $S_q^r(\Delta)$ it is sufficient to check that for any $s \in S_q^r(\Delta)$,

$$s(z_1) = 0, \ldots, s(z_N) = 0, \tag{2.1}$$

implies

$$s(z) = 0\,, \quad \text{for any} \quad z \in \Omega\,. \tag{2.2}$$

To this end we start from T_1 and see that (2.1) implies that

$$s(z) = 0\,, \quad \text{for any} \quad z \in T_1\,,$$

since $\{z_1, \ldots, z_N\} \cap T_1$ is a Lagrange interpolation set for $\widetilde{\Pi}_q$. Then we pass from T_2 to T_n and see that for any $i \in \{2, \ldots, n\}$ such that T_i is a cell of type B,

$$s(z) = 0\,, \quad \text{for any} \quad z \in \bigcup_{j=1}^{i-1} T_j \quad \text{and} \quad z \in \{z_1, \ldots, z_N\} \cap T_i\,,$$

implies

$$s(z) = 0\,, \quad \text{for any} \quad z \in T_i\,,$$

since $\{z_1, \ldots, z_N\} \cap T_i$ is a Lagrange interpolation set for the corresponding space of truncated power functions T_q^r. Similarly, if $T_i, T_{i+1}, \ldots, T_k$ is a family of cells of type C with the same minimal point $m \in \operatorname{int} \Omega$, so that $m(T_{i-1}) < m$, $m(T_i) = m(T_{i+1}) = \cdots = m(T_k) = m$, $m(T_{k+1}) > m$, then

$$s(z) = 0\,, \quad \text{for any} \quad z \in \bigcup_{j=1}^{i-1} T_j \quad \text{and} \quad z \in \{z_1, \ldots, z_N\} \cap \bigcup_{j=i}^{k} T_j\,,$$

implies

$$s(z) = 0\,, \quad \text{for any} \quad z \in \bigcup_{j=i}^{k} T_j\,,$$

since $\{z_1, \ldots, z_N\} \cap \bigcup_{j=i}^{k} T_j$ is a Lagrange interpolation set for the corresponding space of cone splines K_q^r. Thus, (2.2) follows by induction.

3. Approximation Order of Bivariate Spline Interpolation

In this section we consider spaces of bivariate splines with respect to special crosscut partitions Δ^1 and Δ^2. Let a rectangle $\Omega = [a, b] \times [c, d]$ and points $a = x_0 < x_1 < \cdots < x_{n_1} = b$, $c = y_0 < y_1 < \cdots < y_{n_2} = d$ such that $x_i - x_{i-1} = h_1$, $i = 1, \ldots, n_1$; $y_j - y_{j-1} = h_2$, $j = 1, \ldots, n_2$, be given. We set $h = \max\{h_1, h_2\}$. By defining $R_{i,j} = [x_{i-1}, x_i] \times [y_{j-1}, y_j]$, $i = 1, \ldots, n_1$; $j = 1, \ldots, n_2$, we obtain a partition of Ω into subrectangles $R_{i,j}$. If the diagonal from the lower left to the upper right vertex is added to each subrectangle $R_{i,j}$, then we denote the resulting partition by Δ^1. If we add both diagonals to each subrectangle, then the resulting partition is denoted by Δ^2.

Since both Δ^1 and Δ^2 are crosscut partitions, a basis for the spline spaces $S_q^r(\Delta^\mu)$, $\mu = 1, 2$, is given in Theorem 2.1. Similarly, the application of Theorem 2.2 to $S_q^r(\Delta^\mu)$, $\mu = 1, 2$, yields Lagrange interpolation sets of Nürnberger &

Riessinger [13,14]. We first describe these Lagrange interpolation sets for $S_q^r(\Delta^1)$. Then, by "taking limits", some Hermite interpolation sets are obtained, such that interpolation at them yields (nearly) optimal approximation order, under some restrictions on r and q.

For constructing Lagrange interpolation sets for $S_q^r(\Delta^1)$, $q \geq 4$, we describe four basic steps. For an arbitrary subtriangle T of the partition Δ^1, one of the following steps will be applied to T. (If the number of lines in Step C or D below is non-positive, then no points are chosen.)

Step A. (Starting step) Choose $q+1$ disjoint line segments $a_1, \dots, a_{q+1}$ in T. For $i = 1, \dots, q+1$, choose $q+2-i$ distinct points on a_i.

Step B. Choose $q-r$ disjoint line segments $b_1, \dots, b_{q-r}$ in T. For $i = 1, \dots, q-r$, choose $q+1-r-i$ distinct points on b_i.

Step C. Choose $q-2r+[\frac{r}{2}]$ disjoint line segments $c_1, \dots, c_{q-2r+[\frac{r}{2}]}$ in T. For $i = 1, \dots, q-2r$, choose $q+1-r-i$ distinct points on c_i and for $i = q-2r+1, \dots, q-2r+[\frac{r}{2}]$ choose $2(q-i)-3r+1$ distinct points on c_i. (Here $[b] := \max\{a \in \mathbb{Z} : a \leq b\}$.)

Step D. Choose $q-2r-1$ disjoint line segments $d_1, \dots, d_{q-2r-1}$ in T. For $i = 1, \dots, q-2r-1$, choose $q-2r-i$ distinct points on d_i.

Given a partition Δ^1, we apply the above steps to the subtriangles of Δ^1 as indicated in Fig. 3.2, where we choose horizontal, vertical and diagonal line segments as indicated in Fig. 3.1.

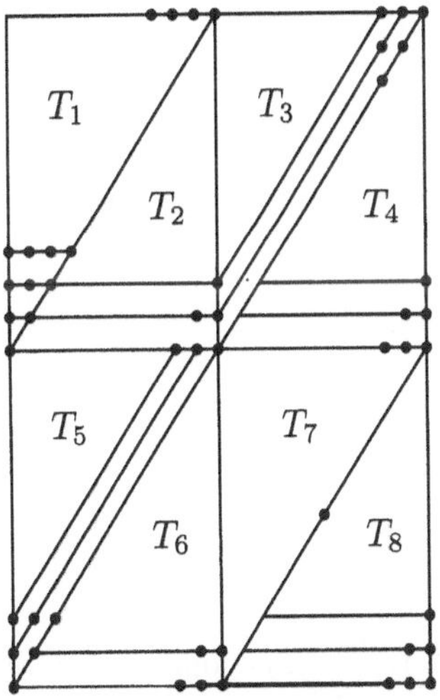

Fig. 3.1.

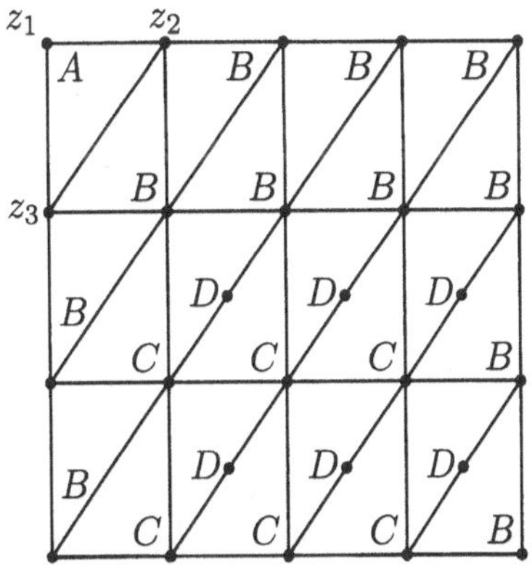

Fig. 3.2.

The following construction of Hermite interpolation sets for $S_q^r(\Delta^1)$, which yields (nearly) optimal order approximation, was given by Nürnberger [12] for $r = 1$ and by Davydov, Nürnberger & Zeilfelder [8] for $r \geq 2$.

Let a sufficiently differentiable function $f \in C(\Omega)$ be given. In order to define Hermite interpolation conditions for a spline $s \in S_q^r(\Delta^1)$, where $q \geq 4$ if $r = 1$, and $q \geq 3.5r + 1$ if $r \geq 2$, we describe four basic conditions. Let T be an arbitrary subtriangle of the partition Δ^1. If T is not the first from the left triangle in the top row, then $\widetilde{T}$ denotes the adjacent subtriangle left of T in the same row if it exists, and up of T otherwise. We impose one of the following four conditions on the polynomial $p = s|_T \in \Pi_q$.

Condition A. (Starting condition) $D^\omega p(z) = D^\omega f(z)$, $\omega = 0, \ldots, q$, where z is a vertex of T.

Condition B. $D^\omega p(z) = D^\omega f(z)$, $\omega = 0, \ldots, q - r - 1$, where z is the vertex of T not belonging to $\widetilde{T}$.

Condition C. $p_{x^\alpha y^\beta}(z) = f_{x^\alpha y^\beta}(z)$, $\alpha \geq 0$, $\beta \geq 0$, $\alpha + \beta \leq q - r - 1$, $\alpha + 2\beta \leq 2q - 3r - 2$, where z is the vertex of T not belonging to $\widetilde{T}$.

Condition D. $D^\omega p(z) = D^\omega f(z)$, $\omega = 0, \ldots, q - 2r - 2$, where z is the midpoint of the diagonal of T.

Note that while Conditions A, B and D are symmetric with respect to x and y, this is not the fact for Condition C. Fig. 3.3 presents the domain in which all integer points (α, β) should be taken in order to define Condition C.

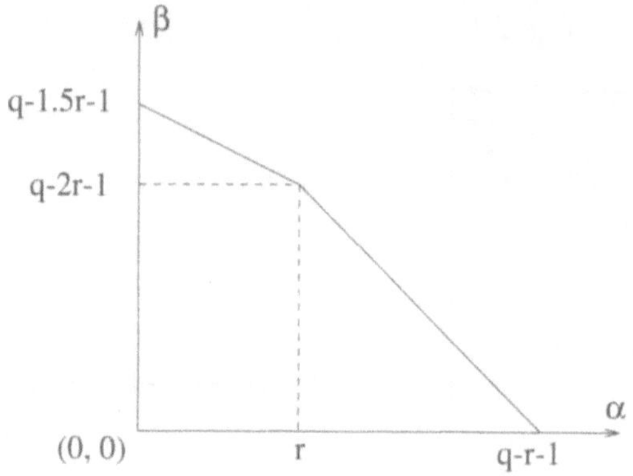

Fig. 3.3. Condition C.

Given a partition Δ^1, the distribution of the Hermite interpolation conditions to the subtriangles is the same as for Lagrange interpolation and is indicated in Fig. 3.2.

In Theorems 3.1, 3.2 and 3.4 below, the norm denotes the maximum of the uniform norm over all subtriangles of the partition (w.r.t. the polynomial pieces).

By using Bernstein-Bezier techniques, several authors proved results similar to those of Theorems 3.1 and 3.4 below for the special spline spaces $S_3^1(\Delta^1)$ (Sha

[16]) and $S_2^1(\Delta^2)$ (Chui & He [4], Sha [17], Zedek [18]). Moreover, Jeeawock-Zedek [9] proved that interpolation by $S_3^1(\Delta^2)$ yields approximation order two.

Theorem 3.1. [12] *For each function $f \in C^{q+1}(\Omega)$, there exists a constant $K > 0$ such that for the unique spline $s \in S_q^1(\Delta^1)$ which satisfies the above Hermite interpolation conditions, the following statements hold: For all $i \in \{0, \dots, \varrho - 1\}$, $\|D^i(f - s)\| \le K h^{\varrho - i}$, where $\varrho = 4$ if $q = 4$, and $\varrho = q + 1$ if $q \ge 5$. (The constant $K > 0$ depends on $\|D^{q+1} f\|$ and is independent of h.)*

Theorem 3.2. [8] *Let integers $r \ge 2$ and $q \ge 3.5r + 1$ be given. For each function $f \in C^{q+1}(\Omega)$, there exists a constant $K > 0$ such that for the unique spline $s \in S_q^r(\Delta^1)$ which satisfies the above Hermite interpolation conditions,*

$$\|D^i(f - s)\| \le K h^{q+1-i}, \quad i = 0, \dots, q.$$

(The constant $K > 0$ depends on $\|D^{q+1} f\|$ and is independent of h.)

In view of Theorem 3.2 it is interesting to note that the approximation order of the spline space $S_q^r(\Delta)$ is optimal (i.e., $q + 1$), if $q \ge 3r + 2$ (see de Boor & Höllig [2], Chui, Hong & Jia [5], and Lai & Schumaker [10]). On the other hand, it was proved by de Boor & Jia [3] that this is not true, if $q < 3r + 2$, even for the Δ^1-partition.

Remark 3.3. The method of proof in [12] can be applied to spline spaces $S_q^1(\widetilde{\Delta}^1)$, $q \ge 5$, where $\widetilde{\Delta}^1$ is a "deformation" of partition Δ^1, by which we mean an arbitrary rectilinear embedding of the same triangulation in $\mathbb{R}^2$, as in Fig. 3.4. Under some restrictions on the angles between adjacent edges, corresponding Hermite interpolation scheme possesses optimal approximation order h^{q+1}, where h is the maximal sidelength of the subtriangles.

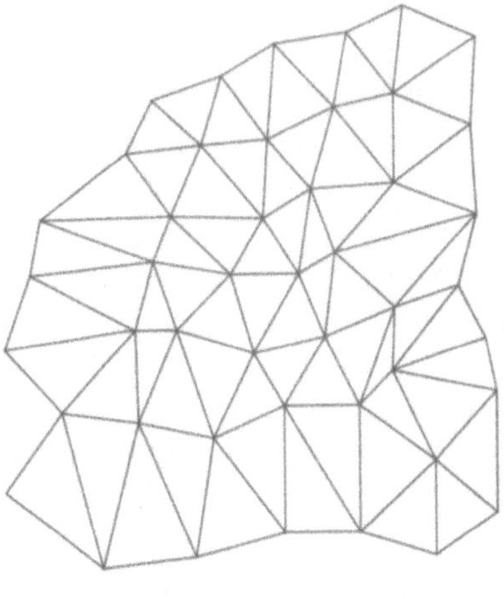

Fig. 3.4. An example of partition $\widetilde{\Delta}^1$.

We now describe in a similar way the construction of Lagrange and Hermite interpolation sets for $S_q^1(\Delta^2)$ (Results for $S_q^r(\Delta^2)$, $r \geq 2$, are not yet available). As above, it turns out that interpolation at these points yields (nearly) optimal approximation order.

For constructing Lagrange interpolation sets for $S_q^1(\Delta^2)$, $q \geq 2$, we again describe four basic steps.

Step A. (Starting step) Choose $q+1$ disjoint line segments $a_1, \ldots, a_{q+1}$ in T. For $i = 1, \ldots, q+1$, choose $q+2-i$ distinct points on a_i.

Step B. Choose $q-1$ disjoint line segments $b_1, \ldots, b_{q-1}$ in T. For $i = 1, \ldots, q-1$, choose $q-i$ distinct points on b_i.

Step C. Choose $q-3$ disjoint line segments $c_1, \ldots, c_{q-3}$ in T. For $i = 1, \ldots, q-3$, choose $q-2-i$ distinct points on c_i.

Step D. Choose $q-2$ disjoint line segments $d_1, \ldots, d_{q-2}$ in T. For $i = 1, \ldots, q-2$, choose $q-i$ distinct points on d_i.

Given a partition Δ^2, we apply the above steps to the subtriangles of Δ^2 as indicated in Fig. 3.6, where we choose horizontal, vertical and diagonal line segments as indicated in Fig. 3.5.

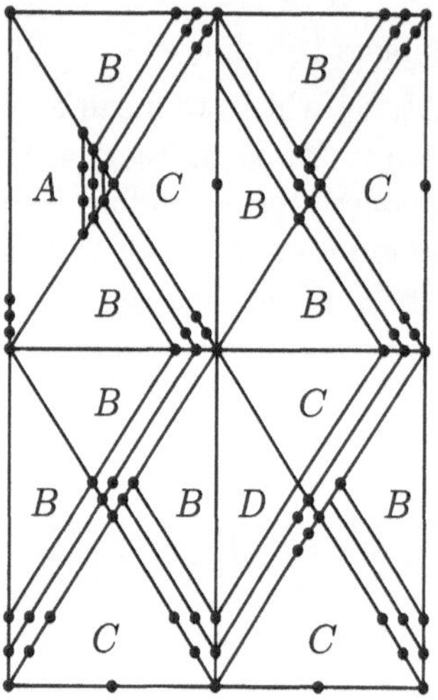

Fig. 3.5.

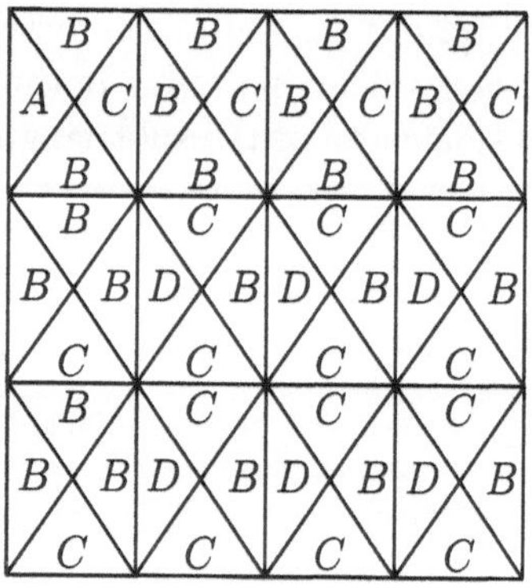

Fig. 3.6.

The following construction of Hermite interpolation sets for $S_q^1(\Delta^2)$, which differs in some parts from the one described above for Δ^1, was given by Nürnberger & Walz [15].

Let a sufficiently differentiable function $f \in C(\Omega)$ be given. We again have to describe four basic conditions. Let T be an arbitrary subtriangle of the partition Δ^2. We impose one of the following four conditions on the polynomial $p = s|_T \in$

$\widetilde{\Pi}_q$, where z is a vertex resp. a midpoint of an edge of T as described below (cf. also Fig. 3.7).

Condition A. (Starting condition) $D^\omega p(z) = D^\omega f(z)$, $\omega = 0, \ldots, q$, where z is a vertex of T.

Condition B. $D^\omega p(z) = D^\omega f(z)$, $\omega = 0, \ldots, q-2$, where z is a vertex of T not adjacent to the subtriangles already considered, e.g. z_4 or z_6.

Condition C. $D^\omega p(z) = D^\omega f(z)$, $\omega = 0, \ldots, q-4$, where z is the midpoint of the edge of T which is not adjacent to the subtriangles already considered, e.g. z_5.

Condition D. $p_{\varrho^\alpha \overline{\varrho}^\beta}(z) = f_{\varrho^\alpha \overline{\varrho}^\beta}(z)$, $\alpha \geq 0$, $\beta \geq 0$, $\alpha + \beta \leq q-2$, $\beta \neq q-2$, where $\varrho = (\varrho_1, \varrho_2)$ is the unit vector in direction of the diagonal of Δ^1, and $\overline{\varrho} = (-\varrho_1, \varrho_2)$, and where z is the crossing point of the two diagonals in one subrectangle, e.g. z_7.

Given a partition Δ^2, we apply the above steps to the subtriangles of Δ^2 as indicated in Fig. 3.7. Note that the configuration of Hermite conditions is different from that of Lagrange conditions (cp. Fig. 3.6 and Fig. 3.7).

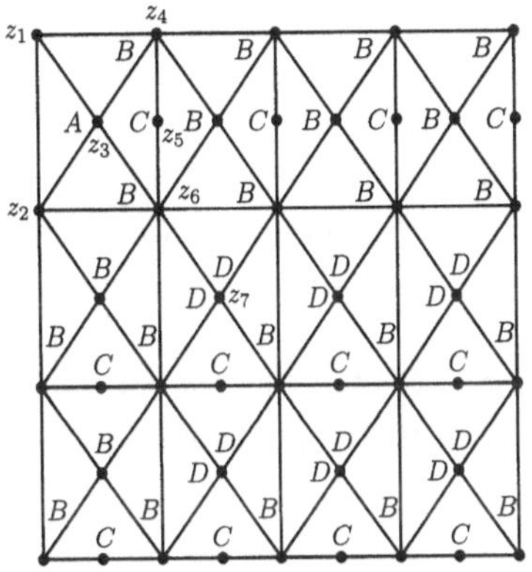

Fig. 3.7.

Theorem 3.4. [15] *For each function $f \in C^{q+1}(\Omega)$, there exists a constant $K > 0$ such that for the unique spline $s \in S_q^1(\Delta^2)$ which satisfies the above Hermite interpolation conditions, the following statements hold: For all $i \in \{0, \ldots, \varrho - 1\}$, $\|D^i(f-s)\| \leq K h^{\varrho - i}$, where $\varrho = q$ if $q \in \{2,3\}$, and $\varrho = q+1$ if $q \geq 4$. (The constant $K > 0$ depends on $\|D^{q+1} f\|$ and is independent of h.)*

We briefly mention that these results can also be used for fitting of scattered data by using a two-step method, originally developed in [12]. The method is as follows: Let a (possibly non-rectangular) domain Ω, points $w_i \in \Omega$ and corresponding data f_i be given. In the first step, we approximate the data f_i by any local method, e.g. interpolation by a piecewise polynomial $\widetilde{s}$ of degree q such that $\|f - \widetilde{s}\| = O(h^{q+1})$ if $f_i = f(w_i)$ and $f \in C^{q+1}(\Omega)$. In general, piecewise polynomial interpolation is a simpler problem than spline interpolation and in any case, this is always possible if the data is regularly distributed over Ω. In the second step, we interpolate the resulting function $\widetilde{s}$ (which may not even be continuous) by a smooth spline s as described in this paper. As in Theorems 3.1, 3.2 and 3.4, it can be shown that $|f_i - s(w_i)| = O(h^q)$ or $O(h^{q+1})$. More details on this and several numerical examples can be found in [12] and [15].

Acknowledgements. The research of O.V.Davydov was supported by the Alexander von Humboldt Foundation, under Research Fellowship.

References

1. Adam M. H., Bivariate Spline-Interpolation auf Crosscut-Partitionen, Doctoral Thesis, Mannheim 1996.
2. de Boor C. and Höllig K., Approximation power of smooth bivariate pp functions, Math. Z. **197** (1988), 343–363.
3. de Boor C. and Jia Q., A sharp upper bound on the approximation order of smooth bivariate pp functions, J. Approx. Theory **72** (1993), 24–33.
4. Chui C. K. and He T. X., On location of sample points in C^1 quadratic bivariate spline interpolation, in *Numerical Methods of Approximation Theory, ISNM 81*, Collatz L., Meinardus G. and Nürnberger G. (eds.), 30–43, Birkhäuser, Basel, 1987.
5. Chui C. K., Hong D., and Jia Q., Stability of optimal-order approximation by bivariate splines over arbitrary triangulations, Trans. Amer. Math. Soc. **347** (1995), 3301–3318.
6. Chui C. K. and Wang R. H., Multivariate spline spaces, J. Math. Anal. Appl. **94** (1983), 197–221.
7. Dahmen W. and Micchelli C. A., Recent progress in multivariate splines, in *Approximation Theory IV*, Chui C. K., Schumaker L. L. and Ward J. D. (eds.), 27–121, Academic Press, New York, 1983.
8. Davydov O. V., Nürnberger G., and Zeilfelder F., Approximation order of bivariate spline interpolation for arbitrary smoothness, submitted for publication.
9. Jeeawock-Zedek F., Interpolation scheme by C^1 cubic splines on a non uniform type-2 triangulation of a rectangular domain, C. R. Acad. Sci. Paris Sér. I Math. **314** (1992), 413–418.
10. Lai M.-J. and Schumaker L., On the approximation power of bivariate splines, preprint.

11. Nürnberger G., *Approximation by Spline Functions*, Springer, Berlin, 1989.
12. Nürnberger G., Approximation order of bivariate spline interpolation, J. Approx. Theory **87** (1996), 117–136.
13. Nürnberger G. and Riessinger Th., Lagrange and Hermite interpolation by bivariate splines, Numer. Funct. Anal. Optim. **13**, (1992), 75–96.
14. Nürnberger G. and Riessinger Th., Bivariate spline interpolation at grid points, Numer. Math. **71** (1995), 91–119.
15. Nürnberger G. and Walz G., Error analysis in bivariate spline interpolation, submitted for publication.
16. Sha Z., On interpolation by $S_3^1(\Delta_{m,n}^1)$, Approx. Theory Appl. **1** (1985), 71–82.
17. Sha Z., On interpolation by $S_2^1(\Delta_{m,n}^2)$, Approx. Theory Appl. **1** (1985), 1–18.
18. Zedek F., Interpolation de Lagrange par de splines quadratique sur un quadrilatere de $\mathbb{R}^2$, RAIRO Model. Math. Anal. Numer. **26** (1992), 575–593.

Oleg V. Davydov
Department of Mechanics and Mathematics
Dnepropetrovsk State University
pr. Gagarina 72
Dnepropetrovsk, GSP 320625, Ukraine
`davydov@euklid.math.uni-mannheim.de`

Günther Nürnberger, Guido Walz, Frank Zeilfelder
Lehrstuhl für Mathematik IV
Universität Mannheim
D-68131 Mannheim, Germany
`{nuernberger}{walz}{zeilfelder}@math.uni-mannheim.de`

Necessary and Sufficient Conditions for Orthonormality of Scaling Vectors

Gerlind Plonka

Abstract. The paper studies necessary and sufficient orthonormality conditions for a scaling vector in terms of its two-scale symbol and its corresponding transfer operator. In particular, it is shown that the conditions of Hogan [10] for the two-scale symbol and the criteria of Shen [21] for the transfer operator are equivalent.

1. Introduction

In this paper we shall discuss orthonormality of compactly supported scaling vectors. These are solutions of functional equations of type

$$\mathbf{\Phi}(x) = \sum_{l=0}^{N} \mathbf{P}_l \, \mathbf{\Phi}(2x - l) \tag{1}$$

with real $r \times r$ coefficient matrices $\mathbf{P}_l$ ($r \in \mathbb{N}, , r \geq 1$) and with an r-dimensional function vector $\mathbf{\Phi} = (\varphi_1, \ldots, \varphi_r)^T$. Equations of the form (1) are called *matrix refinement equations.*

If additionally, the functions $\varphi_\nu(\cdot - l)$ ($l \in \mathbb{Z}$, $\nu = 1, \ldots, r$) form an orthonormal or an L^2-stable basis of their span, then $\mathbf{\Phi}$ is called *multi-scaling function.* In this case, $\mathbf{\Phi}$ can generate a multiresolution analysis with multiplicity r (see [7]). Once, an MRA generated by an orthonormal multi-scaling function $\mathbf{\Phi}$ is given, the construction of an orthonormal multiwavelet $\mathbf{\Psi} = (\psi_1, \ldots, \psi_r)^T$ can be reduced to a problem of matrix extension, as described in [17].
By Fourier transform of (1), we have

$$\hat{\mathbf{\Phi}}(\omega) = \mathbf{P}(\frac{\omega}{2}) \, \hat{\mathbf{\Phi}}(\frac{\omega}{2}), \tag{2}$$

Multivariate Approximation and Splines
G. Nürnberger, J. W. Schmidt, and G. Walz (eds.), pp. 205–218.

ISBN 3-7643-5654-5.

where $\hat{\mathbf{\Phi}}$ is taken componentwisely, *i.e.*, $\hat{\mathbf{\Phi}}(\omega) := (\hat{\varphi}_1(\omega), \ldots, \hat{\varphi}_r(\omega))^T$ with $\hat{\varphi}_\nu(\omega) := \int_{-\infty}^{\infty} \varphi_\nu(x)\, e^{-i\omega x}\, \mathrm{d}x$ $(\nu = 1, \ldots, r)$, and where

$$\mathbf{P}(\omega) := \frac{1}{2} \sum_{l=0}^{N} \mathbf{P}_l\, e^{-i\omega l} \tag{3}$$

denotes the *two-scale symbol* of $\mathbf{\Phi}$.

Hence, we are faced with the problem of how the orthonormality or L^2-stability condition for a solution vector $\mathbf{\Phi}$ can be ensured by appropriate choice of $\mathbf{P}(\omega)$.

As in the scalar case $(r = 1)$, we observe three different methods to express necessary and sufficient stability (orthonormality) conditions in terms of the two-scale symbol $\mathbf{P}(\omega)$.

The first method is based on the so-called *transfer operator* T associated with $\mathbf{P}(\omega)$. Under certain basic conditions on $\mathbf{P}(\omega)$, L^2-stability (and orthonormality, respectively) of $\mathbf{\Phi}$ can be ensured if the transfer operator T associated with $\mathbf{P}(\omega)$ satisfies special spectral conditions. This method can also be applied in the multivariate setting (see [21]). In the meantime, it turned out that the basic conditions assumed by Shen [21] are necessary for stability of $\mathbf{\Phi}$ (see [4,11,14]).

In order to handle the transfer operator T in practice, one has to use its representing matrix, which in fact can be given explicitly in terms of Kronecker products of coefficient matrices $\mathbf{P}_n$ (see [15,20]). The resulting conditions, which are spectral conditions to the representing matrix, can be seen as generalization of Lawton's criteria for scaling functions (see [5,16]).

Second, there are some successful attempts to find necessary and sufficient conditions directly in terms of the trigonometric polynomial matrix $\mathbf{P}(\omega)$ in order to ensure orthonormality, stability or even local linear independence of the solution vector $\mathbf{\Phi}$ (see [10,22]). These results generalize the well-known Cohen criteria [1] and the conditions of Jia and Wang [13]. But this time, the conditions for the two-scale symbol $\mathbf{P}(\omega)$ are much more complicated, since products of matrix polynomials generally do no commute. Moreover, one is faced with a problem which need not to be handled in the scalar case, namely, of how to ensure the algebraic linear independence of the components φ_ν $(\nu = 1, \ldots, r)$ of $\mathbf{\Phi}$ and their translates in terms of $\mathbf{P}(\omega)$.

Third, we want to mention that the stability of scaling vectors is closely related with the convergence of corresponding subdivision schemes and cascade algorithms. In fact, the convergence of the stationary subdivision scheme can be taken as a criteria for stability of $\mathbf{\Phi}$ in $L^p(\mathbb{R})^r$. This subject is addressed in [3,4,12]. In particular, relations between spectral conditions of the transfer operator and the convergence of the cascade algorithm are considered in [21].

We are especially interested in the first two methods. The purpose of this paper is to study the relation between the spectral properties of the transfer operator T and the properties of the two-scale symbol $\mathbf{P}(\omega)$ in the case of orthonormal scaling vectors. In particular, we shall show that the conditions of Hogan [10] and Shen [21] are indeed equivalent.

2. Basic conditions and uniqueness of $\mathbf{\Phi}$

In this section we want to provide some necessary conditions for orthonormality of a compactly supported solution vector $\mathbf{\Phi}$ of a matrix refinement equation (1). These *basic* conditions will also ensure that $\mathbf{\Phi}$ is unique, and moreover that the components of $\mathbf{\Phi}$ are contained in $L^2(\mathbb{R})$.

We say that a function vector $\mathbf{\Phi}$ (with $\varphi_\nu \in L^2(\mathbb{R})$) is L^2-*stable* if the integer translates of φ_ν are algebraically linearly independent and if there are constants $0 < A \leq B < \infty$, such that

$$A \sum_{l=-\infty}^{\infty} \mathbf{c}_l^T \bar{\mathbf{c}}_l \leq \| \sum_{l=-\infty}^{\infty} \mathbf{c}_l^T \mathbf{\Phi}(\cdot - l) \|_{L^2}^2 \leq B \sum_{l=-\infty}^{\infty} \mathbf{c}_l^T \bar{\mathbf{c}}_l \tag{4}$$

for any vector sequence $\{\mathbf{c}_l\}_{l \in \mathbb{Z}} \in l_2^r$. Here l_2^r denotes the set of sequences of vectors $(\mathbf{c}_l)_{l \in \mathbb{Z}}$ $(\mathbf{c}_l \in \mathbb{C}^r)$ with $\sum_{l=-\infty}^{\infty} \mathbf{c}_l^T \overline{\mathbf{c}_l} < \infty$. $\mathbf{\Phi}$ is called *orthonormal* if (4) is satisfied with $A = B = 1$, in other words, if

$$\langle \varphi_\mu, \varphi_\nu(\cdot - n) \rangle_{L^2} = \delta_{0,n} \, \delta_{\mu,\nu}. \tag{5}$$

Introducing the autocorrelation symbol

$$\begin{aligned} \mathbf{\Omega}(\omega) &:= \sum_{n \in \mathbb{Z}} (\langle \varphi_\mu, \varphi_\nu(\cdot - n) \rangle_{L^2})_{\mu,\nu=1}^r \, e^{-i\omega n} \\ &= \sum_{n \in \mathbb{Z}} \left(\int_{-\infty}^{\infty} \mathbf{\Phi}(x) \, \mathbf{\Phi}(x-n)^\star \, \mathrm{d}x \right) e^{-i\omega n} \end{aligned}$$

(with $\mathbf{\Phi}(x)^\star := \overline{\mathbf{\Phi}(x)^T}$) we simply observe that (5) is equivalent with $\mathbf{\Omega}(\omega) = \mathbf{I}$, where $\mathbf{I}$ denotes the unit matrix of size r. Further, the stability condition (4) implies that the autocorrelation symbol is strictly positive definite for all $\omega \in \mathbb{R}$ (see [7]).

From

$$\begin{aligned} \int_{-\infty}^{\infty} \mathbf{\Phi}(x) \, \mathbf{\Phi}(x-n)^\star \, \mathrm{d}x &= \frac{1}{2\pi} \int_{-\infty}^{\infty} \hat{\mathbf{\Phi}}(\omega) \, \hat{\mathbf{\Phi}}(\omega)^\star \, e^{in\omega} \, \mathrm{d}\omega \\ = \frac{1}{2\pi} \int_0^{2\pi} & e^{in\omega} \sum_{l \in \mathbb{Z}} \hat{\mathbf{\Phi}}(\omega + 2\pi l) \, \hat{\mathbf{\Phi}}(\omega + 2\pi l)^\star \, \mathrm{d}\omega \end{aligned}$$

it follows that

$$\mathbf{\Omega}(\omega) = \sum_{l \in \mathbb{Z}} \hat{\mathbf{\Phi}}(\omega + 2\pi l) \, \hat{\mathbf{\Phi}}(\omega + 2\pi l)^\star \qquad a.e.. \tag{6}$$

Substituting (2) leads to

$$\mathbf{\Omega}(\omega) = \sum_{l\in\mathbb{Z}} \mathbf{P}(\frac{\omega}{2}+\pi l)\,\hat{\mathbf{\Phi}}(\frac{\omega}{2}+\pi l)\,\hat{\mathbf{\Phi}}(\frac{\omega}{2}+\pi l)^{\star}\,\mathbf{P}(\frac{\omega}{2}+\pi l)^{\star}.$$

Splitting the sum into even and odd l, it follows that

$$\mathbf{\Omega}(\omega) = \mathbf{P}(\frac{\omega}{2})\,\mathbf{\Omega}(\frac{\omega}{2})\,\mathbf{P}(\frac{\omega}{2})^{\star} + \mathbf{P}(\frac{\omega}{2}+\pi)\,\mathbf{\Omega}(\frac{\omega}{2}+\pi)\,\mathbf{P}(\frac{\omega}{2}+\pi)^{\star} \qquad a.e..$$

Since $\mathbf{P}(\omega)$ and $\mathbf{\Omega}(\omega)$ are assumed to be matrices of trigonometric polynomials, the a.e. can be droped. In particular, for orthonormal L^2-solutions of (1), the condition

$$\mathbf{I} = \mathbf{P}(\frac{\omega}{2})\,\mathbf{P}(\frac{\omega}{2})^{\star} + \mathbf{P}(\frac{\omega}{2}+\pi)\,\mathbf{P}(\frac{\omega}{2}+\pi)^{\star} \tag{7}$$

is necessarily satisfied.

Let $\mathbb{H} = \mathbb{H}_N$ be the space of trigonometric polynomials of degree at most N, *i.e.*, the elements of $\mathbb{H}$ are of the form $h(\omega) = \sum_{n=-N}^{N} h_n\, e^{-i\omega n}$ $(h_n \in \mathbb{C})$. We introduce the following *transfer operator* $T : \mathbb{H}^{r\times r} \to \mathbb{H}^{r\times r}$,

$$T\,\mathbf{H}(\omega) := \mathbf{P}(\frac{\omega}{2})\,\mathbf{H}(\frac{\omega}{2})\,\mathbf{P}(\frac{\omega}{2})^{\star} + \mathbf{P}(\frac{\omega}{2}+\pi)\,\mathbf{H}(\frac{\omega}{2}+\pi)\,\mathbf{P}(\frac{\omega}{2}+\pi)^{\star},$$

acting on $(r \times r)$-matrices $\mathbf{H}(\omega)$ with elements of $\mathbb{H}$ as entries. Observe that the autocorrelation symbol $\mathbf{\Omega}(\omega)$ is an eigenmatrix of the transfer operator T corresponding to the eigenvalue 1.

For a square matrix $\mathbf{M}$ (or a linear operator) let us introduce the following conditions on its eigenvalues:

Condition E. *The spectral radius of* $\mathbf{M}$ *is less than or equal to* 1, *i.e.* $\varrho(\mathbf{M}) \le 1$, *and* 1 *is the only eigenvalue of* $\mathbf{M}$ *on the unit circle. Moreover,* 1 *is a simple eigenvalue.*

Assuming that the components of a solution vector $\mathbf{\Phi}$ of (1) are compactly supported and in $L^2(\mathbb{R})$, it necessarily follows that they are also contained in $L^1(\mathbb{R})$. As shown in [4,11], we have:

Proposition 1. *Let* $\mathbf{\Phi}$ *be a compactly supported,* L^2*-stable solution vector of (1). Then for the corresponding symbol* $\mathbf{P}(\omega)$ *we have:*

a) $\mathbf{P}(0)$ *satisfies Condition E.*

b) *There exists a nonzero vector* $\mathbf{y} \in \mathbb{R}^r$ *such that* $\mathbf{y}^T\,\mathbf{P}(0) = \mathbf{y}^T$ *and* $\mathbf{y}^T\mathbf{P}(\pi) = \mathbf{0}^T$. *Equivalently, the solution vector* $\mathbf{\Phi}$ *provides approximation order* 1, *i.e., we have*

$$\mathbf{y}^T \sum_{l=-\infty}^{\infty} \mathbf{\Phi}(\cdot - l) = c,$$

where $\mathbf{y}$ *is a left eigenvector of* $\mathbf{P}(0)$ *to the eigenvalue* 1, *and* c *is a nonvanishing constant.*

The necessary conditions of Proposition 1 for $\mathbf{P}(\omega)$ are called *basic conditions.*

In the rest of the paper, we want to assume that the basic conditions and the orthonormality condition (7) are satisfied for the two-scale symbol $\mathbf{P}(\omega)$. Indeed, these assumptions already imply the uniqueness of a solution vector of compactly supported L^2-functions. Using the results of Jiang and Shen [14], we find:

Proposition 2. *Let $\mathbf{P}(\omega)$ be of the form (3) satisfying the basic conditions of Proposition 1, and let $\mathbf{a}$ be a right eigenvector of $\mathbf{P}(0)$ corresponding to the eigenvalue 1. Then (1) provides a compactly supported distribution solution $\mathbf{\Phi}$, where*

$$\hat{\mathbf{\Phi}}(\omega) := \lim_{L\to\infty} \prod_{j=1}^{L} \mathbf{P}\left(\frac{\omega}{2^j}\right) \mathbf{a}. \tag{8}$$

This solution vector $\mathbf{\Phi}$ is unique up to multiplication with a constant.

Note that this result is nontrivial; it is based on the observation that the growth of $\hat{\varphi}_1, \ldots, \hat{\varphi}_r$ is at most polynomial on $\mathbb{R}$ (see also [9]). Proposition 2 can be seen as a generalization of an analogous result for scalar refinement equations by Deslauries and Dubuc [6]. Let us mention, that the pointwise convergence of the infinite product in (8) can even be shown, if only $\varrho(\mathbf{P}(0)) < 2$ is satisfied (see [2,8]).

Further, we find:

Proposition 3. *Let $\mathbf{P}(\omega)$ be of the form (3), satisfying the basic conditions of Proposition 1 and the orthonormality condition (7). Further, let $\mathbf{a}$ be a right eigenvector of $\mathbf{P}(0)$ to the eigenvalue 1. Then the function vector $\hat{\mathbf{\Phi}}$ given in (8) is contained in $L^2(\mathbb{R})^r$.*

Proof: We introduce $\mathbf{\Pi}_n(\omega) := \prod_{j=1}^{n} \mathbf{P}(\frac{\omega}{2^j})$ and $\tilde{\mathbf{\Pi}}_n(\omega) := \chi_{[-\pi,\pi]}(2^{-n}\omega)\,\mathbf{\Pi}_n(\omega)$, where $\chi_{[-\pi,\pi]}$ denotes the characteristic function over $[-\pi,\pi]$. Obviously, $\tilde{\mathbf{\Pi}}_n(\omega)$ converges pointwise to $\mathbf{\Pi}(\omega) := \prod_{j=1}^{\infty} \mathbf{P}(\frac{\omega}{2^j})$. By (7), we find that

$$\begin{aligned}
&\frac{1}{2\pi}\int_{-\infty}^{\infty} \tilde{\mathbf{\Pi}}_n(\omega)\,\tilde{\mathbf{\Pi}}_n(\omega)^\star d\omega = \frac{1}{2\pi}\int_{-2^n\pi}^{2^n\pi} \mathbf{\Pi}_n(\omega)\,\mathbf{\Pi}_n(\omega)^\star\, d\omega \\
&= \frac{1}{2\pi}\int_{0}^{2^{n+1}\pi} \mathbf{\Pi}_n(\omega)\,\mathbf{\Pi}_n(\omega)^\star\, d\omega \\
&= \frac{1}{2\pi}\int_{0}^{2^{n}\pi} \mathbf{\Pi}_{n-1}(\omega)\,[\mathbf{P}(\frac{\omega}{2^n})\mathbf{P}(\frac{\omega}{2^n})^\star + \mathbf{P}(\frac{\omega}{2^n}+\pi)\mathbf{P}(\frac{\omega}{2^n}+\pi)^\star]\mathbf{\Pi}_{n-1}(\omega)^\star\, d\omega \\
&= \frac{1}{2\pi}\int_{0}^{2^{n}\pi} \mathbf{\Pi}_{n-1}(\omega)\,\mathbf{\Pi}_{n-1}(\omega)^\star\, d\omega = \ldots \\
&= \frac{1}{2\pi}\int_{0}^{4\pi} \mathbf{\Pi}_1(\omega)\,\mathbf{\Pi}_1(\omega)^\star\, d\omega = \frac{1}{2\pi}\int_{0}^{2\pi} \mathbf{I}\, d\omega = \mathbf{I}
\end{aligned}$$

for all $n \in \mathbb{N}$. Let now $\|\cdot\|$ denote the spectral matrix norm and the Euklidian vector norm, respectively. Using the Lemma of Fatou, it follows that

$$\|\mathbf{\Pi}\|^2_{L^2(\mathbb{R})^{r\times r}} := \frac{1}{2\pi}\int_{-\infty}^{\infty} \|\mathbf{\Pi}(\omega)\|^2 \, d\omega \le \limsup_{n\to\infty} \frac{1}{2\pi}\int_{-\infty}^{\infty} \|\tilde{\mathbf{\Pi}}_n(\omega)\|^2 \, d\omega$$
$$= \limsup_{n\to\infty} \|\frac{1}{2\pi}\int_{-\infty}^{\infty} \tilde{\mathbf{\Pi}}_n(\omega)\, \tilde{\mathbf{\Pi}}_n(\omega)^\star d\omega\| = 1.$$

Finally, observing that

$$\|\hat{\mathbf{\Phi}}\|_{L^2(\mathbb{R})^r} := \left(\frac{1}{2\pi}\int_{-\infty}^{\infty} \|\mathbf{\Pi}(\omega)\, \mathbf{a}\|^2 \, d\omega\right)^{1/2} \le \|\mathbf{\Pi}\|_{L^2(\mathbb{R})^{r\times r}} \|\mathbf{a}\| = \|\mathbf{a}\| < \infty,$$

the assertion follows. □

Remark. This result is a direct generalization of the result of Mallat [19]. For the characterization of L^2-solutions of (1) see also [9,14].

3. Algebraic linear independence of scaling vectors

Let l_0^r denote the set of sequences of vectors $(\mathbf{c}_l)_{l\in\mathbb{Z}}$, where only a finite number of vectors $\mathbf{c}_l$ is different from the zero vector. We say that a compactly supported function vector $\mathbf{\Phi}$ is *(algebraically) linearly independent* if for any $(\mathbf{c}_l)_{l\in\mathbb{Z}} \in l_0^r$

$$\sum_{l\in\mathbb{Z}} \mathbf{c}_l^T \, \mathbf{\Phi}(\cdot - l) = 0 \quad \Rightarrow \quad \mathbf{c}_l = \mathbf{0} \quad \text{for all} \quad l.$$

Equivalently, $\mathbf{\Phi}$ is algebraically linearly independent if for any vector $\mathbf{A}(\omega)$ of trigonometric polynomials

$$\mathbf{A}(\omega)^T \, \hat{\mathbf{\Phi}}(\omega) = 0 \quad \Rightarrow \quad \mathbf{A}^T(\omega) = \mathbf{0}^T.$$

Note that the linear independence is necessary for L^2-stability or orthonormality of $\mathbf{\Phi}$.

In the scalar case, the problem of linear independence of integer translates of a scaling function $\varphi \in L^1(\mathbb{R})$ need not to be handled, since $A(\omega)\, \hat{\varphi}(\omega) = 0$ for some trigonometric polynomial $A(\omega)$ would imply that for all $\omega \in \mathbb{R}$ either $A(\omega) = 0$ or $\hat{\varphi}(\omega) = 0$. But from $\hat{\varphi}(0) \neq 0$ it follows, by continuity, that $\hat{\varphi}(\omega) \neq 0$ in a neighborhood of 0. So, $A(\omega) = 0$ for ω in a neighborhood of 0, and hence for all $\omega \in \mathbb{R}$.

For $r > 1$, we need to investigate this problem. First we observe:

Lemma 4. *Let $\mathbf{\Phi} \in L^2(\mathbb{R})^r$ be a compactly supported function vector, and assume that its autocorrelation symbol $\mathbf{\Omega}(\omega)$ is of the form (6), where the a.e. can be droped. Then we have:*
$\mathbf{\Phi}$ is algebraically linearly dependent if and only if its autocorrelation symbol $\mathbf{\Omega}(\omega)$ satisfies

$$\det \mathbf{\Omega}(\omega) = 0$$

for all $\omega \in [-\pi, \pi)$.

Proof: 1. Assume that $\mathbf{\Phi}$ is algebraically linearly dependent, then, by definition, there exists a nontrivial vector of trigonometric polynomials $\mathbf{A}(\omega)$ such that

$$\mathbf{A}(\omega)^T \,\hat{\mathbf{\Phi}}(\omega) = 0$$

for all $\omega \in [-\pi, \pi)$. Hence, we have

$$\mathbf{A}(\omega)^T \sum_{l=-\infty}^{\infty} \hat{\mathbf{\Phi}}(\omega + 2\pi l)\, \hat{\mathbf{\Phi}}(\omega + 2\pi l)^\star \,\overline{\mathbf{A}(\omega)} = 0,$$

that means, $\det \mathbf{\Omega}(\omega) = 0$ for all $l \in \mathbb{Z}$.

2. Assume that $\mathbf{\Omega}(\omega)$ is singular for all $\omega \in [-\pi, \pi)$. Then there exists a vector $\mathbf{A}(\omega)$ of 2π-periodic functions such that

$$\mathbf{A}(\omega)^T \,\mathbf{\Omega}(\omega)\, \overline{\mathbf{A}(\omega)} = \mathbf{A}(\omega)^T \sum_{l=-\infty}^{\infty} \hat{\mathbf{\Phi}}(\omega + 2\pi l)\, \hat{\mathbf{\Phi}}(\omega + 2\pi l)^\star \,\overline{\mathbf{A}(\omega)} = 0.$$

Moreover, since $\mathbf{\Omega}(\omega)$ is a matrix of trigonometric polynomials, we can also find a suitable $\mathbf{A}(\omega)$ with trigonometric polynomials as entries. Observing that each summand $\hat{\mathbf{\Phi}}(\omega+2\pi l)\, \hat{\mathbf{\Phi}}(\omega+2\pi l)^\star$ is positive semidefinite, it follows that $\mathbf{A}(\omega)^T \,\hat{\mathbf{\Phi}}(\omega+2\pi l) = 0$ for all $l \in \mathbb{Z}$, *i.e.*, $\mathbf{\Phi}$ is linearly dependent. □

Note that Lemma 4 is not restricted to scaling vectors.

For $r > 1$ the following conditions on the two-scale symbol $\mathbf{P}(\omega)$ imply linear dependence of the solution vector $\mathbf{\Phi}$ of (1):

Theorem 5. *Let $\mathbf{P}(\omega)$ be of the form (3), satisfying the basic conditions of Proposition 1 and (7). Let $\mathbf{a}$ be a right eigenvector of $\mathbf{P}(0)$ to the eigenvalue 1. Then the following assertions are equivalent:*

(a) *The solution vector $\mathbf{\Phi}$ of (1), determined in (8), is algebraically linearly dependent.*
(b) *There exist an $(r - s) \times r$ matrix $\tilde{\mathbf{M}}(\omega)$ of trigonometric polynomials with $\text{rank}(\tilde{\mathbf{M}}(0)) = r - s$ and an $r \times s$ matrix $\mathbf{M}(\omega)$ of trigonometric polynomials such that*

$$\begin{aligned} \tilde{\mathbf{M}}(\omega)\, \mathbf{M}(\omega) &= \mathbf{0}, \\ \tilde{\mathbf{M}}(0)\, \mathbf{a} &= \mathbf{0}, \\ \tilde{\mathbf{M}}(2\omega)\, \mathbf{P}(\omega)\, \mathbf{M}(\omega) &= \mathbf{0} \end{aligned} \tag{9}$$

with zero matrices of suitable size.

(c) There exists a positive semidefinite, hermitian matrix $\mathbf{F}(\omega) \in \mathbb{H}^{r\times r}$ with $\det \mathbf{F}(\omega) = 0$ for all $\omega \in [-\pi, \pi)$ and satisfying $T\mathbf{F}(\omega) = \mathbf{F}(\omega)$, *i.e.*,

$$\mathbf{P}(\omega)\,\mathbf{F}(\omega)\,\mathbf{P}(\omega) + \mathbf{P}(\omega+\pi)\,\mathbf{F}(\omega+\pi)\,\mathbf{P}(\omega+\pi) = \mathbf{F}(2\omega). \qquad (10)$$

Proof: The equivalence of (a) and (b) was already shown by Hogan [10].

We only need to show the equivalence of (a) and (c).

1. Let $\mathbf{\Phi}$ be linearly dependent. Then, by Lemma 4, its autocorrelation symbol $\mathbf{\Omega}(\omega)$ satisfies $\det \mathbf{\Omega}(\omega) = 0$ for all $\omega \in \mathbb{R}$. Moreover, $\mathbf{\Omega}(\omega)$ is an eigenmatrix of T corresponding to the eigenvalue 1.

2. Assume that $\mathbf{\Phi}$ is linearly independent, and that $\mathbf{F}(\omega)$ is a positive semidefinite, hermitian matrix of trigonometric polynomials satisfying (10) and with $\det \mathbf{F}(\omega) = 0$ for all ω. Then, there exists a nontrivial vector $\mathbf{A}(\omega)$ of trigonometric polynomials such that

$$\mathbf{A}(\omega)^T\,\mathbf{F}(\omega)\,\overline{\mathbf{A}(\omega)} = 0 \quad \text{for all } \omega \in \mathbb{R}.$$

Hence, (10) implies that also $\mathbf{A}(\omega)^T\,\mathbf{P}(\frac{\omega}{2})\,\mathbf{F}(\frac{\omega}{2})\,\mathbf{P}(\frac{\omega}{2})^\star\,\overline{\mathbf{A}(\omega)} = 0$ and $\mathbf{A}(\omega)^T\,\mathbf{P}(\frac{\omega}{2}+\pi)\,\mathbf{F}(\frac{\omega}{2}+\pi)\,\mathbf{P}(\frac{\omega}{2}+\pi)^\star\,\overline{\mathbf{A}(\omega)} = 0$. Using the notion $\tilde{\mathbf{\Pi}}_n(\omega) := \chi_{[-\pi,\pi]}(2^{-n}\omega)\prod_{l=1}^{n}\mathbf{P}(\frac{\omega}{2^l})$, it follows, by repeated application of (10), that

$$\mathbf{A}(\omega)^T\,\tilde{\mathbf{\Pi}}_n(\omega)\,\mathbf{F}(\frac{\omega}{2^n})\,\tilde{\mathbf{\Pi}}_n(\omega)^\star\,\overline{\mathbf{A}(\omega)} = 0$$

and finally for $n \to \infty$, for all $\omega \in \mathbb{R}$

$$\mathbf{A}(\omega)^T\,\mathbf{\Pi}(\omega)\,\mathbf{F}(0)\,\mathbf{\Pi}(\omega)^\star\,\overline{\mathbf{A}(\omega)} = 0 \qquad (11)$$

with $\mathbf{\Pi}(\omega) = \prod_{l=1}^{\infty}\mathbf{P}(\frac{\omega}{2^l})$. (Observe that, by basic conditions, $\tilde{\mathbf{\Pi}}_n$ converges pointwise to $\mathbf{\Pi}(\omega)$ for all ω.)

3. Let $\mathbf{a}$ be a right eigenvector, and let $\mathbf{y}$ be a left eigenvector of $\mathbf{P}(0)$ to the simple eigenvalue 1, then $\mathbf{y}^T\,\mathbf{a} \neq 0$. We show that $\mathbf{y}^T\,\mathbf{F}(0)\,\overline{\mathbf{y}} = 0$:

We can assume that $\mathbf{F}(0)$ is of the form $c\,\mathbf{a}\mathbf{a}^\star + \mathbf{F}_0(0)$, where c is a suitable nonnegative constant and $\mathbf{F}_0(0)$ is a positive semidefinite matrix satisfying $\mathbf{y}^T\,\mathbf{F}_0(0)\,\overline{\mathbf{y}} = 0$. Then (11) implies that $c\,\mathbf{A}(\omega)^T\,\mathbf{\Pi}(\omega)\,\mathbf{a}\mathbf{a}^\star\,\mathbf{\Pi}(\omega)^\star\,\overline{\mathbf{A}(\omega)} = 0$ for all ω, and hence

$$\begin{aligned} &c\,\mathbf{A}(\omega)^T \left(\sum_{l\in\mathbb{Z}} \mathbf{\Pi}(\omega+2\pi l)\,\mathbf{a}\mathbf{a}^\star\,\mathbf{\Pi}(\omega+2\pi l)^\star\right)\overline{\mathbf{A}(\omega)} \\ &= c\,\mathbf{A}(\omega)^T \left(\sum_{l\in\mathbb{Z}} \hat{\mathbf{\Phi}}(\omega+2\pi l)\,\hat{\mathbf{\Phi}}(\omega+2\pi l)^\star\right)\overline{\mathbf{A}(\omega)} = c\,\mathbf{A}(\omega)^T\,\mathbf{\Omega}(\omega)\,\overline{\mathbf{A}(\omega)} = 0. \end{aligned}$$

But, since $\mathbf{\Phi}$ is assumed to be linearly independent, its autocorrelation symbol is nonsingular a.e., and hence $c = 0$. So, we find $\mathbf{y}^T\,\mathbf{F}(0)\,\overline{\mathbf{y}} = 0$.

4. We introduce the space $V_1 := \{\mathbf{H} \in \mathbb{H}^{r\times r} : \mathbf{y}^T \mathbf{H}(0)\,\bar{\mathbf{y}} = 0\}$, which was already considered in [21]. Observe that $\mathbf{F} \in V_1$. Using Proposition 3.5 in [21], it follows that the transfer operator T, restricted to V_1 has spectral radius < 1, contradicting (10). Note that Proposition 3.5. in [21] (see also [18])) can be used since the integrability of $\tilde{\mathbf{\Pi}}_n\,\tilde{\mathbf{\Pi}}_n^\star$ is ensured by (7). □

Remark. 1. Note, that condition (7) can be replaced by a weaker condition in this theorem. We only need to ensure that the solution vector $\mathbf{\Phi}$ is contained in $L^2(\mathbb{R})^r$.

2. Note that for a given $(r-s)\times r$ matrix $\tilde{\mathbf{M}}(\omega)$ of trigonometric polynomials we can find an $r\times s$ matrix $\mathbf{M}(\omega)$ with $\tilde{\mathbf{M}}(\omega)\,\mathbf{M}(\omega) = \mathbf{0}$, everytimes. Introducing the determinants $\Delta_{d_1,\dots,d_{r-s}}(\omega)$ of $(r-s)\times(r-s)$ submatrices of $\tilde{\mathbf{M}}(\omega)$ consisting of the d_1th, d_2th,..., and the d_{r-s}th column of $\tilde{\mathbf{M}}(\omega)$, and letting $\mathbf{M}(\omega) = (m_{l,k}(\omega))_{l=1,\dots,r,k=1,\dots,s}$, choose

$$m_{l,k}(\omega) = \begin{cases} 0 & l<k \text{ or } l>k+r-s,\\ \Delta_{l+1,\dots,l+r-s}(\omega) & l=k,\\ (-1)^{r-s}\,\Delta_{l,\dots,l+r-s-1}(\omega) & l=k+r-s,\\ (-1)^{k+l}\,\Delta_{l,\dots,k-1,k+1,\dots,l+r-s}(\omega) & k<l<k+r-s. \end{cases}$$

4. Necessary and sufficient conditions for orthonormality

Let us introduce the following definition. We say that an $\omega \in (0, 2\pi)$ is cyclic, if there exists an integer $m \geq 2$ such that $2^m\omega \equiv \omega \,(\text{mod } 2\pi)$. Equivalently, ω is cyclic, if and only if it is of the form

$$\omega = \frac{2\pi\mu}{2^m - 1}$$

for some $m \in \mathbb{N}$, $m \geq 2$ and $\mu \in \{1,\dots,2^m-1\}$. Considering a cyclic $\omega_1 \in (0,\,2\pi)$, we can associate a cycle of numbers $\{\omega_1,\dots,\omega_m\}$, where $\omega_k := 2\,\omega_{k+1}\,(\text{mod } 2\pi)$ and $\omega_m := 2\,\omega_1\,(\text{mod } 2\pi)$. With $\omega_1 = \frac{2\pi\mu}{2^m-1}$ we obtain

$$\omega_k = \frac{2^{m-k+2}\pi\mu}{2^m-1}\,(\text{mod } 2\pi) \qquad (k = 1,\dots,m).$$

It can easily be shown that ω and $\omega + \pi\,(\text{mod } 2\pi)$ can not both be cyclic.

We are now ready to state the following theorem giving the relation between orthonormality conditions in terms of the transfer operator T (see [21]) and direct conditions to the two-scale symbol $\mathbf{P}$ (see [10,22]).

Theorem 6. *Let $\mathbf{P}(\omega)$ be of the form (3), satisfying the basic conditions of Proposition 1 and (7). Further, let $\mathbf{a}$ be a suitable right eigenvector of $\mathbf{P}(0)$ corresponding to the eigenvalue 1. Then the following assertions are equivalent:*

(A) *The solution vector $\mathbf{\Phi}$ of (1) determined by (8) is orthonormal (up to multiplication with a constant).*

(B) *(i) The two-scale symbol $\mathbf{P}(\omega)$ does not satisfy the linear dependence condition (9), and*

(ii) for all cycles $\{\omega_1, \dots, \omega_m\}$ in $(0, 2\pi)$ for the operation $\omega \to 2\omega \pmod{2\pi}$, and for all $\mathbf{x} \in \mathbb{R}^r$ there exists an $n \in \mathbb{N}_0$ and an $k \in \{0, \dots, m-1\}$ such that

$$\mathbf{x}^T \mathbf{Q}_m^n \mathbf{P}(\omega_1) \dots \mathbf{P}(\omega_k) \mathbf{P}(\omega_{k+1} + \pi) \neq \mathbf{0}^T,$$

where $\mathbf{Q}_m := \mathbf{P}(\omega_1) \dots \mathbf{P}(\omega_m)$.

(C) *The transfer operator T possesses a simple eigenvalue 1.*

Proof: We show that (A) $\Rightarrow$ (B) $\Rightarrow$ (C) $\Rightarrow$ (A).

1. (A) $\Rightarrow$ (B): Let $\mathbf{\Phi}$ be an orthonormal solution vector of (1). We show the necessity of the conditions (B). Here, we only need to consider condition (ii), the condition (i) is already proved to be necessary in Theorem 5.

We show that: If (B) (ii) is not satisfied, then the solution vector $\mathbf{\Phi}$ is not L^2-stable (and hence not orthonormal). Assume that there is a cycle $\{\omega_1, \dots, \omega_m\}$ and a nonzero vector $\mathbf{x} \in \mathbb{R}^r$ satisfying $\mathbf{x}^T \mathbf{Q}_m^n \mathbf{P}(\omega_1) \dots \mathbf{P}(\omega_k) \mathbf{P}(\omega_{k+1} + \pi) = \mathbf{0}^T$ for all $n \in \mathbb{N}_0$ and all $k \in \{0, \dots, m-1\}$. We show that $\mathbf{x}^T \mathbf{\Omega}(\omega_m) \overline{\mathbf{x}} = 0$:
By definition of a cycle, ω_k equals $2\omega_{k+1} \pmod{2\pi}$, and $\omega_m = 2\omega_1 \pmod{2\pi}$. For an arbitrary, fixed $l \in \mathbb{Z}$, choose l_1, such that $\omega_m + 2\pi l = 2\omega_1 + 2\pi l_1$. Hence, the refinement equation implies that $\mathbf{x}^T \hat{\mathbf{\Phi}}(\omega_m + 2\pi l) = \mathbf{x}^T \hat{\mathbf{\Phi}}(2\omega_1 + 2\pi l_1)$ is either (for odd l_1) equal to $\mathbf{x}^T \mathbf{P}(\omega_1 + \pi) \hat{\mathbf{\Phi}}(\omega_1 + \pi l) = 0$ or (for even l_1) to $\mathbf{x}^T \mathbf{P}(\omega_1) \hat{\mathbf{\Phi}}(\omega_1 + \pi l_1) = \mathbf{x}^T \mathbf{P}(\omega_1) \hat{\mathbf{\Phi}}(2\omega_2 + 2\pi l_2)$ for some l_2 with $|l_2| \leq 1 + |l_1|/2$. Now, we can use the same argument again, and find that $\mathbf{x}^T \mathbf{P}(\omega_1) \hat{\mathbf{\Phi}}(2\omega_2 + 2\pi l_2)$ either equals to $\mathbf{x}^T \mathbf{P}(\omega_1) \mathbf{P}(\omega_2 + \pi) \hat{\mathbf{\Phi}}(\omega_2 + \pi l_2) = 0$ or to $\mathbf{x}^T \mathbf{P}(\omega_1) \mathbf{P}(\omega_2) \hat{\mathbf{\Phi}}(2\omega_3 + 2\pi l_3)$ and so on. If $\mathbf{x}^T \hat{\mathbf{\Phi}}(\omega_m + 2\pi l)$ is of the form $\mathbf{x}^T \mathbf{P}(\omega_1) \dots \mathbf{P}(\omega_m) \hat{\mathbf{\Phi}}(\omega_m + \pi l_m) = \mathbf{x}^T \mathbf{Q}_m \hat{\mathbf{\Phi}}(\omega_m + \pi l_m)$, we can keep going through the cycle as before. The procedure comes to an end, since the sequence $\{|l_k|\}_{k \geq 1}$ is monotonly decreasing; in fact we have $|l_{k+1}| \leq 1 + |l_k|/2$.

Hence, it follows that $\mathbf{x}^T \hat{\mathbf{\Phi}}(\omega_m + 2\pi l) = 0$ for all $l \in \mathbb{Z}$ implying that

$$\mathbf{x}^T \sum_{l \in \mathbb{Z}} \hat{\mathbf{\Phi}}(\omega_m + 2\pi l) \hat{\mathbf{\Phi}}(\omega_m + 2\pi l)^\star \overline{\mathbf{x}} = \mathbf{x}^T \mathbf{\Omega}(\omega_m) \overline{\mathbf{x}} = 0$$

and contradicting the positive definiteness of the autocorrelation symbol $\mathbf{\Omega}(\omega)$.

2. In the next steps we show that (B) $\Rightarrow$ (C).

Let $\mathbf{H}_0 \in \mathbb{H}^{r \times r}$ be an eigenmatrix of T corresponding to the eigenvalue 1. Since, by (7), $\mathbf{I}$ is already an eigenmatrix of T to 1, we have to show that, under the assumptions (B), the assertion $\mathbf{H}_0(\omega) \neq c\mathbf{I}$ leads to a contradiction.

So, let us suppose that $\mathbf{H}(\omega) \neq c\mathbf{I}$. Observe that $T\mathbf{H}_0(\omega) = \mathbf{H}_0(\omega)$ implies that $T\mathbf{H}_0(\omega)^\star = \mathbf{H}_0(\omega)^\star$. Since each matrix is representable as a sum of a hermitian and an antihermitian part, we can restrict us to the case that $\mathbf{H}_0(\omega)$ is hermitian. Hence, the eigenvalues of $\mathbf{H}_0(\omega)$ are real. Introducing the minimum and the maximum eigenvalue of $\mathbf{H}_0$:

$$\lambda_{\min} := \min_{\omega} \lambda_1(\omega) \quad \text{with} \quad \lambda_1(\omega) := \min\{\lambda : \mathbf{H}_0(\omega)\mathbf{x} = \lambda\mathbf{x},\, \mathbf{x} \neq 0\},$$
$$\lambda_{\max} := \min_{\omega} \lambda_2(\omega) \quad \text{with} \quad \lambda_2(\omega) := \max\{\lambda : \mathbf{H}_0(\omega)\mathbf{x} = \lambda\mathbf{x},\, \mathbf{x} \neq 0\},$$

we consider the matrices

$$\mathbf{F}_1(\omega) := \mathbf{H}_0(\omega) - \lambda_{\min}\mathbf{I}, \qquad \mathbf{F}_2(\omega) := \lambda_{\max}\mathbf{I} - \mathbf{H}_0(\omega).$$

Assuming that $\lambda_{\min} = \lambda_1(\omega_0)$, $\lambda_{\max} = \lambda_2(\tilde{\omega}_0)$, it follows that the matrices $\mathbf{F}_1(\omega)$, $\mathbf{F}_2(\omega)$, both are hermitian and positive semidefinite with $\det \mathbf{F}_1(\omega_0) = 0$ and $\det \mathbf{F}_2(\tilde{\omega}_0) = 0$. Further, observe that both, $\mathbf{F}_1(\omega)$ and $\mathbf{F}_2(\omega)$, are eigenmatrices of T to the eigenvalue 1.

3. We show that there is an $\omega_1 \in (0, 2\pi)$ such that either $\det \mathbf{F}_1(\omega_1) = 0$ or $\det \mathbf{F}_2(\omega_1) = 0$: If this assertion were not true, then we would have $\det \mathbf{F}_1(\omega) \neq 0$ and $\det \mathbf{F}_2(\omega) \neq 0$ for all $\omega \in (0, 2\pi)$. But the determinants of $\mathbf{F}_1$ and $\mathbf{F}_2$ have at least one zero by construction, hence $\det \mathbf{F}_1(0) = \det \mathbf{F}_2(0) = 0$. That means, there exist nonzero vectors $\mathbf{x}$ and $\mathbf{y}$ with $\mathbf{x}^T\mathbf{F}_1(0)\overline{\mathbf{x}} = \mathbf{y}^T\mathbf{F}_2(0)\overline{\mathbf{y}} = 0$. In case of $\mathbf{x} = c\mathbf{y}$, it follows from the definition of $\mathbf{F}_1$ and $\mathbf{F}_2$ that

$$\mathbf{x}^T(\mathbf{H}_0(0) - \lambda_{\min}\mathbf{I})\overline{\mathbf{x}} = \mathbf{x}^T(\lambda_{\max}\mathbf{I} - \mathbf{H}_0(0))\overline{\mathbf{x}} = 0,$$

i.e., $\mathbf{x}^T(\lambda_{\max} - \lambda_{\min})\mathbf{I}\overline{\mathbf{x}} = 0$, implying $\lambda_{\min} = \lambda_{\max}$. Hence, all eigenvalues of $\mathbf{H}_0(\omega)$ are equal, and $\mathbf{H}_0(\omega) = c\mathbf{I}$. This contradicts our assumption.

So, we only need to consider the case that $\mathbf{x}$ and $\mathbf{y}$ are linearly independent. Using that $T\mathbf{F}_\nu(\omega) = \mathbf{F}_\nu(\omega)$ $(\nu = 1, 2)$, it follows that

$$0 = \mathbf{x}^T\mathbf{F}_1(0)\overline{\mathbf{x}} = \mathbf{x}^T\mathbf{P}(0)\mathbf{F}_1(0)\mathbf{P}(0)^\star\overline{\mathbf{x}} + \mathbf{x}^T\mathbf{P}(\pi)\mathbf{F}_1(\pi)\mathbf{P}(\pi)^\star\overline{\mathbf{x}}$$
$$0 = \mathbf{y}^T\mathbf{F}_2(0)\overline{\mathbf{y}} = \mathbf{y}^T\mathbf{P}(0)\mathbf{F}_2(0)\mathbf{P}(0)^\star\overline{\mathbf{y}} + \mathbf{y}^T\mathbf{P}(\pi)\mathbf{F}_2(\pi)\mathbf{P}(\pi)^\star\overline{\mathbf{y}}$$

implying that

$$\mathbf{x}^T\mathbf{P}(\pi)\mathbf{F}_1(\pi)\mathbf{P}(\pi)^\star\overline{\mathbf{x}} = \mathbf{y}^T\mathbf{P}(\pi)\mathbf{F}_2(\pi)\mathbf{P}(\pi)^\star\overline{\mathbf{y}} = 0.$$

Since $\mathbf{F}_1(\pi)$, $\mathbf{F}_2(\pi)$ were supposed to be nonsingular, it follows that $\mathbf{x}^T\mathbf{P}(\pi) = \mathbf{y}^T\mathbf{P}(\pi) = \mathbf{0}^T$. Hence, (7) implies that $\mathbf{x}^T\mathbf{P}(0)\mathbf{P}(0)^\star\overline{\mathbf{x}} = \mathbf{x}^T\overline{\mathbf{x}}$ and $\mathbf{y}^T\mathbf{P}(0)\mathbf{P}(0)^\star\overline{\mathbf{y}} = \mathbf{y}\overline{\mathbf{y}}$. But this a contradiction to the basic condition that $\mathbf{P}(0)$ possesses a simple eigenvalue 1.

4. Let $\mathbf{F}$ be one of the matrices $\mathbf{F}_1, \mathbf{F}_2$ satisfying $\det \mathbf{F}(\omega_1) = 0$ for some $\omega_1 \in (0, 2\pi)$. Let $\mathbf{x}_1$ be a right eigenvector corresponding to the eigenvalue 0, *i.e.*, $\mathbf{x}_1^T \mathbf{F}(\omega_1) \overline{\mathbf{x}}_1 = 0$. Hence,

$$\begin{aligned} 0 &= \mathbf{x}_1^T \mathbf{F}(\omega_1) \overline{\mathbf{x}}_1 = \mathbf{x}_1^T (T\mathbf{F})(\omega_1) \overline{\mathbf{x}}_1 \\ &= \mathbf{x}_1^T \mathbf{P}(\frac{\omega_1}{2}) \mathbf{F}(\frac{\omega_1}{2}) \mathbf{P}(\frac{\omega_1}{2})^\star \overline{\mathbf{x}}_1 + \mathbf{x}_1^T \mathbf{P}(\frac{\omega_1}{2} + \pi) \mathbf{F}(\frac{\omega_1}{2} + \pi) \mathbf{P}(\frac{\omega_1}{2} + \pi)^\star \overline{\mathbf{x}}_1 \end{aligned}$$

implying that

$$\mathbf{x}_1^T \mathbf{P}(\frac{\omega_1}{2}) \mathbf{F}(\frac{\omega_1}{2}) \mathbf{P}(\frac{\omega_1}{2})^\star \overline{\mathbf{x}}_1 = 0$$

as well as

$$\mathbf{x}_1^T \mathbf{P}(\frac{\omega_1}{2} + \pi) \mathbf{F}(\frac{\omega_1}{2} + \pi) \mathbf{P}(\frac{\omega_1}{2} + \pi)^\star \overline{\mathbf{x}}_1 = 0.$$

Observing that, by (7), never both $\mathbf{x}_1^T \mathbf{P}(\frac{\omega_1}{2})$ and $\mathbf{x}_1^T \mathbf{P}(\frac{\omega_1}{2} + \pi)$ can be zero vectors at the same time, it follows that either $\det \mathbf{F}(\frac{\omega_1}{2}) = 0$ or $\det \mathbf{F}(\frac{\omega_1}{2} + \pi) = 0$ with corresponding left eigenvectors $\mathbf{x}_1^T \mathbf{P}(\frac{\omega_1}{2})$ and $\mathbf{x}_1^T \mathbf{P}(\frac{\omega_1}{2} + \pi)$ to the eigenvalue 0, respectively. Let ω_2 be equal to $\frac{\omega_1}{2}$ or to $\frac{\omega_1}{2} + \pi$ such that $\det \mathbf{F}(\omega_2) = 0$. With $\mathbf{x}_2^T := \mathbf{x}_1^T \mathbf{P}(\omega_2)$ we then have $\mathbf{x}_2^T \mathbf{F}(\omega_2) \overline{\mathbf{x}}_2 = 0$. We again apply the transfer operator and find a further zero of $\det \mathbf{F}$, namely either $\frac{\omega_2}{2}$ and $\frac{\omega_2}{2} + \pi$.

Continuing this process, to each $\omega_1 \in (0, 2\pi)$ with $\mathbf{x}_1^T \mathbf{F}(\omega_1) \overline{\mathbf{x}}_1 = 0$ we can find a chain of zeros of $\det \mathbf{F}$, $\omega_1, \omega_2, \ldots$ in $(0, 2\pi)$, where ω_{k+1} is either $\frac{\omega_k}{2}$ or $\frac{\omega_k}{2} + \pi$ (or, equivalently, $\omega_k = 2\,\omega_{k+1} \pmod{2\pi}$) with $\mathbf{x}_k^T \mathbf{F}(\omega_k) \overline{\mathbf{x}}_k = 0$ and $\mathbf{x}_{k+1}^T := \mathbf{x}_k^T \mathbf{P}(\omega_{k+1})$ $(k = 1, 2, \ldots)$. Since $\det \mathbf{F}(\omega)$ is a trigonometric polynomial, it can only have a finite number of different zeros. The case $\det \mathbf{F}(\omega) = 0$ for all ω can not happen, since, by Theorem 5 (c), this assertion contradicts the assumption (i) of (B). Hence, the chain $\omega_1, \omega_2, \ldots$ is finite, *i.e.*, there is an $l \in \mathbb{N}$, such that $\omega_l = \omega_k$ for some $k < l$, that means, $\omega_k = 2^{l-k} \omega_l \pmod{2\pi} = 2^{l-k}\, \omega_k \pmod{2\pi}$. Thus, ω_k is cyclic, and moreover, by

$$\omega_1 2^{l-k} \pmod{2\pi} = 2^k\, \omega_k\, 2^{l-k} \pmod{2\pi} = 2^k\, \omega_k \pmod{2\pi} = \omega_1 \pmod{2\pi},$$

ω_1 is cyclic. The same procedure can be applied to all zeros of $\det \mathbf{F}(\omega)$ in $(0, 2\pi)$ yielding a finite number of cycles. In particular, $\det \mathbf{F}(\omega)$ only has cyclic zeros.

5. Since ω and $\omega + \pi$ can not both be cyclic at the same time, it follows from $\det \mathbf{F}(\omega) = 0$ that $\det \mathbf{F}(\omega + \pi) \neq 0$. Using again that $\mathbf{F}(\omega) = T\mathbf{F}(\omega)$, we find for a cycle $\{\omega_1, \ldots, \omega_m\}$ with $\mathbf{x}_k^T \mathbf{F}(\omega_k) \overline{\mathbf{x}}_k = 0$ and $\det \mathbf{F}(\omega_{k+1} + \pi) \neq 0$ from

$$0 = \mathbf{x}_k^T \mathbf{P}(\omega_{k+1}) \mathbf{F}(\omega_{k+1}) \mathbf{P}(\omega_{k+1})^\star \overline{\mathbf{x}}_k + \mathbf{x}_k^T \mathbf{P}(\omega_{k+1} + \pi) \mathbf{F}(\omega_{k+1} + \pi) \mathbf{P}(\omega_{k+1} + \pi)^\star \overline{\mathbf{x}}_k$$

that $\mathbf{x}_k^T \mathbf{P}(\omega_{k+1} + \pi) = \mathbf{0}^T$ $(k = 0, \ldots, m-1)$. This process can be continued by going again through the cycle. Hence, there is a cycle $\{\omega_1, \ldots, \omega_m\}$ and a nonzero vector $\mathbf{x}$, which does not satisfy the condition (B) (ii), and we have found the contradiction.

6. We finally observe that (C) $\Rightarrow$ (A): Let the eigenvalue 1 of T be simple. By (7), I is already an eigenmatrix of T to the eigenvalue 1. Since the autocorrelation symbol $\mathbf{\Omega}(\omega)$ is also an eigenmatrix of T corresponding to 1, it has to be a multiple of $\mathbf{I}$. Hence, $\mathbf{\Phi}$ is orthonormal (up to multiplication with a constant). □

Remarks. 1. The above Theorem 6 can be seen as a generalization of Theorem 6.3.5 in [5] showing the equivalence between Lawton's condition [16] and Cohen's criteria [1].
2. As shown in [10], condition (B) (ii) is already satisfied if there is no cycle $\{\omega_1, \ldots, \omega_m\}$ with det $\mathbf{P}(\omega_k + \pi) = 0$ $(k = 0, \ldots, n)$.

References

1. Cohen A., Ondelettes, analyses multirésolutions et filtres miroir en quadrature, Ann. Inst. H. Poincaré, Anal. non linéaire **7** (1990), 439–459.
2. Cohen A., Daubechies I., and Plonka G., Regularity of refinable function vectors, J. Fourier Anal. Appl., to appear.
3. Cohen A., Dyn N., and Levin D., Matrix subdivision schemes, preprint, 1995.
4. Dahmen W., and Micchelli C. A., Biorthogonal wavelet expansions, Constr. Approx., to appear.
5. Daubechies I., *Ten Lectures on Wavelets*, SIAM, Philadelphia, 1992.
6. Deslauriers G. and Dubuc S., Interpolation dyadique, in: Fractals, dimensions non entières et applications, G. Cherbit (ed.), Masson, Paris, 1987, 44–55.
7. Goodman T. N. T. and Lee S. L., Wavelets with multiplicity r, Trans. Amer. Math. Soc. **342(1)** (1994), 307–324.
8. Heil C. and Colella D., Matrix refinement equations: Existence and uniqueness, J. Fourier Anal. Appl. **2** (1996), 363–377.
9. Hervé L., Multi-resolution analysis of multiplicity d: Application to dyadic interpolation, Appl. Comput. Harmonic Anal. **1** (1994), 199–315.
10. Hogan T. A., Stability and independence of shifts of finitely many refinable functions, J. Fourier Anal. Appl., to appear.
11. Hogan T. A., A note on matrix refinement equations, preprint, 1996.
12. Jia R. Q., Riemenschneider S. D., and Zhou D. X., Vector subdivision schemes and multiple wavelets, preprint, 1996.
13. Jia R. Q. and Wang J. Z., Stability and linear independence associated with wavelet decompositions, Proc. Amer. Math. Soc. **117** (1993), 1115–1124.
14. Jiang Q. and Shen Z., On existence and weak stability of matrix refinable functions, preprint, 1996.
15. Lian J., Orthogonality criteria for multi-scaling functions, preprint 1996.

16. Lawton W., Necessary and sufficient conditions for constructing orthonormal wavelet bases, J. Math. Phys. **32** (1991), 57–61.
17. Lawton W., Lee S. L., and Shen Z., An algorithm for matrix extension and wavelet construction, Math. Comp. **65** (1996), 723–737.
18. Long R., Chen W., and Yuan S., Wavelets generated by vector multiresolution analysis, preprint, 1995.
19. Mallat S., Multiresolution approximation and wavelets, Trans. Amer. Math. Soc. **315** (1989), 69–88.
20. Plonka G., On stability of scaling vectors, in *Surface Fitting and Multiresolution Methods*, Le Mehauté, A., Rabut, C. and Schumaker, L. L. (eds.), Vanderbilt University Press, Nashville, 1997, 293–300.
21. Shen Z., Refinable function vectors, SIAM J. Math. Anal., to appear.
22. Wang J. Z., Stability and linear independence associated with scaling vectors, SIAM J. Math. Anal., to appear.

Gerlind Plonka
Fachbereich Mathematik,
Universität Rostock,
D-18051 Rostock, Germany
gerlind.plonka@mathematik.uni-rostock.de

Trigonometric Preconditioners for Block Toeplitz Systems

Daniel Potts, Gabriele Steidl, and Manfred Tasche

Abstract. This paper is concerned with the solution of a system of linear equations $\mathbf{T}_{M,N}\mathbf{x} = \mathbf{b}$, where $\mathbf{T}_{M,N}$ denotes a positive definite doubly symmetric block-Toeplitz matrix with Toeplitz blocks arising from a generating function f of the Wiener class. We derive optimal and Strang–type trigonometric preconditioners $\mathbf{P}_{M,N}$ of $\mathbf{T}_{M,N}$ from the Fejér and Fourier sum of f, respectively. Using relations between trigonometric transforms and Toeplitz matrices, we prove that for all $\varepsilon > 0$ and sufficiently large M, N, at most $\mathcal{O}(M) + \mathcal{O}(N)$ eigenvalues of $\mathbf{P}_{M,N}^{-1}\mathbf{T}_{M,N}$ lie outside the interval $(1-\varepsilon, 1+\varepsilon)$ such that the preconditioned conjugate gradient method converges in at most $\mathcal{O}(M)+\mathcal{O}(N)$ steps.

1. Introduction

Systems of linear equations

$$\mathbf{T}_{M,N}\mathbf{x} = \mathbf{b}$$

where $\mathbf{T}_{M,N}$ denotes a positive definite doubly symmetric block-Toeplitz matrix with Toeplitz blocks (BTTB matrices) arise in a variety of applications in mathematics and engineering (see [8] and the references therein). Along with stabilization techniques for direct fast and superfast Toeplitz solvers, preconditioned conjugate gradient methods (PCG–methods) have attained much attention during the last years. Two types of so-called "level–2" preconditioners are mainly exploited for linear BTTB systems, namely optimal (Cesáro) doubly circulant preconditioners [4] and more simple so-called "Strang" doubly circulant preconditioners [16]. One reason for the choice of doubly circulant preconditioners is the fact that doubly circulant matrices can be diagonalized by tensor products of Fourier matrices such that the multiplication with doubly circulant matrices requires only $\mathcal{O}(MN\log(MN))$

Multivariate Approximation and Splines
G. Nürnberger, J. W. Schmidt, and G. Walz (eds.), pp. 219–234.

ISBN 3-7643-5654-5.

arithmetical operations. In this paper, we restrict our attention to *real* BTTB matrices. Here, it seems to be natural, to replace the doubly circulant matrices by matrices which are diagonalizable by tensor products of some orthogonal matrices. We construct "level-2" trigonometric preconditioners of BTTB matrices with respect to various trigonometric transforms and show that for doubly symmetric BTTB matrices arising from a generating function f of the Wiener class the corresponding PCG-method converges in at most $\mathcal{O}(M) + \mathcal{O}(N)$ steps.
The special case of preconditioning of doubly symmetric BTTB matrices with respect to a discrete sine transform was studied in [11]. However, in many examples discrete cosine transform based preconditioning leads to better convergence results.

2. Trigonometric Transforms and Toeplitz Matrices

We introduce four discrete cosine transforms (DCT) and four discrete sine transforms (DST) as classified in [17]:

$$\text{DCT}-\text{I}: \quad \mathbf{C}^{I}_{N+1} := \left(\frac{2}{N}\right)^{1/2} \left(\varepsilon_j^N \varepsilon_k^N \cos \frac{jk\pi}{N}\right)_{j,k=0}^{N} \in \mathbb{R}^{N+1,N+1},$$

$$\text{DCT}-\text{II}: \quad \mathbf{C}^{II}_{N} := \left(\frac{2}{N}\right)^{1/2} \left(\varepsilon_j^N \cos \frac{j(2k+1)\pi}{2N}\right)_{j,k=0}^{N-1} \in \mathbb{R}^{N,N},$$

$$\text{DCT}-\text{III}: \quad \mathbf{C}^{III}_{N} := \left(\mathbf{C}^{II}_{N}\right)' \in \mathbb{R}^{N,N},$$

$$\text{DCT}-\text{IV}: \quad \mathbf{C}^{IV}_{N} := \left(\frac{2}{N}\right)^{1/2} \left(\cos \frac{(2j+1)(2k+1)\pi}{4N}\right)_{j,k=0}^{N-1} \in \mathbb{R}^{N,N},$$

$$\text{DST}-\text{I}: \quad \mathbf{S}^{I}_{N-1} := \left(\frac{2}{N}\right)^{1/2} \left(\sin \frac{(j+1)(k+1)\pi}{N}\right)_{j,k=0}^{N-2} \in \mathbb{R}^{N-1,N-1},$$

$$\text{DST}-\text{II}: \quad \mathbf{S}^{II}_{N} := \left(\frac{2}{N}\right)^{1/2} \left(\varepsilon_{j+1}^N \sin \frac{(j+1)(2k+1)\pi}{2N}\right)_{j,k=0}^{N-1} \in \mathbb{R}^{N,N},$$

$$\text{DST}-\text{III}: \quad \mathbf{S}^{III}_{N} := \left(\mathbf{S}^{II}_{N}\right)' \in \mathbb{R}^{N,N},$$

$$\text{DST}-\text{IV}: \quad \mathbf{S}^{IV}_{N} := \left(\frac{2}{N}\right)^{1/2} \left(\cos \frac{(2j+1)(2k+1)\pi}{4N}\right)_{j,k=0}^{N-1} \in \mathbb{R}^{N,N},$$

where $\varepsilon_k^N := 2^{-1/2}$ $(k = 0, N)$ and $\varepsilon_k^N := 1$ otherwise. We refer to the corresponding transforms as *trigonometric transforms*. It is well-known that the above matrices are orthogonal and that the multiplication of such matrix with a vector takes only $\mathcal{O}(N \log N)$ arithmetical operations. There exist implementations of fast algorithms for the multiplication of above sine and cosine matrices with a vector, for example a C-implementation based on [15] and [1].

Moreover, we use the slightly modified DCT–I and DST–I matrices

$$\tilde{\mathbf{C}}_{N+1}^{I} := \left((\varepsilon_k^N)^2 \cos \frac{jk\pi}{N} \right)_{j,k=0}^{N}, \quad \tilde{\mathbf{S}}_{N-1}^{I} := \left(\sin \frac{jk\pi}{N} \right)_{j,k=1}^{N-1}$$

and the slightly modified DCT–III and DST–III matrices

$$\tilde{\mathbf{C}}_{N}^{III} := \left((\varepsilon_k^N)^2 \cos \frac{(2j+1)k\pi}{2N} \right)_{j,k=0}^{N-1},$$
$$\tilde{\mathbf{S}}_{N}^{III} := \left((\varepsilon_{k+1}^N)^2 \sin \frac{(2j+1)(k+1)\pi}{2N} \right)_{j,k=0}^{N-1}.$$

It holds that

$$\tilde{\mathbf{C}}_{N+1}^{I}\, \tilde{\mathbf{C}}_{N+1}^{I} = \frac{N}{2} \mathbf{I}_{N+1}.$$

Let

$$\mathbf{Z}'_{N,1} := \begin{pmatrix} 0 & 1 & & 0 \\ \vdots & & \ddots & \vdots \\ 0 & 0 & \dots & 1 \end{pmatrix} \in \mathbb{R}^{N,N+1}, \quad \mathbf{Z}'_{N,2} := \begin{pmatrix} 1 & \dots & 0 & 0 \\ \vdots & \ddots & \vdots & \vdots \\ 0 & \dots & 1 & 0 \end{pmatrix} \in \mathbb{R}^{N,N+1},$$

$$\mathbf{R}'_N := \begin{pmatrix} 0 & 1 & & 0 & 0 \\ \vdots & & \ddots & \vdots & \vdots \\ 0 & 0 & \dots & 1 & 0 \end{pmatrix} \in \mathbb{R}^{N-1,N+1}.$$

By $\operatorname{stoep} \mathbf{a}'$ we denote the symmetric Toeplitz matrix with first row $\mathbf{a}'$ and by $\operatorname{atoep} \mathbf{a}'$ the antisymmetric Toeplitz matrix with first row $\mathbf{a}'$, where $a_0 = 0$. Similarly, let $\operatorname{shank} \mathbf{a}'$ be the persymmetric Hankel matrix with first row $\mathbf{a}'$ and let $\operatorname{ahank} \mathbf{a}'$ be the antipersymmetric Hankel matrix with first row $\mathbf{a}'$, where $a_{N-1} = 0$. Let $\operatorname{diag} \mathbf{a}$ be the diagonal matrix with diagonal $\mathbf{a}$ and let $\delta(\mathbf{A}) := \operatorname{diag}(a_{k,k})_{k=0}^{N-1}$, where $a_{k,k}$ is the (k,k)–th entry of $\mathbf{A} \in \mathbb{R}^{N,N}$.

Theorem 1. *There exist the following relations between trigonometric transforms and Toeplitz–plus–Hankel matrices:*

i) DCT–I *and* DST–I:

$$\begin{aligned}
2\,\mathbf{R}'_N \mathbf{C}_{N+1}^{I}\, \mathbf{D}\, \mathbf{C}_{N+1}^{I} \mathbf{R}_N &= \operatorname{stoep}(a_0, \dots, a_{N-2}) + \operatorname{shank}(a_2, \dots, a_{N-2}, 0, 0), \\
2\,\mathbf{S}_{N-1}^{I}\, \mathbf{R}'_N\, \mathbf{D}\, \mathbf{R}_N \mathbf{S}_{N-1}^{I} &= \operatorname{stoep}(a_0, \dots, a_{N-2}) - \operatorname{shank}(a_2, \dots, a_{N-2}, 0, 0), \\
2\,\mathbf{R}'_N \mathbf{C}_{N+1}^{I}\, \tilde{\mathbf{D}}\, \mathbf{R}_N \mathbf{S}_{N-1}^{I} &= \operatorname{atoep}(0, a_1, \dots, a_{N-2}) + \operatorname{ahank}(a_2, \dots, a_{N-1}, 0), \\
2\,\mathbf{S}_{N-1}^{I}\, \mathbf{R}'_N\, \tilde{\mathbf{D}}\, \mathbf{C}_{N+1}^{I} \mathbf{R}_N &= -\operatorname{atoep}(0, a_1, \dots, a_{N-2}) + \operatorname{ahank}(a_2, \dots, a_{N-1}, 0)
\end{aligned}$$

with

$$\mathbf{D} := \operatorname{diag}(d_0, \ldots, d_N)\,,\ \tilde{\mathbf{D}} := \operatorname{diag}(0, \tilde{d}_1, \ldots, \tilde{d}_{N-1}, 0)\,,$$
$$(d_0, \ldots, d_N)' := \tilde{\mathbf{C}}^I_{N+1}\,(a_0, \ldots, a_{N-2}, 0, 0)'\,,$$
$$(\tilde{d}_1, \ldots, \tilde{d}_{N-1})' := \tilde{\mathbf{S}}^I_{N-1}\,(a_1, \ldots, a_{N-1})'\,.$$

ii) DCT–II *and* DST–II:

$$2\left(\mathbf{C}_N^{II}\right)'\,\mathbf{Z}'_{N,2}\,\mathbf{D}\,\mathbf{Z}_{N,2}\mathbf{C}_N^{II} = \operatorname{stoep}(a_0, \ldots, a_{N-1}) + \operatorname{shank}(a_1, \ldots, a_{N-1}, 0)\,,$$
$$2\left(\mathbf{S}_N^{II}\right)'\,\mathbf{Z}'_{N,1}\,\mathbf{D}\,\mathbf{Z}_{N,1}\mathbf{S}_N^{II} = \operatorname{stoep}(a_0, \ldots, a_{N-1}) - \operatorname{shank}(a_1, \ldots, a_{N-1}, 0)\,,$$
$$2\left(\mathbf{C}_N^{II}\right)'\,\mathbf{Z}'_{N,2}\,\tilde{\mathbf{D}}\,\mathbf{Z}_{N,1}\mathbf{S}_N^{II} = \operatorname{atoep}(0, a_1, \ldots, a_{N-1}) + \operatorname{ahank}(a_1, \ldots, a_{N-1}, 0),$$
$$2\left(\mathbf{S}_N^{II}\right)'\,\mathbf{Z}'_{N,1}\,\tilde{\mathbf{D}}\mathbf{Z}_{N,2}\mathbf{C}_N^{II} = -\operatorname{atoep}(0, a_1, \ldots, a_{N-1}) + \operatorname{ahank}(a_1, \ldots, a_{N-1}, 0)$$

with

$$\mathbf{D} := \operatorname{diag}(d_0, \ldots, d_N)\,,\ \tilde{\mathbf{D}} := \operatorname{diag}(0, \tilde{d}_1, \ldots, \tilde{d}_{N-1}, 0)\,,$$
$$(d_0, \ldots, d_N)' := \tilde{\mathbf{C}}^I_{N+1}\,(a_0, \ldots, a_{N-1}, 0)'\,,$$
$$(\tilde{d}_1, \ldots, \tilde{d}_{N-1})' := \tilde{\mathbf{S}}^I_{N-1}\,(a_1, \ldots, a_{N-1})'\,.$$

iii) DCT–IV *and* DST–IV:

$$2\,\mathbf{C}_N^{IV}\,\mathbf{D}\,\mathbf{C}_N^{IV} = \operatorname{stoep}(a_0, \ldots, a_{N-1}) + \operatorname{ahank}(a_1, \ldots, a_{N-1}, 0)\,,$$
$$2\,\mathbf{S}_N^{IV}\,\mathbf{D}\,\mathbf{S}_N^{IV} = \operatorname{stoep}(a_0, \ldots, a_{N-1}) - \operatorname{ahank}(a_1, \ldots, a_{N-1}, 0)\,,$$
$$2\,\mathbf{C}_N^{IV}\,\tilde{\mathbf{D}}\,\mathbf{S}_N^{IV} = \operatorname{atoep}(0, a_1, \ldots, a_{N-1}) + \operatorname{shank}(a_1, \ldots, a_{N-1}, 0)\,,$$
$$2\,\mathbf{S}_N^{IV}\,\tilde{\mathbf{D}}\,\mathbf{C}_N^{IV} = -\operatorname{atoep}(0, a_1, \ldots, a_{N-1}) + \operatorname{shank}(a_1, \ldots, a_{N-1}, 0)$$

with

$$\mathbf{D} := \operatorname{diag}(d_0, \ldots, d_{N-1})\,,\ \tilde{\mathbf{D}} := \operatorname{diag}(\tilde{d}_0, \ldots, \tilde{d}_{N-1}),$$
$$(d_0, \ldots, d_{N-1})' := \tilde{\mathbf{C}}_N^{III}\,(a_0, \ldots, a_{N-1})'\,,$$
$$(\tilde{d}_0, \ldots, \tilde{d}_{N-1})' := \tilde{\mathbf{S}}_N^{III}\,(a_1, \ldots, a_{N-1}, 0)'\,.$$

Typewriting this paper, we got a ps–file of a paper of G. Heinig and K. Rost [12], which contains similar results as presented in Theorem 1.

Corollary 2. *Let* $\mathbf{T} = \mathbf{T}_N := (t_{j-k})_{j,k=0}^{N-1}$ *be given and let* $\mathbf{C} := \mathbf{C}_N^{IV}$, $\mathbf{S} := \mathbf{S}_N^{IV}$. *Then*

$$\mathbf{T} = \frac{1}{2}(\mathbf{T}+\mathbf{T}') + \frac{1}{2}(\mathbf{T}-\mathbf{T}') = \mathbf{CDC} + \mathbf{SDS} + \mathbf{C\tilde{D}S} - \mathbf{S\tilde{D}C},$$

where

$$\mathbf{D} := \operatorname{diag}(d_0, \ldots, d_{N-1}), \ \tilde{\mathbf{D}} := \operatorname{diag}(0, \tilde{d}_1, \ldots, \tilde{d}_{N-1}),$$

$$(d_0, \ldots, d_{N-1})' := \tilde{\mathbf{C}}_N^{III} \left(t_0, \frac{t_1 + t_{-1}}{2}, \ldots, \frac{t_{N-1} + t_{-(N-1)}}{2}\right)',$$

$$(\tilde{d}_0, \ldots, \tilde{d}_{N-1})' := \tilde{\mathbf{S}}_N^{III} \left(\frac{t_{-1} - t_1}{2}, \ldots, \frac{t_{-(N-1)} - t_{N-1}}{2}, 0\right)'.$$

Corollary 2, which can also be formulated for the other trigonometric transforms, provides a new method for the fast multiplication of real Toeplitz matrix with a vector that avoids the complex arithmetic which comes into the play if we exploit FFT–based methods.

We are interested in real BTTB matrices

$$\mathbf{T}_{M,N} := (\mathbf{T}_{r-s})_{r,s=0}^{M-1} \quad \text{with} \quad \mathbf{T}_r := (t_{r,j-k})_{j,k=0}^{N-1}. \tag{1}$$

If $t_{r,j} = t_{|r|,|j|}$, then $\mathbf{T}_{M,N}$ is called *doubly symmetric.* Corollary 2 can be generalized to BTTB matrices. For given doubly symmetric BTTB matrix $\mathbf{T}_{M,N}$ we obtain the representation

$$\mathbf{T}_{M,N} = (\mathbf{C}_M^{IV} \otimes \mathbf{C}_N^{IV})\mathbf{D}(\mathbf{C}_M^{IV} \otimes \mathbf{C}_N^{IV}) + (\mathbf{S}_M^{IV} \otimes \mathbf{C}_N^{IV})\mathbf{D}(\mathbf{S}_M^{IV} \otimes \mathbf{C}_N^{IV})$$
$$+ (\mathbf{C}_M^{IV} \otimes \mathbf{S}_N^{IV})\mathbf{D}(\mathbf{C}_M^{IV} \otimes \mathbf{S}_N^{IV}) + (\mathbf{S}_M^{IV} \otimes \mathbf{S}_N^{IV})\mathbf{D}(\mathbf{S}_M^{IV} \otimes \mathbf{S}_N^{IV})$$

with $\mathbf{D} := \operatorname{diag}((\tilde{\mathbf{C}}_M^{III} \otimes \tilde{\mathbf{C}}_N^{III}) \operatorname{col}(t_{r,j})_{j,r=0}^{N-1,M-1})$. Here col: $\mathbb{R}^{N,M} \to \mathbb{R}^{MN}$ and its inverse col^{-1} are defined by

$$\operatorname{col}\,(x_{j,k})_{j=0,k=0}^{N-1,M-1} := (x_r)_{r=0}^{MN-1} \quad \text{with} \quad x_{kN+j} := x_{j,k},$$
$$\operatorname{col}^{-1}\,(x_r)_{r=0}^{MN-1} := (x_{j,k})_{j=0,k=0}^{N-1,M-1} \quad \text{with} \quad x_{j,k} := x_{kN+j}.$$

Algorithm 3. (*Fast multiplication of* BTTB *matrix with a vector*)
Input: $\mathbf{x} \in \mathbb{R}^{MN}$, $t_{r,j} \in \mathbb{R}$ $(j = -(N-1), \ldots, N-1; r = -(M-1), \ldots, M-1)$.
Output: $\mathbf{y} := \mathbf{T}_{M,N}\mathbf{x}$ *with* $\mathbf{T}_{M,N}$ *given by* (1).
Precomputation:
For $j = -(N-1), \ldots, N-1$ *compute*

$$(d_{r,j}^c)_{r=0}^{M-1} := \tilde{\mathbf{C}}_M^{III} \left(t_{0,j}, \frac{t_{-1,j} + t_{1,j}}{2}, \ldots, \frac{t_{-(M-1),j} + t_{M-1,j}}{2}\right)',$$

$$(d_{r,j}^s)_{r=0}^{M-1} := \tilde{\mathbf{S}}_M^{III} \left(\frac{t_{-1,j} - t_{1,j}}{2}, \ldots, \frac{t_{-(M-1),j} - t_{M-1,j}}{2}, 0\right)'.$$

For $r = -(M-1), \ldots, M-1$ and $\alpha \in \{c, s\}$ compute

$$\left(d_{r,j}^{\alpha,c}\right)_{j=0}^{N-1} := \tilde{\mathbf{C}}_N^{III} \left(d_{r,0}^{\alpha}, \frac{d_{r,-1}^{\alpha} + d_{r,1}^{\alpha}}{2}, \ldots, \frac{d_{r,-(N-1)}^{\alpha} + d_{r,N-1}^{\alpha}}{2}\right)',$$

$$\left(d_{r,j}^{\alpha,s}\right)_{j=0}^{N-1} := \tilde{\mathbf{S}}_N^{III} \left(\frac{d_{r,-1}^{\alpha} - d_{r,1}^{\alpha}}{2}, \ldots, \frac{d_{r,-(N-1)}^{\alpha} - d_{r,N-1}^{\alpha}}{2}, 0\right)'.$$

Let

$$\mathbf{X} := \mathrm{col}^{-1}\mathbf{x},$$

$$\mathbf{D}^{\alpha,\beta} := \left(d_{r,j}^{\alpha,\beta}\right)_{j=0,r=0}^{N-1,M-1} \qquad (\alpha, \beta \in \{c, s\}).$$

1. $\mathbf{X}^c := \mathbf{X}\mathbf{C}_M^{IV}$, $\mathbf{X}^s := \mathbf{X}\mathbf{S}_M^{IV}$.
2. $\mathbf{X}^{\alpha,c} := \mathbf{C}_N^{IV}\mathbf{X}^{\alpha}$, $\mathbf{X}^{\alpha,s} := \mathbf{S}_N^{IV}\mathbf{X}^{\alpha}$ $(\alpha \in \{c, s\})$.
3. $\mathbf{Z}^{\beta,\alpha} := \mathbf{C}_N^{IV}(\mathbf{D}^{\beta,c} \circ \mathbf{X}^{\alpha,c} + \mathbf{D}^{\beta,s} \circ \mathbf{X}^{\alpha,s}) + \mathbf{S}_N^{IV}(\mathbf{D}^{\beta,c} \circ \mathbf{X}^{\alpha,s} - \mathbf{D}^{\beta,s} \circ \mathbf{X}^{\alpha,c})$ $(\alpha, \beta \in \{c, s\})$. *Here* $\circ$ *denotes the componentwise matrix product.*
4. $\mathbf{Y} := (\mathbf{Z}^{c,c} + \mathbf{Z}^{s,s})\mathbf{C}_M^{IV} + (\mathbf{Z}^{c,s} - \mathbf{Z}^{s,c})\mathbf{S}_M^{IV}$.
Set $\mathbf{y} := \mathrm{col}\,\mathbf{Y}$.

Algorithm 3 requires $\mathcal{O}(MN\,\log(MN))$ real arithmetical operations.

3. Trigonometric Preconditioners

We are concerned with the solution of a system of linear equations

$$\mathbf{T}_{M,N}\mathbf{x} = \mathbf{b} \tag{2}$$

with doubly symmetric positive definite BTTB matrix

$$\mathbf{T} = \mathbf{T}_{M,N} := \mathrm{stoep}(\mathbf{T}_0, \ldots, \mathbf{T}_{M-1}), \quad \mathbf{T}_j = \mathrm{stoep}(t_{j,0}, \ldots, t_{j,N-1}) \tag{3}$$

by the PCG-method. We will see that with a good preconditioner at hand, this can be realized in a fast way. There are several requirements on a preconditioner $\mathbf{P}_{M,N}$ of (2) resulting from the construction and the convergence behaviour of the CG-method as well as from the fact that the multiplication of $\mathbf{T}_{M,N}$ with a vector requires only $\mathcal{O}(MN\,\log(MN))$ arithmetical operations. Therefore, we are looking for a preconditioner with the following properties:
(P1) $\mathbf{P}_{M,N}$ is symmetric and positive definite.
(P2) For all $\varepsilon > 0$ and sufficiently large M, N, at most $\mathcal{O}(M) + \mathcal{O}(N)$ eigenvalues of $\mathbf{P}_{M,N}^{-1}\mathbf{T}_{M,N}$ lie outside the interval $(1-\varepsilon, 1+\varepsilon)$ such that the PCG-method converges in at most $\mathcal{O}(M) + \mathcal{O}(N)$ steps.

(P3) The multiplication of $\mathbf{P}_{M,N}$ with a vector can be computed with $\mathcal{O}(MN\log(MN))$ arithmetical operations.
(P4) The construction of $\mathbf{P}_{M,N}$ takes only $\mathcal{O}(MN\ \log(MN))$ arithmetical operations.
Having property (P3) in mind, a straightforward idea consists in choosing $\mathbf{P}_{M,N}$ from an algebra

$$\mathcal{A}_O := \{\mathbf{O}'\,(\operatorname{diag}\mathbf{d})\,\mathbf{O} : \mathbf{d} \in \mathbb{R}^{MN}\}$$

of all matrices which are diagonalizable by some orthogonal matrix $\mathbf{O} \in \mathbb{R}^{MN,MN}$, where $\mathbf{O}$ has the additional property that its fast multiplication with a vector requires only $\mathcal{O}(MN\ \log(MN))$ arithmetical operations. As orthogonal matrices, we will use tensor products of the trigonometric matrices of the previous section. For $\mathbf{A} \in \mathbb{R}^{MN,MN}$, the matrix $\mathbf{P}(\mathbf{A}) \in \mathcal{A}_O$ is called an *optimal preconditioner* of $\mathbf{A}$ [4] if

$$||\mathbf{P}(\mathbf{A}) - \mathbf{A}||_F = \min\{||\mathbf{B} - \mathbf{A}||_F : \mathbf{B} \in \mathcal{A}_O\}\,.$$

Here $||\mathbf{A}||_F$ denotes the *Frobenius norm*

$$||\mathbf{A}||_F := (\sum_{j,k=0}^{MN-1} a_{jk}^2)^{1/2}\,.$$

The choice of the Frobenius norm results from the fact that this norm is induced by an inner product of $\mathbb{R}^{MN,MN}$

$$\langle \mathbf{A}, \mathbf{B}\rangle := \operatorname{tr}(\mathbf{A}'\mathbf{B}) = \sum_{j,k=0}^{MN-1} a_{j,k} b_{j,k}.$$

In particular, we have

$$||\mathbf{OAO}'||_F^2 = \operatorname{tr}(\mathbf{OA}'\mathbf{O}'\mathbf{OAO}') = \operatorname{tr}(\mathbf{A}'\mathbf{A}) = ||\mathbf{A}||_F^2\,.$$

For $\mathbf{B} := \mathbf{O}'\,(\operatorname{diag}\mathbf{d})\,\mathbf{O} \in \mathcal{A}_O$ this implies

$$||\mathbf{B} - \mathbf{A}||_F = ||\operatorname{diag}\mathbf{d} - \mathbf{OAO}'||_F$$

such that the optimal preconditioner of $\mathbf{A}$ is given by

$$\mathbf{P}(\mathbf{A}) = \mathbf{O}'\,\delta(\mathbf{OAO}')\,\mathbf{O}\,. \tag{4}$$

If $\mathbf{O}$ is the tensor product of any two orthogonal matrices which correspond to the DST–I, DST–II, DCT–II, DST–IV and DCT–IV, respectively, then $\mathbf{P}(\mathbf{A})$ is said to be an *optimal trigonometric preconditioner* of $\mathbf{A}$.
In the following, we restrict our attention to

$$\mathbf{O} = \mathbf{O}_{M,N} := \mathbf{C}_M^{IV} \otimes \mathbf{C}_N^{IV}\,.$$

The approach for the other transforms follows exactly the same lines. Numerical examples are included for different transforms.
By Theorem 1, $\mathcal{A}_O$ consists of block–Toeplitz–plus–Hankel matrices with Toeplitz–plus–Hankel blocks, more precisely

$$2\,\mathbf{ODO} = \operatorname{stoep}(\mathbf{A}_0,\ldots,\mathbf{A}_{M-1}) + \operatorname{ahank}(\mathbf{A}_1,\ldots,\mathbf{A}_{M-1},\mathbf{0}) \tag{5}$$

where

$$2\,\mathbf{A}_r := \operatorname{stoep}(a_{r,0},\ldots,a_{r,N-1}) + \operatorname{ahank}(a_{r,1},\ldots,a_{r,N-1},0)\,,$$

$$\mathbf{D} := \operatorname{diag}(\tilde{\mathbf{C}}_M^{III} \otimes \tilde{\mathbf{C}}_N^{III})\,\mathbf{a}, \quad \mathbf{a} := \operatorname{col}\,(a_{r,k})_{k=0,r=0}^{N-1,M-1}\,.$$

The optimal trigonometric preconditioner $\mathbf{P} := \mathbf{P}(\mathbf{T}) \in \mathcal{A}_O$ of a doubly symmetric BTTB matrix $\mathbf{T}$ can be computed as follows:

Theorem 4. *Let a doubly symmetric* BTTB *matrix* $\mathbf{T}$ *of form* (3) *be given. Then its optimal trigonometric preconditioner* $\mathbf{P} \in \mathcal{A}_O$ *reads*

$$\mathbf{P} = 4\,\mathbf{O}\left(\operatorname{diag}(\tilde{\mathbf{C}}_M^{III} \otimes \tilde{\mathbf{C}}_N^{III})\,(\mathbf{w}^F \circ \mathbf{t})\right)\mathbf{O}$$

where $\mathbf{t} := \operatorname{col}(t_{r,k})_{k=0,r=0}^{N-1,M-1}$ *and* $\mathbf{w}^F := \operatorname{col}\left((1-\frac{r}{M})(1-\frac{k}{N})\right)_{k=0,r=0}^{N-1,M-1}$.

Proof: By (4), we are interested in $\delta(\mathbf{OTO}) = \operatorname{diag}\,(d_r)_{r=0}^{MN-1}$.
For $k = 0,\ldots,M-1$ and $j = 0,\ldots,N-1$, we obtain by straightforward calculation that

$$d_{kN+j} = \frac{4}{MN} \sum_{u,v=0}^{M-1} \sum_{r,s=0}^{N-1} t_{|u-v|,|r-s|} \cos\frac{(2k+1)(2u+1)\pi}{4M} \cos\frac{(2j+1)(2r+1)\pi}{4N}$$
$$\cdot \cos\frac{(2v+1)(2k+1)\pi}{4M} \cos\frac{(2s+1)(2j+1)\pi}{4N}\,.$$

Using

$$\cos x \cos y = \frac{1}{2}\cos(x-y) + \frac{1}{2}\cos(x+y)\,,$$

we can simplify the above expression in

$$d_{kN+j} = \frac{1}{MN} \sum_{u,v=0}^{M-1} \sum_{r,s=0}^{N-1} t_{|u-v|,|r-s|} \cos\frac{(2k+1)(u-v)\pi}{2M} \cos\frac{(2j+1)(r-s)\pi}{2N}\,,$$

since the remaining three terms vanish. Summation over all (u,v) with $m := |u-v| = 0,\ldots,M-1$ leads to

$$d_{kN+j} = \frac{2}{N} \sum_{r,s=0}^{N-1} \cos\frac{(2j+1)(r-s)\pi}{2N} \sum_{m=0}^{M-1} (\varepsilon_m^M)^2 (1-\frac{m}{M}) t_{m,|r-s|} \cos\frac{(2k+1)m\pi}{2M}$$

and summation over all (r,s) with $n := |r-s| = 0,\dots,N-1$ finally to

$$\begin{aligned} d_{kN+j} &= 4\sum_{m=0}^{M-1}\sum_{n=0}^{N-1}(\varepsilon_m^M)^2(\varepsilon_n^N)^2(1-\frac{m}{M})(1-\frac{n}{N})\, t_{m,n} \\ &\qquad\qquad \cdot \cos\frac{(2k+1)m\pi}{2M}\cos\frac{(2j+1)n\pi}{2N}\,. \end{aligned} \tag{6}$$

Consequently, we get

$$(d_r)_{r=0}^{MN-1} = 4\left((\tilde{\mathbf{C}}_M^{III}\otimes\tilde{\mathbf{C}}_N^{III})\,(\mathbf{w}^F\circ\mathbf{t})\right). \qquad \square$$

Let $t_{j,k}\in\mathbb{R}$ $(j,k\in\mathbb{N}_0)$ with $\sum_{j,k=0}^{\infty}|t_{j,k}|<\infty$ be given. Then the function

$$f(x,y) := 4\sum_{j,k=0}^{\infty}\lambda_j\lambda_k\, t_{j,k}\, \cos(jx)\cos(ky) \tag{7}$$

with $\lambda_0 := 1/2$ and $\lambda_j := 1$ for $j\in\mathbb{N}$ belongs to the Wiener class. Note that f is 2π–periodic and even in both variables. Under these assumptions, f is called *generating function* of the doubly symmetric BTTB matrix (3).
Using the Fourier sums

$$(s_{m,n}f)(x,y) := 4\sum_{j=0}^{m}\sum_{k=0}^{n}\lambda_j\lambda_k\, t_{j,k}\, \cos(jx)\cos(ky)\,, \tag{8}$$

the Fejér mean [3] is defined by

$$\begin{aligned} (\sigma_{M-1,N-1}f)(x,y) &:= \frac{1}{MN}\sum_{m=0}^{M-1}\sum_{n=0}^{N-1}(s_{m,n}f)(x,y) \\ &= 4\sum_{m=0}^{M-1}\sum_{n=0}^{N-1}\lambda_j\lambda_k w_{m,n}^F t_{m,n}\cos(mx)\cos(ny) \end{aligned}$$

with the corresponding *Fejér window*

$$w_{m,n}^F := \begin{cases} \left(1-\frac{m}{M}\right)\left(1-\frac{n}{N}\right) & m=0,\dots,M-1\,; n=0,\dots,N-1\,, \\ 0 & \text{otherwise}\,. \end{cases}$$

Defining the Fejér kernel by

$$F_{N-1}(x) := 1+2\sum_{k=1}^{N-1}\left(1-\frac{k}{N}\right)\cos(kx) = \begin{cases} \dfrac{1}{N}\left(\dfrac{\sin\frac{xN}{2}}{\sin\frac{x}{2}}\right)^2 & x\neq 0\,, \\ N & x=0 \end{cases}$$

and the *even shift* by

$$(\tau_{u,v}\, f)\,(x,y) := \frac{1}{4} \sum_{j,k=0}^{1} f(x+(-1)^j u, y+(-1)^k v)\,,$$

we obtain the known integral representation of the Fejér mean

$$(\sigma_{M-1,N-1}f)\,(x,y) = \frac{1}{\pi^2} \int_0^\pi \int_0^\pi (\tau_{u,v}f)(x,y)\, F_{M-1}(u)\, F_{N-1}(v)\; \mathrm{d}u\mathrm{d}v\,.$$

For $M, N \to \infty$, the Fejér mean $\sigma_{M-1,N-1}f$ tends uniformly to f on $[0,\pi]^2$. Furthermore, we know that for all $(x,y) \in [0,\pi]^2$

$$f_{\min} \leq (\sigma_{M-1,N-1}f)(x,y) \leq f_{\max}\,, \tag{9}$$

where $f_{\min}$ and $f_{\max}$ denote the minimal and maximal values of f, respectively. By (9) and Theorem 4, we obtain:

Corollary 5. *Let f be given by (7) and let $\mathbf{T} := \mathbf{T}_{M,N}$ be the associated doubly symmetric* BTTB *matrix* (3). *Then the eigenvalues of the optimal preconditioner $\mathbf{P} \in \mathcal{A}_O$ of $\mathbf{T}$ are special values of the Fejér mean, i.e.*

$$\mathrm{col}\left((\sigma_{M-1,N-1}f)(\frac{(2j+1)\pi}{2M}, \frac{(2k+1)\pi}{2N}) \right)_{k,j=0}^{N-1,M-1} = 4\left(\tilde{\mathbf{C}}_M^{III} \otimes \tilde{\mathbf{C}}_N^{III}\right)(\mathbf{w}^F \circ \mathbf{t})\,.$$

If f is positive, then $\mathbf{P}$ is positive definite.

Using the simple *rectangular window*

$$w_{m,n}^S := \begin{cases} 1 & m=0,\ldots M-1,\; n=0,\ldots N-1\,, \\ 0 & \text{otherwise,} \end{cases}$$

instead of the Fejér window in Theorem 4, we obtain the definition of the *Strang–type preconditioner* [10]

$$\mathbf{S} = \mathbf{S}(\mathbf{T}) := 4\,\mathbf{O}\left(\mathrm{diag}(\tilde{\mathbf{C}}_M^{III} \otimes \tilde{\mathbf{C}}_N^{III})(\mathbf{w}^S \circ \mathbf{t})\right)\mathbf{O}\,.$$

The eigenvalues of $\mathbf{S}$ are special values of the Fourier sum $s_{M-1,N-1}f$. For positive f it could happen that not all eigenvalues of $\mathbf{S}$ are positive. Since f is from the Wiener class, the Fourier sum $s_{M-1,N-1}f$ tends uniformly to f on $[0,\pi]^2$. Hence, for all $\varepsilon > 0$, there exist $\mu > 0$ such that for $M, N > \mu$ all eigenvalues of $\mathbf{S}$ lie in the interval $[f_{\min} - \varepsilon, f_{\max} + \varepsilon]$. Thus, for positive f and sufficiently large M and N, the Strang–type preconditioner $\mathbf{S}$ is positive definite.

4. Convergence Analysis

Let f be given by (7). We use f as generating function of doubly symmetric BTTB matrices $\mathbf{T}_{M,N}$ of form (3). The eigenvalues of $\mathbf{T}_{M,N}$ are distributed as f [13] and

$$f_{\min} \leq \lambda_{\min}(\mathbf{T}_{M,N}) \leq \lambda_{\max}(\mathbf{T}_{M,N}) \leq f_{\max},$$

where $\lambda_{\min}(\mathbf{T}_{M,N})$ and $\lambda_{\max}(\mathbf{T}_{M,N})$ denote the smallest and largest eigenvalues of $\mathbf{T}_{M,N}$, respectively. We consider again the case $\mathbf{O}_{M,N} := \mathbf{C}_M^{IV} \otimes \mathbf{C}_N^{IV}$.

Theorem 6. *Let f in (7) be a positive function from the Wiener class with $f_{\min} \geq \gamma > 0$. Further, let $\mathbf{T}_{M,N}$ be the associated doubly symmetric BTTB matrix (3). By $\mathbf{S}_{M,N}$, $\mathbf{P}_{M,N} \in \mathcal{A}_{O_{M,N}}$, we denote the Strang-type preconditioner and the optimal trigonometric preconditioner of $\mathbf{T}_{M,N}$, respectively. Then, for all $\varepsilon > 0$, and for sufficiently large M, N at most $\mathcal{O}(M) + \mathcal{O}(N)$ eigenvalues of $\mathbf{S}_{M,N}^{-1}\mathbf{T}_{M,N}$ and $\mathbf{P}_{M,N}^{-1}\mathbf{T}_{M,N}$ lie outside the interval $(1-\varepsilon, 1+\varepsilon)$.*

Proof: 1. First we consider $\mathbf{S}_{M,N}^{-1}\mathbf{T}_{M,N}$. By (5), it holds that

$$\begin{aligned}\mathbf{S}_{M,N} - \mathbf{T}_{M,N} =& \operatorname{ahank}(\mathbf{T}_1, \ldots, \mathbf{T}_{M-1}, \mathbf{0}) + \operatorname{stoep}(\mathbf{H}_0, \ldots, \mathbf{H}_{M-1}) \\ &+ \operatorname{ahank}(\mathbf{H}_1, \ldots, \mathbf{H}_{M-1}, \mathbf{0}),\end{aligned} \tag{10}$$

where $\mathbf{H}_r := \operatorname{ahank}(t_{r,1}, \ldots, t_{r,N-1}, 0)$. Since f belongs to the Wiener class, for all $\varepsilon > 0$, there exist $m, n > 0$ such that

$$\sum_{s=m+1}^{\infty} \sum_{j=0}^{\infty} |t_{s,j}| < \varepsilon\gamma/6 \quad , \quad \sum_{s=0}^{\infty} \sum_{j=n+1}^{\infty} |t_{s,j}| < \varepsilon\gamma/6. \tag{11}$$

For $N > 2n$ and $M > 2m$, we decompose (10) as

$$\mathbf{S}_{M,N} - \mathbf{T}_{M,N} = \mathbf{U} + \mathbf{V},$$

where

$$\begin{aligned}\mathbf{U} :=& \operatorname{ahank}(\mathbf{0}_m, \mathbf{T}_{m+1}, \ldots, \mathbf{T}_{M-1}, \mathbf{0}) + \operatorname{stoep}(\mathbf{H}_0^B, \ldots, \mathbf{H}_{M-1}^B) \\ &+ \operatorname{ahank}(\mathbf{0}_m, \mathbf{H}_{m+1}, \ldots, \mathbf{H}_{M-1}, \mathbf{0}), \\ \mathbf{V} :=& \operatorname{ahank}(\mathbf{T}_1, \ldots, \mathbf{T}_m, \mathbf{0}_{M-m}) + \operatorname{stoep}(\mathbf{H}_0^E, \ldots, \mathbf{H}_{M-1}^E) \\ &+ \operatorname{ahank}(\mathbf{H}_1, \ldots, \mathbf{H}_m, \mathbf{0}_{M-m})\end{aligned}$$

and

$$\mathbf{H}_r^B := \operatorname{ahank}(\mathbf{0}_n, t_{r,n+1}, \ldots, t_{r,N-1}, 0), \qquad \mathbf{H}_r^E := \operatorname{ahank}(t_{r,1}, \ldots, t_{r,n}, \mathbf{0}_{N-n}).$$

Here $\mathbf{0}_j$ denotes either the zero vector of length j or the vector of length j with zero matrices of size (N,N) as entries. By (11), it holds that

$$||\mathbf{U}||_2 \leq ||\mathbf{U}||_1 \leq \varepsilon\gamma.$$

Moreover, $\mathbf{V}$ is a matrix of low rank $\leq 2mn + 2nM$. Next, we verify that

$$\mathbf{I} - \mathbf{S}_{M,N}^{-1}\mathbf{T}_{M,N} = \mathbf{S}_{M,N}^{-1}(\mathbf{S}_{M,N} - \mathbf{T}_{M,N}) = \mathbf{S}_{M,N}^{-1}\mathbf{U} + \mathbf{S}_{M,N}^{-1}\mathbf{V},$$

where $\mathbf{S}_{M,N}^{-1}\mathbf{V}$ is of low rank $\leq 2mN + 2nM$ and where by $f_{\min} \geq \gamma > 0$

$$||\mathbf{S}_{M,N}^{-1}\mathbf{U}||_2 \leq ||\mathbf{S}_{M,N}^{-1}||_2||\mathbf{U}||_2 \leq \varepsilon.$$

Hence, by Cauchy's interlace theorem [18] (applied to the matrix $\mathbf{S}_{M,N}^{-1/2}\mathbf{U}\mathbf{S}_{M,N}^{-1/2}$ $+ \mathbf{S}_{M,N}^{-1/2}\mathbf{U}\mathbf{S}_{M,N}^{-1/2}$) at most $2mN + 2nM$ eigenvalues of $\mathbf{I} - \mathbf{S}_{M,N}^{-1}\mathbf{T}_{M,N}$ have absolute values larger than ε.

2. To prove the assertion for $\mathbf{P}_{M,N}^{-1}\mathbf{T}_{M,N}$, we use that

$$\begin{aligned}\mathbf{P}_{M,N} - \mathbf{T}_{M,N} &= \mathbf{S}_{M,N} - \mathbf{T}_{M,N} - \mathbf{O}_{M,N}\,\delta\,(\mathbf{O}_{M,N}(\mathbf{S}_{M,N} - \mathbf{T}_{M,N})\mathbf{O}_{M,N})\,\mathbf{O}_{M,N}\\ &= \mathbf{U} + \mathbf{V} + \mathbf{O}_{M,N}\,\delta(\mathbf{O}_{M,N}\mathbf{U}\mathbf{O}_{M,N})\,\mathbf{O}_{M,N} + \mathbf{O}_{M,N}\,\delta(\mathbf{O}_{M,N}\mathbf{V}\mathbf{O}_{M,N})\,\mathbf{O}_{M,N}.\end{aligned}$$

Regarding that the entries of $\mathbf{O}_{M,N}$ have absolute values $\leq 2/\sqrt{MN}$, we get for sufficiently large M, N that

$$||\delta(\mathbf{O}_{M,N}\mathbf{V}\mathbf{O}_{M,N})||_2 \leq \varepsilon\gamma.$$

On the other hand, we have

$$||\delta(\mathbf{O}_{M,N}\mathbf{U}\mathbf{O}_{M,N})||_2 \leq ||\mathbf{U}||_2 \leq \varepsilon\gamma$$

such that

$$\mathbf{C}_{M,N} - \mathbf{T}_{M,N} = \mathbf{V} + \tilde{\mathbf{U}}$$

with $||\tilde{\mathbf{U}}||_2 \leq 3\varepsilon\gamma$. The rest of the proof follows the lines of the first part. □

Using standard error analysis, we conclude that the CG-method converges in at most $\mathcal{O}(M) + \mathcal{O}(N)$ steps for sufficiently large M, N.

4. Numerical Results

In this section we illustrate the efficiency of trigonometric preconditioning by various numerical examples. The algorithms were realized for the optimal preconditioners with respect to the DCT–II, DST–II, DCT–IV and DST–IV, respectively. The fast computation of the preconditioners and the PCG–method were implemented in Matlab and tested on a Sun SPARCstation 20. The fast trigonometric transforms appearing both in the computation of the preconditioner and in the PCG–steps were taken from the C–implementation based on [15] and [1] by using the cmex–program.

As transform length we choose $N = 2^n$. The right–hand side $\mathbf{b}$ of (2) is the vector consisting of N entries „1". The PCG–method starts with the zero vector and stops if $\|\mathbf{r}^{(j)}\|_2/\|\mathbf{r}^{(0)}\|_2 < 10^{-7}$, where $\mathbf{r}^{(j)}$ denotes the residual vector after j iterations. We begin with symmetric Toeplitz matrices $\mathbf{T}_{1,N}$. Our test matrices correspond to the following generating functions:

(i) (see [8])

$$f(x) := x^4 + 1 \quad (-\pi \leq x \leq \pi)\,,$$

(ii) (see [9])

$$f(x) := x^2 \quad (-\pi \leq x \leq \pi)\,.$$

The second column of each table contains the number of iteration steps of the CG–method without preconditioning. The last four columns of tables 1 – 2 show the number of iterations required by the PCG–method for different optimal trigonometric preconditioners.

n	I	$\mathbf{C}_N^{II}$	$\mathbf{S}_N^{II}$	$\mathbf{C}_N^{IV}$	$\mathbf{S}_N^{IV}$
8	67	5	5	7	7
9	70	5	5	7	7
10	71	5	5	7	7
11	70	5	5	7	7
12	68	5	5	7	7
13	68	5	5	7	7
14	65	5	5	7	7

Table 1. Number of iterations for example (i).

n	I	$\mathbf{C}_N^{II}$	$\mathbf{S}_N^{II}$	$\mathbf{C}_N^{IV}$	$\mathbf{S}_N^{IV}$
8	179	23	5	25	25
9	375	29	5	33	33
10	771	38	5	41	41
11	*	51	5	55	55
12	*	68	5	59	59

Table 2. Number of iterations for example (ii).

Further extensive numerical tests with matrices from [2] and [9] lead to the following general observations:

– For symmetric Toeplitz matrices, the DST–I based PCG–method shows a similar convergence behaviour as the DST–II based PCG–method.

– The use of a Strang–type preconditioner with respect to DCT–II or DST–II yields similar results as the DCT–II or DST–II based PCG–method in all numerical tests.

– Preconditioning with respect to the DST–II and DCT–II, respectively, leads to a faster convergence of the PCG-method than preconditioning with respect to the DST–IV or DCT–IV.

Next we consider (2) with doubly symmetric BTTB matrices $\mathbf{T}_{N,N}$. By our observations, we restrict our attention to $\mathbf{O}_1 := \mathbf{C}_N^{II} \otimes \mathbf{C}_N^{II}$ and $\mathbf{O}_2 := \mathbf{S}_N^{II} \otimes \mathbf{S}_N^{II}$. Our test matrices are the four doubly symmetric BTTB matrices with the entries:

(iii) (see [6] and [11])

$$t_{j,k} := \frac{1}{(j+1)\,(k+1)^{1+0.1(j+1)}} \quad (j,k = 0,\ldots,N-1)\,,$$

(iv) (see [6] and [11])

$$t_{j,k} := \frac{1}{(j+1)^{1.1} + (k+1)^{1.1}} \quad (j,k = 0,\ldots,N-1)\,,$$

(v) (see [11] and [14])

$$t_{j,k} := 0.7^{\sqrt{j^2+k^2}} + 0.5^{\sqrt{j^2+k^2}} + 0.3^{\sqrt{j^2+k^2}} \quad (j,k = 0,\ldots,N-1)$$

(vi) and with the generating function (see [11])

$$f(x,y) := x^2 + y^2 + x^2 y^2 \quad ((x,y) \in [0,\pi]^2)\,.$$

N	I	$\mathbf{O}_1$	$\mathbf{O}_2$
8	15	8	10
16	28	9	12
32	38	10	13
64	45	11	14
128	49	12	14
256	51	13	14
512	50	13	15

N	I	$\mathbf{O}_1$	$\mathbf{O}_2$
8	11	7	8
16	27	8	10
32	43	9	13
64	71	9	15
128	103	10	16
256	144	10	18
512	198	11	20

Table 3. Number of iterations for examples (iii) and (iv).

N	I	$\mathbf{O}_1$	$\mathbf{O}_2$
8	12	7	8
16	30	8	10
32	48	9	12
64	68	9	13
128	79	8	13
256	*	7	13
512	*	7	12

N	I	$\mathbf{O}_1$	$\mathbf{O}_2$
8	10	10	9
16	52	18	9
32	164	25	10
64	*	36	10
128	*	56	10
256	*	90	10
512	*	152	9

Table 4. Number of iterations for example (v) and (vi).

In all examples from [6], our "level–2" trigonometric preconditioners work significantly better than the "level–2" circulant preconditioners and also often better than the "level–1" circulant preconditioners. Only in example (vi), the preconditioning based on $\mathbf{O}_2$ as proposed in [11] leads to a faster convergence than the preconditioning based on $\mathbf{O}_1$. In most of our examples with BTTB matrices the $\mathbf{O}_1$ based PCG–method requires fewer iterations than the PCG–method based on $\mathbf{C}_N^{IV} \otimes \mathbf{S}_N^{IV}$ or $\mathbf{O}_2$.
Moreover, in all examples the optimal trigonometric preconditioner with respect to both $\mathbf{O}_1$ and $\mathbf{O}_2$ yields a little better convergence of the PCG–method than the Strang–type preconditioner based on $\mathbf{O}_1$ and $\mathbf{O}_2$, respectively.

References

1. Baszenski G. and Tasche M., Fast polynomial multiplication and convolution related to the discrete cosine transform, Linear Algebra Appl. **252** (1997), 1–25.
2. Boman E. and Koltracht I., Fast transform based preconditioners for Toeplitz equations, SIAM J. Matrix Anal. Appl. **16** (1995), 628–645.
3. Butzer P. L. and Stens R. L., The operational properties of the Chebyshev transform I: General properties, Funct. Approx. Comment. Math. **5** (1977), 129–160.
4. Chan T. F., An optimal circulant preconditioner for Toeplitz systems, SIAM J. Sci. Statist. Comput. **9** (1988), 766–771.
5. Chan T. F. and Olkin J. A., Circulant preconditioners for Toeplitz-block matrices, Numer. Algorithms **6** (1994), 89 - 101.
6. Chan R. and Jin X. Q., A family of block preconditioners for block systems, SIAM J. Sci. Statist. Comput. **13** (1992), 1218 - 1235.
7. Chan R., Nagy J. G., and Plemmons R. J., FFT-based preconditioners for Toeplitz-block least squares problems, SIAM J. Numer. Anal. **30** (1993), 1740 - 1768.
8. Chan R. and Ng M. K., Conjugate gradient methods for Toeplitz systems, SIAM Review **38** (1996), 427 - 482.
9. Chan R., Ng M. K., and Wong C.K., Sine transform based preconditioners for symmetric Toeplitz systems, Linear Algebra Appl. **232** (1996), 237–259.
10. Chan R. and Yeung M. , Circulant preconditioners constructed from kernels, SIAM J. Numer. Anal. **29** (1992), 1093–1103.
11. Di Benedetto F., Preconditioning of block matrices by sine transforms, Preprint, Univ. Genoa, 1993.
12. Heinig G. and Rost K., Representations of Toeplitz–plus–Hankel matrices using trigonometric transforms with application to fast matrix-vector multiplication, Preprint, December 1996.

13. Jin X. Q., A note on preconditioned Toeplitz matrices, SIAM J. Scient. Computing **16** (1995), 951 - 955.

14. Ku T. and Kuo C. C. J., On the spectrum of a family of preconditioned block Toeplitz matrices, SIAM J. Sci. Statist. Comput. **13** (1992), 948 - 966.

15. Steidl G. and Tasche M., A polynomial approach to fast algorithms for discrete Fourier–cosine and Fourier–sine transforms, Math. Comp. **56** (1991), 281–296.

16. Strang, G., A proposal for Toeplitz matrix calculations, Studies in Appl. Math. **74** (1986), 171–176.

17. Wang Z., Fast algorithms for the discrete W transform and for the discrete Fourier transform, IEEE Trans. Acoustic, Speech, and Signal Processing **32** (1984), 803–816.

18. Wilkinson J., *The Algebraic Eigenvalue Problem*, Clarendon Press, Oxford, 1965.

Daniel Potts
Institut für Mathematik, Medizinische Universität zu Lübeck
Wallstr. 40
D - 23560 Lübeck
Germany
potts@informatik.mu-luebeck.de

Gabriele Steidl
Fakultät für Mathematik und Informatik
Universität Mannheim
D–68131 Mannheim
Germany
steidl@kiwi.math.uni-mannheim.de

Manfred Tasche
Fachbereich Mathematik
Universität Rostock
D–18051 Rostock
Germany
tasche@mathematik.uni-rostock.d400.de

The Average Size of Certain Gram-Determinants and Interpolation on Non-Compact Sets

Manfred Reimer

Abstract. The determinants of Gram-matrices defined by reproducing kernels can be averaged with respect to the arguments occuring. The results can be used to prove that, in a very general situation, interpolation points exist which furnish small Lagrange elements. The results are complementary to Auerbach's Theorem. They apply even in cases where interpolation takes place on non-compact sets.

1. Introduction and Results

Let X denote an algebra of functions $D \to \mathbb{R}$ (necessarily commutative) with unity $\underline{1}$, and let I denote an integral on X, i.e. a positive linear functional, which is normalized by

$$I\underline{1} = 1. \tag{1}$$

I induces on X the inner product

$$< f, g >:= I(fg), \quad f, g \in X. \tag{2}$$

As a standard example we may think of the case $X = C(D)$, D compact, which we call the "continuous case".

Next let V be a linear subspace of X with

$$N := \dim V < \infty. \tag{3}$$

V possesses a uniquely determined reproducing kernel $G : D \times D \to \mathbb{R}$ with regard to $< \cdot, \cdot >$. For arbitrary orthonormal systems $\{S_1, \ldots, S_N\}$

$$G(x, y) = \sum_{j=1}^{N} S_j(x) S_j(y), \quad x, y \in D, \tag{4}$$

Multivariate Approximation and Splines
G. Nürnberger, J. W. Schmidt, and G. Walz (eds.), pp. 235–243.

ISBN 3-7643-5654-5.

holds, implying

$$I_x G(x,x) = N. \tag{5}$$

Recall that the main property of G is

$$< G(x,\cdot), f >= f(x), \tag{6}$$

which is valid for arbitrary $f \in V$ and $x \in D$.

Now let $T = (t_1, \ldots, t_N) \in D^N$ be a system of nodes t_j. Then we may define the functions $g_1, \ldots, g_n \in V$ by

$$g_j := G(t_j, \cdot), \; j = 1, \ldots, N, \tag{7}$$

and the $N \times N$-Gram-matrix $\mathbf{G}$ by

$$\mathbf{G} := (G_{jk}) = (< g_j, g_k >), \tag{8}$$

where $G_{jk} = G(t_j, t_k), \; j,k = 1, \ldots, N$.

T is called a fundamental system for V if the corresponding evaluation functionals are linearly independent in the dual of V. In this case the related interpolation problem is unisolvable, in particular the Lagrange-elements $l_1, \ldots, l_N \in V$ are defined by

$$l_j(t_k) = \delta_{jk}, \quad j,k = 1, \ldots, N. \tag{9}$$

It follows from (7) and (9) that

$$< g_j, l_k >= \delta_{j,k}, \quad j,k = 1, \ldots, N. \tag{10}$$

The systems $\{g_j\}$ and $\{l_k\}$ are biorthogonal, both are bases in V, and $\mathbf{G}$ is regular (if and only if T is fundamental). $\mathbf{G}$ is always positively semi-definite with eigenvalues

$$0 \leq \lambda_1 \leq \lambda_2 \leq \cdots \leq \lambda_N, \tag{11}$$

where $\lambda_1 > 0$ holds if and only if T is fundamental. If T is fundamental, then (10) implies

$$\begin{aligned} g_j &= \sum_{k=1}^{N} < g_j, g_k > l_k, \\ l_j &= \sum_{j=1}^{N} < l_j, l_k > g_k \end{aligned} \tag{12}$$

and hence

$$\mathbf{G}^{-1} = (< l_j, l_k >), \tag{13}$$

such that $\mathbf{G}^{-1}$ is also a Gram-matrix. The importance of $\mathbf{G}$ and $\mathbf{G}^{-1}$ for the numerical treatment of the corresponding interpolation problem is well known, see [4], e. g.

In what follows let

$$\Delta := \Delta(T) := \det \mathbf{G}(T) \tag{14}$$

and

$$\Delta^* := \Delta_V^* := \sup\{\det \mathbf{G}(T) : T \in D^N\} \tag{15}$$

In the continuous case the supremum is attained for some $T = T^* \in D^N$, i. e. with

$$\Delta(T^*) = \Delta^*.$$

In this case

$$\| l_1 \|_\infty = \| l_2 \|_\infty = \cdots = \| l_N \|_\infty = 1$$

holds in the uniform norm. The outstanding situation can be characterized by each l_j being best approximated by zero in $span\{l_k : k \neq j\}$. In addition the corresponding interpolation norm (Lebesgue-constant) is bounded in this case by the value of N (Auerbach's Theorem, [6], [4], e. g.). This makes us ask for the value of Δ^*, or at least for realisations of a great value of Δ, in the general case, where X is an arbitrary function algebra. We are going to prove

Theorem 1. *Let X be an arbitrary function algebra with unity and with inner product as above, and let V be a subspace with $\dim V = N < \infty$. Then a fundamental system T for V exists, such that*

$$N! \leq \Delta(T) \leq \Delta_V^* \tag{16}$$

holds.

If $X = C(S^{r-1}), r \in \mathbb{N}\backslash\{1\}$, where S^{r-1} is the unit sphere in $\mathbb{R}^r$, and if V is a rotation-invariant subspace of X, then G takes form

$$G(x,y) = g(x \cdot y), \;\; g \in C([-1,1]),$$

[4]. In this case, (4) and (1) imply

$$g(1) = N, \tag{17}$$

if the integral is defined as the normalized surface-integral

$$If := \frac{1}{\omega_{r-1}} \int\limits_{S^{r-1}} f(x) d\omega(x),$$

ω_{r-1} volume of S^{r-1}. It follows that the diagonal of $\mathbf{G}$ consists of entries all equal to N, such that

$$\lambda_1 + \lambda_2 + \cdots + \lambda_N = trace(\mathbf{G}) = N^2. \tag{18}$$

Because of this side condition,

$$\Delta(T) = \lambda_1 \cdot \lambda_2 \cdots \lambda_N \leq N^N \tag{19}$$

is valid for arbitrary T, such that (16) reads like

$$\frac{N}{e} \sim \sqrt[N]{N!} \leq \sqrt[N]{\lambda_1 \lambda_2 \cdots \lambda_N} \leq \sqrt[N]{\Delta_V^*} \leq N, \tag{20}$$

i. e. the upper and the lower bound for the geometric mean are rather close to each other.

In the general case this need not be true, since $G(x,x)$ and hence $\Delta(T)$ may be even unbounded. In this case, a better strategy to obtain a valuable fundamental system consists in looking for

$$\Lambda^* = \Lambda_V^* := \inf\{\varphi(\lambda_1, \cdots, \lambda_N) : T \in D^N\}$$

where

$$\varphi(\lambda_1, \cdots, \lambda_N) := \sum_{j=1}^{N} \lambda_j^{-1} = trace(\mathbf{G}^{-1}(T)).$$

Note that, because of (13),

$$\Lambda_V^* = \inf\{\sum_{j=1}^{N} < l_j, l_j >: T \in D^N, T \ fundamental for \ V\} \tag{21}$$

is valid. We are going to prove

Theorem 2. *Under the assumptions of Theorem 1, a fundamental system T for V exists such that*

$$\Lambda_V^* \leq \sum_{j=1}^{N} < l_j, l_j > \leq N \tag{22}$$

holds.

Lemma 3. *Let X and V be as in Theorem 1. For $k \in \{0, 1, \cdots, N-1\}$ and arbitrary $T \in D^N$ let $\Delta^{(k)} = \Delta^{(k)}(T)$ denote any minor determinant of $\Delta = \Delta(T)$ which originates by cutting k different rows and the corresponding columns. Then the N-fold integral*

$$I^N \Delta^{(k)} := I_{t_N} \cdots I_{t_1} \Delta^{(k)}(t_1, \cdots, t_N)$$

has the value of

$$I^N \Delta^{(k)} = \frac{N!}{k!}.$$

The proofs are given in Section 2. Note that Lemma 3 indicates, in some sense, stability of the determinants against degeneration. Note also, that for complex functions similar holds if the inner product has form $< f, g >= I(f\overline{g})$.

2. Proofs

We begin with Lemma 3. By $\Delta^{jp\cdots}_{kq\cdots}$ we denote the adjoint of Δ where the rows/ columns with the indices $j, p, \cdots / k, q, \cdots$ are cut.

Obviously, without restriction of generality it suffices to prove the statements for the adjoints

$$\Delta, \Delta^1_1, \Delta^{12}_{12}, \cdots .$$

But, expanding first Δ by its first column, and after this all originating adjoints by their first row, we obtain

$$\Delta = G(t_1, t_1)\Delta^1_1 - \sum_{j=2}^{N}\sum_{k=2}^{N} G(t_j, t_1)\Delta^{1j}_{1k} G(t_1, t_k),$$

where Δ^1_1 and the Δ^{1j}_{1k} do not depend on t_1. Because of (5) and (1), integration with respect to t_1 yields

$$I_{t_1}\Delta = N\Delta^1_1 - \sum_{j=2}^{N}\sum_{k=2}^{N} \Delta^{1j}_{1k} G(t_j, t_k),$$

where we used the reproducing property (6) of the kernel. The inner sum can be regarded to be the expansion of Δ^1_1 with respect to its j-th row, so the whole sum equals to the value of $(N-1)\Delta^1_1$ and we obtain

$$I_{t_1}\Delta = \Delta^1_1 \ . \tag{23}$$

Similarly we expand

$$\Delta^1_1 = G(t_2, t_2)\Delta^{12}_{12} - \sum_{j=3}^{N}\sum_{k=3}^{N} G(t_j, t_2)\Delta^{12j}_{12k} G(t_2, t_k)$$

where Δ^{12}_{12} and the Δ^{12j}_{12k} do not depend on t_2. So we obtain likewise

$$I_{t_2}\Delta^1_1 = N\Delta^{12}_{12} - \sum_{j=3}^{N}\sum_{k=3}^{N} \Delta^{12j}_{12k} G(t_j, t_k)$$

where the inner sums equal the value of Δ^{12}_{12}. Hence we obtain

$$I_{t_2}\Delta^1_1 = 2 \cdot \Delta^{12}_{12}.$$

So we go on, with the result that

$$I_{t_k}\Delta^{1,\cdots,k-1}_{1,\cdots,k-1} = k\Delta^{1,\cdots,k}_{1,\cdots,k} \tag{24}$$

is valid for $k = 2, \cdots, N-1$, where we obtain, in the end,

$$I_{t_N} \Delta_{1,\cdots,N-1}^{1,\cdots,N-1} = I_{t_N} G(t_N, t_N) = N.$$

Putting this together we obtain

$$I_{t_N} \cdots I_{t_{k+1}} \Delta_{1,\cdots,k}^{1,\cdots,k} = \frac{N!}{k!} \, .$$

But $\Delta_{1,\cdots,k}^{1,\cdots,k}$ does not depend on $t_1, \cdots, t_k$, such that

$$\Delta_{1,\cdots k}^{1,\cdots,k} = I_{t_k} \cdots I_{t_1} \Delta_{1,\cdots,k}^{1,\cdots,k}$$

holds because of (1). Together with (23) and (24) this yields

$$I^N \Delta_{1,\cdots,k}^{1,\cdots,k} = \frac{N!}{k!} \, , \quad k = 0, 1, \cdots, N-1,$$

with

$$\Delta_{\cdots,0}^{\cdots,0} := \Delta \, ,$$

and Lemma 3 is proved.

Next we turn to Theorem 1. By the definitions of (15) and (14)

$$\Delta(T) \leq \Delta^* \tag{25}$$

holds for arbitrary $T \in D^N$. If $\Delta^* = +\infty$, nothing is to be proved. In case of $\Delta^* < \infty$, we obtain by the positivity of the integral and the result of Lemma 3 by integration of (25).

$$N! \leq \Delta^*,$$

where we again used (1). In addition, if

$$\Delta(T) < N!$$

would hold for arbitrary T, integration would lead to a contradiction, again because of the positivity of the integral. So there exists a $T \in D^N$ with $N! \leq \Delta(T)$.

Note that we did not use continuity arguments in this proof.

Theorem 2 can be proved as follows: Let T be a fundamental system. Because of (13) we have

$$< l_j, l_j >= \frac{\Delta_j^j}{\Delta} \, ,$$

where we do not notify the dependency of T. From (21) it follows that

$$\Lambda^* \leq \frac{1}{\Delta} \sum_{j=1}^{N} \Delta_j^j$$

holds for arbitrary fundamental systems T. Hence we obtain

$$\Lambda^* \Delta(T) \leq \sum_{j=1}^{N} \Delta_j^j(T),$$

and because of Lemma 3, integration yields

$$\Lambda^* N! \leq N \cdot N!, \quad i.\ e.\ \Lambda^* \leq N.$$

If

$$\sum_{j=1}^{N} \frac{\Delta_j^j(T)}{\Delta(T)} > N$$

would hold for arbitrary fundamental systems T, then we would obtain

$$N \cdot \Delta(T) - \sum_{j=1}^{N} \Delta_j^j(T) < 0,$$

except for non-fundamental T, where the left-hand side is vanishing or less than zero. Since a fundamental system always exists, integration would yield

$$N \cdot N! - N \cdot N! < 0,$$

which is a contradiction. Hence a fundamental system T with $\sum_{j=1}^{N} < l_j, l_j >= \sum_{j=1}^{N} \frac{\Delta_j^j}{\Delta} \leq N$ exists.

3. Applications

Let T be the fundamental system of Theorem 2. Then it follows

$$\begin{aligned}\left(\sum_{j=1}^{N} |l_j(x)|\right)^2 &\leq N \cdot \sum_{j=1}^{N} l_j^2(x) = N \sum_{j=1}^{N} < l_j, G(x, \cdot) >^2 \\ &\leq N^2 \cdot G(x,x),\end{aligned}$$

i. e. the Lebesgue-function can be estimated by

$$\sum_{j=1}^{N} |l_j(x)| \leq N \cdot \sqrt{G(x,x)}, \quad x \in D. \tag{26}$$

Apart of the factor $\sqrt{G(x,x)}$, this estimate holds in the continuous case if T realizes $\Delta(T) = \Delta^*$. But it is quite more general, especially it applies also to the case where D is non-compact.

Interpolation on non-compact intervals was discussed, for instance, by [2], though in a different context. We investigate the following

Example. Let $D := \mathbb{R}$,

$$X := L^2_w(\mathbb{R}) \cap C^*(\mathbb{R})$$

where $C^*(\mathbb{R})$ consists of the bounded continuous functions on $\mathbb{R}$, while $L^2_w(\mathbb{R})$ is the space of functions which are square-integrable with respect to the strictly positive continuous weight-function w. The integral is to be defined by

$$If := \int_{-\infty}^{\infty} f(x)w(x)dx, \quad f \in X.$$

In order to guarantee $1 \in X$, it suffices to assume that

$$\int_{-\infty}^{\infty} w(x)dx = 1$$

holds. Then the integral fits also to the normalisation (1). For instance we could choose $w(x) = \pi^{-1}(1+x^2)^{-1}$.

It is well known that the functions

$$S_n(x) = \pi^{-\frac{1}{4}}(2^n n!)^{-\frac{1}{2}} e^{-\frac{1}{2}x^2} H_n(x), \tag{27}$$

$n \in \mathbb{N}_0$, H_n Hermite-polynomial, are orthonormal on $\mathbb{R}$ with respect to the weight-function 1. It follows that the functions

$$\tilde{S}_n(x) := S_n(x)w(x)^{-\frac{1}{2}}, \quad n \in \mathbb{N}_0,$$

are orthonormal with respect to w, i. e. to the inner product induced by I. Now let

$$V_N := span\{\tilde{S}_0, \cdots, \tilde{S}_{N-1}\} = \{p(x) \cdot e^{-\frac{1}{2}x^2} \cdot w^{-\frac{1}{2}}(x) | p \in \mathbb{P}_{N-1}\}.$$

Then V_N is a subspace of X with dimension N and reproducing kernel

$$G_N(x,y) = \sum_{n=0}^{N-1} S_n(x)S_n(y)w(x)^{-\frac{1}{2}}w(y)^{-\frac{1}{2}}, \quad x, y \in \mathbb{R}.$$

Moreover, following Sansone [5], p. 3.31, we have

$$|\tilde{S}_n(x)|^2 \le 4 \cdot \left\{ \frac{2^{\frac{3}{4}}}{\sqrt{\pi}} + \frac{|x|^{\frac{5}{2}}}{\sqrt[4]{2n+1}} \right\}^2 \cdot \frac{1}{\sqrt{2n+1}} \cdot w(x)^{-1}$$

for arbitrary $x \in \mathbb{R}$, $n \in \mathbb{N}_0$. It follows by (4) that the reproducing kernel $G = G_N$ of V_N satisfies

$$G_N(x,x) = O(N^{\frac{1}{2}}), \ N \to \infty, \tag{28}$$

on compact sets, where the O-constant does not depend on x.

Now assume that for arbitrary $N \in \mathbb{N}$ we choose $T = T_N = \{t_1^N, \cdots, t_N^N\}$ according to Theorem 2, while $l_1^{(N)}, \cdots, l_N^{(N)}$ are the corresponding Lagrange elements in V_N. Then (26) and (28) imply that

$$\sum_{j=1}^{N} |l_j^N(x)| = O(N^{\frac{5}{4}}) \tag{29}$$

holds uniformly on compact subsets of $\mathbb{R}$ as $N \to \infty$. Hence it follows by well-known arguments that

$$\lim_{N\to\infty} \sum_{j=1}^{N} f(t_j^N) l_j^N(x) = f(x) \tag{30}$$

is valid uniformly on all compact sets, if $f \in X$ and $E(f, V_N) = o(N^{-\frac{5}{4}})$, $N \to \infty$. I. e. we are able to interpolate a class of bounded continuous functions on the open real line by means of a fixed matrix of nodes, and this convergently in the sense of uniform convergence on compact sets.

For a detailed discussion of weighted approximations on $\mathbb{R}$ see [3].

References

1. König H., Norms of minimal projections, J. Funct. Anal. **119** (1994), 253–280.
2. Lau T.-S., On an extremal problem of Fejér, J. Approx. Theory **53** (1988), 184–194.
3. Lubinsky D. S., Ideas of weighted approximations on $(-\infty, \infty)$, in *Approximation Theory VIII* (C. K. Chui, L. L. Schumaker, eds.), World Scientific, Singapore, 1995, 371–396.
4. Reimer M., *Constructive theory of multivariate functions.* BI-Wisenschaftsverlag, Mannheim, 1990.
5. Sansone G., *Orthogonal functions.* Interscience Publ., London, 1959, 322, 323.
6. Taylor A. E., A geometric theorem and its application to biorthogonal systems, Bull. Amer. Math. Soc. **53** (1947), 614–616.

Manfred Reimer
Lehrstuhl für Mathematik III
Universität Dortmund
44221 Dortmund
`reimer@euler.math.uni-dortmund.de`

Radial Basis Functions Viewed From Cubic Splines

Robert Schaback

Abstract. In the context of radial basis function interpolation, the construction of native spaces and the techniques for proving error bounds deserve some further clarification and improvement. This can be described by applying the general theory to the special case of cubic splines. It shows the prevailing gaps in the general theory and yields a simple approach to local error bounds for cubic spline interpolation.

1. Optimal Recovery

The theoretical starting point for both cubic splines and radial basis functions is provided by optimal recovery of functions f from scattered data in a set $X = \{x_1, x_2, \ldots, x_M\}$ of *centers* or *knots*. The functions must be in some space $\mathcal{F}$ of real–valued functions on some domain $\Omega \subset \mathbb{R}^d$ containing $X = \{x_1, x_2, \ldots, x_M\}$. The space carries a semi-inner product $(.,.)_\mathcal{F}$ that has a finite–dimensional nullspace $\mathcal{P} \subset \mathcal{F}$ such that $\mathcal{F}/\mathcal{P}$ is a Hilbert space. Once the space $\mathcal{F}$ and its semi–inner product $(.,.)_\mathcal{F}$ are fixed, the recovery problem consists in finding a function $s \in \mathcal{F}$ that minimizes $|s|^2_\mathcal{F} = (s, s)_\mathcal{F}$ under the constraints $s(x_j) = f(x_j)$, $1 \leq j \leq M$.

For cubic splines, the space $\mathcal{F}$ is known beforehand as Sobolev space

$$\mathcal{F} = W_2^2[a,b] = \left\{f \ : \ f'' \in L_2[a,b]\right\}, \quad (f,g)_\mathcal{F} = (f'',g'')_{L_2[a,b]} \tag{1}$$

while the case of radial basis functions usually requires a fairly abstract construction of the space $\mathcal{F}$ from some given radial basis function Φ (see e.g.: [5,6,9] for details). We shall mainly consider the special function $\Phi(x) = |x|^3$ on $\mathbb{R}^1$ here and show how this construction works to recover the theory of natural cubic splines. We note in passing that it is possible (see [8] for a simple presentation) to go the other way round, i.e.: to construct Φ from $\mathcal{F}$, but this is not the standard procedure.

Multivariate Approximation and Splines
G. Nürnberger, J. W. Schmidt, and G. Walz (eds.), pp. 245–258.

ISBN 3-7643-5654-5.

2. Native Spaces

The radial basis function setting starts out with the space $\mathbb{P}_m^d$ of d-variate m-th order polynomials, a subset Ω of $\mathbb{R}^d$, and defines the linear space $(\mathbb{P}_m^d)_\Omega^\perp$ of all linear functionals

$$f \mapsto \lambda_{X,M,\alpha}(f) := \sum_{j=1}^{M} \alpha_j f(x_j)$$

that vanish on $\mathbb{P}_m^d$ and are finitely supported in $\Omega \subseteq \mathbb{R}^d$. The functionals $\lambda_{X,M,\alpha}$ depend each on a finite support set $X = \{x_1, x_2, \ldots, x_M\} \subseteq \Omega$ and a vector $\alpha \in \mathbb{R}^M$, where M is unrestricted, but where α and X are subject to the condition $\lambda_{X,M,\alpha}(\mathbb{P}_m^d) = \{0\}$. We shall write $\lambda^x f(x)$ to indicate the action of a functional λ with respect to the variable x on a function f.

Definition 1. *An even continuous function Φ on $\mathbb{R}^d$ is conditionally positive definite of order m on $\Omega \subseteq \mathbb{R}^d$, iff the symmetric bilinear form*

$$(\lambda, \mu)_\Phi := \lambda^x \mu^y \Phi(x - y) \tag{2}$$

is positive definite on $(\mathbb{P}_m^d)_\Omega^\perp$.

Definition 2. *With*

$$s_\lambda := \lambda^y \Phi(\cdot - y) \tag{3}$$

the native space associated to the conditionally positive definite function Φ of order m on $\mathbb{R}^d$ is

$$\mathcal{F} := \mathbb{P}_m^d + \operatorname{clos}_{(\cdot,\cdot)_\Phi} \{ s_\lambda : \lambda \in (\mathbb{P}_m^d)_\Omega^\perp \} \tag{4}$$

The above definition of the native space involves an abstract closure with respect to a somewhat unusual inner product. At this point it is not even clear that it consists of well–defined functions on Ω or $\mathbb{R}^d$ (see [9] for a discussion). Altogether, it is a major problem to characterize the native space as a space of functions with one of the usual differentiability properties. This seems to be the major obstacle for understanding the theory of radial basis functions.

Though quite abstractly defined, the native space carries a useful Hilbert space structure that is induced via continuity arguments by

$$(s_\lambda, s_\mu)_\Phi := (\lambda, \mu)_\Phi = \lambda(s_\mu) = \mu(s_\lambda),$$

which is an inner product on the second summand of $\mathcal{F}$. There are several ways to define an inner product on all of $\mathcal{F}$, but for most purposes it suffices to extend the above bilinear form to $\mathcal{F}$ by setting it to zero on $\mathbb{P}_m^d$, and to work in the Hilbert space $\mathcal{F}/\mathbb{P}_m^d$ instead of $\mathcal{F}$. The Hilbert space structure is the major tool for analysis of the native space.

These things will be somewhat more transparent if we specialize to the cubic spline case. Here we have $d = 1$, $\Phi(x) = |x|^3$ and we will see soon that Φ is conditionally positive definite of order 2 on $\mathbb{R}$, which is not immediately clear. We first connect the two bilinear forms in

Theorem 3. *For $\Phi(x) = |x|^3$, $d = 1$, $\Omega = [a, b] \subset \mathbb{R}$, and $m = 2$ we have*

$$(\lambda, \mu)_\Phi = (s_\lambda, s_\mu)_\Phi = \frac{1}{12}(s_\lambda'', s_\mu'')_{L_2[a,b]}$$

for all $\lambda, \mu \in (\mathbb{P}_m^d)_\Omega^\perp$.

Proof: We can assume $X = \{x_1, x_2, \ldots, x_M\} \subset \Omega = [a, b] \subset \mathbb{R}$ to be ordered as $a \le x_1 < x_2 < \ldots < x_M \le b$. For any $\lambda_{X,M,\alpha} \in (\mathbb{P}_m^d)_\Omega^\perp$ we have

$$s_{\lambda_{X,M,\alpha}}(x) = \sum_{j=1}^{M} \alpha_j |x - x_j|^3$$

due to (3), and with $|x|^3 = 2x_+^3 - x^3$ we find

$$s_{\lambda_{X,M,\alpha}}(x) = 2\sum_{j=1}^{M} \alpha_j (x - x_j)_+^3 + \sum_{j=1}^{M} \alpha_j (x_j^3 - 3x x_j^2) + 0$$

because $\lambda_{X,M,\alpha}$ annihilates linear polynomials. Then

$$s''_{\lambda_{X,M,\alpha}}(x) = 12\sum_{j=1}^{M} \alpha_j (x - x_j)_+^1 \tag{5}$$

is a piecewise linear function. Its support lies in $[x_1, x_M]$, again because $\lambda_{X,M,\alpha}$ annihilates linear polynomials. If two functionals

$$\lambda_{X,M,\alpha}(f) = \sum_{j=1}^{M} \alpha_j f(x_j), \quad \lambda_{Y,N,\beta}(f) = \sum_{k=1}^{N} \beta_k f(y_k)$$

from $(\mathbb{P}_m^d)_\Omega^\perp$ are given, then

$$(\lambda_{X,M,\alpha}, \lambda_{Y,N,\beta})_\Phi = \sum_{j=1}^{M}\sum_{k=1}^{N} \alpha_j \beta_k |x_j - y_k|^3 \tag{6}$$

by (2), and we want to compare this to

$$\begin{aligned}
&(s''_{\lambda_{X,M,\alpha}}(x), s''_{\lambda_{Y,N,\beta}}(x))_{L_2(\mathbb{R})} \\
&:= \int_{-\infty}^{\infty} s''_{\lambda_{X,M,\alpha}}(x) s''_{\lambda_{Y,N,\beta}}(x) dx. \\
&= \int_a^b s''_{\lambda_{X,M,\alpha}}(x) s''_{\lambda_{Y,N,\beta}}(x) dx.
\end{aligned}$$

Using $x_+^1 = (-x)_+^1 + x$ we rewrite $s''_{\lambda_{X,M,\alpha}}$ as

$$s''_{\lambda_{X,M,\alpha}}(x) = 12\sum_{j=1}^{M}\alpha_j(x_j - x)_+^1 + 12\sum_{j=1}^{M}\alpha_j(x - x_j)$$
$$= 12\sum_{j=1}^{M}\alpha_j(x_j - x)_+^1.$$

with a remarkable swap of x with x_j when compared to (5). We now use Taylor's formula

$$f(x) = f(a) + (x-a)f'(a) + \int_a^b f''(u)(x-u)_+^1 du$$

for functions $f \in C^2[a,b]$ and $a \le x \le b$. Fixing $y \in [a,b]$, we insert $f_y(x) := (y-x)_+^3/3!$ and get

$$\frac{(y-x)_+^3}{3!} = \frac{(y-a)^3}{3!} - (x-a)\frac{(y-a)^2}{2!} + \int_a^b (y-u)_+^1(x-u)_+^1 du$$
$$= \frac{1}{2\cdot 3!}(|y-x|^3 + (y-x)^3).$$

To the two right–hand sides of this identity we now apply functionals $\lambda_{X,M,\alpha}$ and $\lambda_{Y,N,\beta}$. This yields

$$\begin{aligned}
&\sum_{j=1}^{M}\sum_{k=1}^{N}\alpha_j\beta_k|x_j - y_k|^3 &&+ \sum_{j=1}^{M}\sum_{k=1}^{N}\alpha_j\beta_k(y_k - x_j)^3\\
=\ &\sum_{j=1}^{M}\sum_{k=1}^{N}\alpha_j\beta_k|x_j - y_k|^3 &&+\ 0\\
=\ &2\sum_{j=1}^{M}\sum_{k=1}^{N}\alpha_j\beta_k(y_k - a)^3 &&-\ 6\sum_{j=1}^{M}\sum_{k=1}^{N}\alpha_j\beta_k(x_j - a)(y_k - a)^2\\
& &&+\ 12\sum_{j=1}^{M}\sum_{k=1}^{N}\alpha_j\beta_k\int_a^b (y_k - u)_+^1(x_j - u)_+^1 du\\
=\ &0 - 0 &&+\ 12\int_a^b \sum_{j=1}^{M}\alpha_j(x_j - u)_+^1 \sum_{k=1}^{N}\beta_k(y_k - u)_+^1 du\\
=\ & && \frac{1}{12}\left(s''_{\lambda_{X,M,\alpha}}, s''_{\lambda_{Y,N,\beta}}\right)_{L_2(\mathbb{R})},
\end{aligned}$$

where the functions $s''_{\lambda_{X,M,\alpha}}$ and $s''_{\lambda_{Y,N,\beta}}$ are supported in $[x_1, x_M]$ and $[y_1, y_M]$, respectively, such that the L_2 integral can be restricted to $[a,b]$. □

Corollary 4. *The function $\Phi(x) := |x|^3$ is conditionally positive definite of order 2 on $\mathbb{R}$.*

Proof: Theorem 3 yields that the quadratic form (2) is positive semidefinite for $\Phi(x) = |x|^3$. If $\|\lambda_{X,M,\alpha}\|_\Phi$ vanishes, then $s''_{\lambda_{X,M,\alpha}} = 0$ holds, and the representation (5) as a piecewise linear function implies that all coefficients α_j must vanish. □

Theorem 5. *The spaces $\mathcal{F}$ of (1) and (4) coincide for $\Phi(x) = |x|^3$.*

Proof: By standard arguments, taking the L_2 closure of the space of piecewise linear continuous functions. □

We have cut the above proof short because we want to illustrate the use of the abstract space

$$\mathcal{G}_\Omega = \{f : \Omega \to \mathbb{R} : |\lambda(f)| \le C_f \|\lambda\|_\Phi \text{ for all } \lambda \in (\mathbb{P}^d_m)^\perp_\Omega\} \tag{7}$$

that occurs in the fundamental papers of Madych and Nelson [5,6] and provides another possibility to define a native space for Φ.

Theorem 6. *The spaces $\mathcal{F}$ of (1) and (4) coincide with $\mathcal{G}_{[a,b]}$ of (7) for $\Phi(x) = |x|^3$.*

Proof: We shall be somewhat more explicit this time, and start with

Lemma 7. *We have the inclusion $W^2_2[a,b] \subseteq \mathcal{G}_{[a,b]}$.*

Proof: Generalizing Taylor's formula for $f \in W^2_2[a,b]$, we find the identity

$$\begin{aligned}
\lambda_{X,M,\alpha}(f) &= \sum_{j=1}^M \alpha_j f(x_j) = 0 + \int_a^b f''(u) \sum_{j=1}^M \alpha_j (x_j - u)^1_+ du \\
&= \frac{1}{12}(f'', s''_{\lambda_{X,M,\alpha}})_{L_2[a,b]} \\
&\le \frac{1}{12}\|f''\|_{L_2[a,b]} \cdot \|s''_{\lambda_{X,M,\alpha}}\|_{L_2[a,b]} \\
&\le \frac{1}{\sqrt{12}}\|f''\|_{L_2[a,b]} \cdot \|\lambda_{X,M,\alpha}\|_\Phi.
\end{aligned}$$

for all $\lambda_{X,M,\alpha} \in (\mathbb{P}^d_m)^\perp_\Omega$. □

Lemma 8. *The other inclusion is $\mathcal{G}_{[a,b]} \subseteq W^2_2[a,b]$.*

Proof: Define the subspace $\mathcal{F}_0 := \{s''_\lambda : \lambda \in (\mathbb{P}^d_m)^\perp_\Omega\}$ of $L_2[a,b]$. It carries an inner product $(s''_\lambda, s''_\mu)_{L_2[a,b]} = 12(\lambda,\mu)_\Phi$ constructed from the inner product $(\cdot,\cdot)_\Phi$, and we define $\mathcal{F} := \overline{\mathcal{F}_0}$ to be the L_2 closure of $\mathcal{F}_0$ with respect to $(\cdot,\cdot)_{L_2[a,b]}$. Any $g \in \mathcal{G}_{[a,b]}$ defines a linear functional on $\mathcal{F}_0$ by

$$s''_\lambda \mapsto \lambda(g), \qquad \lambda \in (\mathbb{P}^d_m)^\perp_\Omega.$$

Here, we used that the map $\lambda \mapsto s''_\lambda$ is one–to–one. The above functional is continuous on $\mathcal{F}_0$ by definition of $\mathcal{G}_{[a,b]}$. Thus there is some $h_g \in \mathcal{F}_{[a,b]} = \overline{\mathcal{F}_0} \subseteq L_2[a,b]$ such that $\lambda(g) = (h''_g, s''_\lambda)_{L_2[a,b]}$ for all $\lambda \in (\mathbb{P}^d_m)^\perp_\Omega$, and we can assume $h_g \in W_2^2[a,b]$ because we can start with $h''_g = f_g \in L_2[a,b]$ and do integration. Now Taylor's formula for h_g yields

$$\lambda(h_g) = 0 + \frac{1}{12}(h''_g, s''_\lambda)_{L_2[a,b]} = \lambda(g)$$

for all $\lambda \in (\mathbb{P}^d_m)^\perp_\Omega$. By considering a fixed $\mathbb{P}^d_m$–unisolvent set $X = \{x_1, x_2, \ldots, x_M\}$ and functionals supported on $\{x, x_1, \ldots, x_M\}$ defined by interpolation in Lagrange form

$$\lambda(p) := p(x) - \sum_{j=1}^{M} u_j(x)p(x_j) = 0 \text{ for all } p \in \mathbb{P}^d_m,$$

we can form the interpolating polynomial

$$p_g(x) := \sum_{j=1}^{M} u_j(x)(g - h_g)(x_j)$$

to $g - h_g$ and use $\lambda(g - h_g) = 0$ to see that $g = h_g + p_g$ holds. □

Thus we have proven $\mathcal{G}_{[a,b]} = W_2^2[a,b]$ in detail. A full proof of Theorem 5 can be given by similar techniques. Furthermore, the abstractly defined spaces $\mathcal{G}_{[a,b]}$ and $\mathcal{F}$ can be proven to coincide using abstract Hilbert space arguments. □

This finishes the proof of Theorem 6, and we can reconstruct functions from $\mathcal{G}_{[a,b]} = W_2^2[a,b]$ from data at locations $a \leq x_1 < x_2 < \ldots < x_M \leq b$ uniquely by cubic splines of the form

$$s(x) = \sum_{j=1}^{M} \alpha_j |x - x_j|^3 + \sum_{k=0}^{1} \beta_k x^k \tag{8}$$

under the two additional conditions

$$\sum_{j=1}^{M} \alpha_j x_j^k = 0, \; k = 0, 1. \tag{9}$$

The representation (5) shows that these conditions imply linearity of s outside of $[x_1, x_M]$. Thus the solution is a *natural cubic spline*, defined on all of $\mathbb{R}$.

3. Working Equations

Note that the standard approach of radial basis functions cannot use the piecewise polynomial structure of the underlying basis function. Thus it has to start with the linear system

$$\begin{pmatrix} 0 & |x_1-x_2|^3 & \dots & |x_1-x_M|^3 & 1 & x_1 \\ |x_2-x_1|^3 & 0 & \dots & |x_2-x_M|^3 & 1 & x_2 \\ \vdots & \vdots & \ddots & \vdots & \vdots & \vdots \\ |x_M-x_1|^3 & |x_M-x_2|^3 & \dots & 0 & 1 & x_M \\ 1 & 1 & \dots & 1 & 0 & 0 \\ x_1 & x_2 & \dots & x_M & 0 & 0 \end{pmatrix} \begin{pmatrix} \alpha_1 \\ \alpha_2 \\ \vdots \\ \alpha_M \\ \beta_0 \\ \beta_1 \end{pmatrix} = \begin{pmatrix} f(x_1) \\ f(x_2) \\ \vdots \\ f(x_M) \\ 0 \\ 0 \end{pmatrix}$$

when interpolating a function f in $X = \{x_1, x_2, \dots, x_M\}$ by a function represented as in (8) with the conditions (9). This is a non-sparse system with entries increasing when moving away from the diagonal. The usual systems for cubic splines, however, are tridiagonal and diagonally dominant, allowing a solution at $\mathcal{O}(M)$ computational cost. By going over to a new basis of second divided differences of $|\,.\,|^3$, and by taking second divided differences of the above equations, one can bring them down to tridiagonal form. This was already pointed out in the early paper [7]. In modern terminology this is a preconditioning process, and it was thoroughly investigated for general radial basis functions by Jetter and Stöckler [3].

4. Power Functions

We now want to explain the technique of error analysis for radial basis functions in terms of cubic splines; the results will yield explicit local error bounds that seem to be new.

Definition 9. *For any general quasi–interpolant of the form*

$$s_{u,f}(x) := \sum_{j=1}^{M} u_j(x) f(x_j) \tag{10}$$

reproducing $\mathbb{P}_m^d$ *the power function*

$$P_u(x) := \sup_{|f|_\Phi \neq 0} \frac{|f(x) - s_{u,f}(x)|}{|f|_\Phi}.$$

is the pointwise norm of the error functional in $(\mathbb{P}_m^d)_\Omega^\perp$.

Introducing the Lagrange formulation

$$s_{u^*,f}(x) := \sum_{j=1}^{M} u_j^*(x) f(x_j)$$

of the radial basis function interpolant on $X = \{x_1, x_2, \dots, x_M\}$ to some function f, we have $u_j^*(x_k) = \delta_{jk}$, $1 \le j, k \le M$ by interpolation, but the functions u_j in (10) need not satisfy these conditions. The crucial fact for error analysis is the optimality principle

Theorem 10. *[5,11] For all x, the power function for radial basis function interpolation minimizes*

$$P_{u^*}(x) = \inf_u P_u(x),$$

where all quasi–interpolants (10) are admitted, provided that they reproduce $\mathbb{P}_m^d$.

This allows to insert piecewise polynomial quasi–interpolation processes in order to get explicit error bounds. It is a major problem to do this for scattered multivariate data. If we consider the univariate cubic spline case, things are much easier. We have to assume reproduction of linear polynomials and can apply Taylor's formula to the error:

$$\begin{aligned}
f(x) - s_{u,f}(x) &= f(x) - \sum_{j=1}^{M} u_j(x) f(x_j) \\
&= \int_a^b f''(t) \left((x-t)_+^1 - \sum_{j=1}^{M} u_j(x)(x_j - t)_+^1 \right) dt \\
&\le \|f''\|_{L_2[a,b]} \|(x-\cdot)_+^1 - \sum_{j=1}^{M} u_j(x)(x_j-\cdot)_+^1\|_{L_2[a,b]}.
\end{aligned}$$

Since the Cauchy–Schwarz inequality is sharp, we have

$$\begin{aligned}
P_u(x)^2 &= \|(x-\cdot)_+^1 - \sum_{j=1}^{M} u_j(x)(x_j-\cdot)_+^1\|_{L_2[a,b]}^2 \\
&= \int_a^b \left((x-t)_+^1 - \sum_{j=1}^{M} u_j(x)(x_j-t)_+^1 \right)^2 dt
\end{aligned}$$

as an explicit representation of any upper bound of the power function. We now fix an index k such that $x \in [x_{k-1}, x_k]$ and use the piecewise linear interpolant defined by

$$\begin{aligned}
u_{k-1}(x) &= \frac{x_k - x}{x_k - x_{k-1}} \\
u_k(x) &= \frac{x - x_{k-1}}{x_k - x_{k-1}} \\
u_j(x) &= 0, \quad j \neq k, \quad j \neq k-1.
\end{aligned}$$

This simplifies the integral to

$$P_u(x)^2 = \int_{x_{k-1}}^{x_k} \left((x-t)_+^1 - \frac{(x_k - x)(x_{k-1}-t)_+^1 + (x - x_{k-1})(x_k - t)_+^1}{x_k - x_{k-1}} \right)^2 dt,$$

because the integrand vanishes outside $[x_{k-1}, x_k]$. Furthermore, the bracketed function is a piecewise linear B–spline with knots x_{k-1}, x, x_k and absolute value $(x - x_{k-1})(x_k - x)/(x_k - x_{k-1})$ at $t = x$. This suffices to evaluate the integral as

$$P_u^2(x) = \frac{1}{3}(x_k - x)^3 \left(\frac{x - x_{k-1}}{x_k - x_{k-1}}\right)^2 + \frac{1}{3}(x - x_{k-1})^3 \left(\frac{x_k - x}{x_k - x_{k-1}}\right)^2,$$

$$P_u(x) = \frac{1}{\sqrt{3}} \frac{(x_k - x)(x - x_{k-1})}{\sqrt{x_k - x_{k-1}}}$$

to get

Theorem 11. *The natural cubic spline s_f interpolating a function f with $f'' \in L_2$ has the local error bound*

$$|f(x) - s_f(x)| \le \frac{1}{\sqrt{3}} \frac{(x_k - x)(x - x_{k-1})}{\sqrt{x_k - x_{k-1}}} \|f''\|_{L_2[x_{k-1}, x_k]} \tag{11}$$

for all x between two adjacent knots $x_{k-1} < x_k$.

If we define the local meshwidth $h_k := x_k - x_{k-1}$, we thus get local convergence of order 3/2 by

$$|f(x) - s_f(x)| \le \frac{h_k^{3/2}}{4\sqrt{3}} \|f''\|_{L_2[x_{k-1}, x_k]},$$

and 3/2 is known to be the optimal approximation order for functions with smoothness at most of the type $f'' \in L_2$. Taking the Chebyshev norm of f'' instead of the L_2 norm we similarly get

$$|f(x) - s_f(x)| \le \frac{1}{2}(x_k - x)(x - x_{k-1}) \|f''\|_{L_\infty[x_{k-1}, x_k]}$$

proving that natural cubic splines satisfy the standard error bound for piecewise linear interpolation.

Note that these bounds have the advantage to be local and of optimal order, but the disadvantage to be no better than those for piecewise linear interpolation. This is due to their derivation via local linear interpolation. To improve the bounds one must add more regularity to the function f, and this is the topic of the next section. Piecewise linear interpolation is optimal in its native space of functions f with $f' \in L_2$ and has optimal order 1/2 there, but at the same time it is used to provide bounds of the optimal order 3/2 for cubic spline interpolation of functions with $f'' \in L_2[a, b]$, forming the native space of cubic splines.

5. Improved Error Bounds

For interpolation of sufficiently smooth functions by cubic splines, the following facts about improved approximation orders in the L_∞ norm are well–known [1]:

a) The approximation order can reach four, but

b) order four is a saturation order in the sense that any higher order can only occur for the trivial case of $f \in \mathbb{P}_4^1$.

c) For orders greater than 2 on all of $[a, b]$ one needs additional boundary conditions. These can be imposed in several ways:

   c1) Conditions that force f to be linear outside of $[a, b]$ while keeping splines natural (i.e.: with linearity outside of $[a, b]$), or

   c2) additional interpolation conditions for derivatives at the boundary, or

   c3) periodicity requirements for both f and the interpolant.

d) General Hilbert space techniques including boundary conditions lead to orders up to 7/2 for functions with $f^{(4)} \in L_2$, but

e) techniques for order four (as known so far) require additional stability arguments (diagonal dominance of the interpolation matrix or bounds on elements of the inverse matrix providing exponential off-diagonal decay) and need $f \in C^4[a, b]$.

In this area the theory of radial basis functions still lags behind the theory of cubic splines, as far as interpolation of finitely may scattered data on a compact set is concerned (see [2] for better results in case of data on infinite grids). In particular, when specializing to cubic splines,

a) current convergence orders reach only 7/2 at most, and

b) there is no saturation result at all.

c) For orders greater than 2 on all of $[a, b]$ one needs additional boundary conditions which still deserve clarification in the general setting.

d) There is no proper theory for C^k functions.

Let us look at these problems from the classical cubic spline point of view. There, the standard technique [1] uses the orthogonality relation following from the minimum–norm property and applies integration by parts for functions with $f^{(4)} \in L_2[a, b]$. This yields

$$\begin{aligned} \|f'' - s_f''\|_2^2 &= (f'' - s_f'', f'' - s_f'')_2 = (f'' - s_f'', f'')_2 \\ &= (f - s_f, f^{(4)})_2 + f'' \cdot (f' - s_f')|_a^b \end{aligned}$$

and now it is clear that the various possibilities of imposing additional boundary conditions are just boiling down to the condition $f'' \cdot (f' - s_f')|_a^b = 0$ that we assume from now on. Then by Cauchy-Schwarz

$$\|f'' - s_f''\|_2^2 \leq \|f - s_f\|_2 \|f^{(4)}\|_2$$

and one has to evaluate the L_2 norm of the error. This is done by summing up the bound provided by Theorem 11 in a proper way:

$$\begin{aligned}
\|f - s_f\|_2^2 &\leq \sum_{k=2}^{M} \frac{\|f''\|^2_{L_2[x_{k-1},x_k]}}{3(x_k - x_{k-1})} \int_{x_{k-1}}^{x_k} (x_k - x)^2 (x - x_{k-1})^2 dx \\
&\leq \sum_{k=2}^{M} \frac{\|f''\|^2_{L_2[x_{k-1},x_k]}}{3h_k} \frac{h_k^5}{30} \\
&\leq \frac{h^4}{90} \|f''\|^2_{L_2[x_1,x_M]}
\end{aligned}$$

with

$$h := \max_{2\leq k\leq M} h_k = \max_{2\leq k\leq M} (x_k - x_{k-1}).$$

This step used that the right–hand side of (11) has a *localized* norm, and this fact is missing in general radial basis function cases. The technique works for norms that can be localized properly, and this was clearly pointed out in the recent paper [4]. In particular, it works for spaces that are norm–equivalent to Sobolev spaces, and this covers the case of Wendland's compactly supported unconditionally positive definite functions [10].

Using $s_{f-s_f} = 0$, one can replace $\|f''\|_2$ by $\|f'' - s_f''\|_2$ in the right–hand side and combine the above inequalities into

$$\begin{aligned}
\|f'' - s_f''\|_2^2 &\leq \frac{h^2}{\sqrt{90}} \|f'' - s_f''\|_2 \|f^{(4)}\|_2 \\
\|f'' - s_f''\|_2 &\leq \frac{h^2}{\sqrt{90}} \|f^{(4)}\|_2 \\
\|f - s_f\|_2 &\leq \frac{h^4}{90} \|f^{(4)}\|_2 \\
\|f - s_f\|_\infty &\leq \frac{h^{7/2}}{12\sqrt{30}} \|f^{(4)}\|_2.
\end{aligned}$$

This is how far we can get by Hilbert space techniques for cubic splines.

To see the problems occurring for general radial basis functions, we try to repeat the above argument, starting from the necessary and sufficient variational equation

$$(s_f, v)_\Phi = 0 \text{ for all } v \in \mathcal{F},\ v(X) = \{0\} \tag{12}$$

for any minimum–norm interpolant s_f based on data in $X = \{x_1, x_2, \ldots, x_M\}$. We want to apply this to $v := f - s_f$ and use integration by parts in some way or other. In view of Theorem 3 it is reasonable to impose the condition

$$(f, g)_\Phi = (Lf, Lg)_2 \text{ for all } f,\ g \in \mathcal{F}$$

with a linear operator L that maps $\mathcal{F}$ into some L_2 space, which we would prefer to be $L_2(\Omega)$, not $L_2(\mathbb{R}^d)$. But both the definition of L and the "localization" of the space are by no means trivial, since section 2 made use of very special properties of Φ that are not available in general. Looking back at cubic splines, we see that the variational equation (12) in its special form

$$(s_f'', v'')_{L_2(\mathbb{R})} = 0 \text{ for all } v'' \in L_2(\mathbb{R}^d),\ v(x_j) = 0$$

contains a lot of information:

1) the function s_f must be a cubic polynomial in all real intervals containing no knot (i.e. also in $(-\infty, x_1)$ and (x_M, ∞)), and
2) because of $v'' \in L_2(\mathbb{R}^d)$, the outer cubic pieces must be linear.

This follows from standard arguments in the calculus of variations, but it does not easily generalize, because the general form (12) of the variational equation does not admit the same analytic tools as in the cubic case.

If (generalized) direct and inverse Fourier transforms are properly introduced, the radial basis function literature [6,9,11] uses the operator

$$Lf := \left(\frac{f^\wedge}{\sqrt{\Phi^\wedge}}\right)^\vee$$

that (up to a constant factor) is a second derivative operator in the cubic case, as expected. This is due to the fact that $\Phi^\wedge = \|.\|_2^{-4}$ for $\Phi = \|.\|_2^3$ in one dimension, and up to a constant factor. Unconditionally positive definite smooth functions Φ like Gaussians $\Phi(x) = \exp(-\|x\|_2^2)$ or Wendland's function $\Phi(x) = (1 - \|x\|_2)_+^4(1 + 4\|x\|_2)$ have classical Fourier transforms with proper decay at infinity, and then the above operator L is well–defined even without resorting to generalized Fourier transforms. However, it does not represent a classical differential operator, but rather an awkward pseudodifferential operator, making the analysis of the variational equation (12) and the corresponding boundary conditions a hard task. It would be nice to conclude from (12) that a *weak boundary condition* of the form

$$L^*Ls_f = 0 \text{ outside of } \Omega$$

holds, or even a *strong boundary condition*

$$Ls_f = 0 \text{ outside of } \Omega$$

as in the cubic spline case, where L^* is the L_2 adjoint of L. So far, the proper characterization of necessary boundary conditions following from (12) seems to be an open question whose answer would improve our understanding of the error behavior of radial basis functions considerably.

The paper [9] avoids these problems by simply assuming that the function f satisfies the conditions

$$L^*Lf := \frac{f^\wedge}{\Phi^\wedge} \in L_2(\mathbb{R}^d),\ \ \operatorname{supp} L^*Lf = \operatorname{supp}\left(\frac{f^\wedge}{\Phi^\wedge}\right) \subset \Omega \subset \mathbb{R}^d.$$

This allows to mimic part of the cubic spline argument in the form

$$\begin{aligned}|f - s_f|_\Phi^2 &= (f - s_f, f - s_f)_\Phi = (f - s_f, f)_\Phi \\ &= (L(f - s_f), Lf)_{L_2(\mathbb{R}^d)} = (f - s_f, L^*Lf)_{L_2(\mathbb{R}^d)} \\ &= (f - s_f, L^*Lf)_{L_2(\Omega)} \leq \|f - s_f\|_{L_2(\Omega)} \|L^*Lf\|_{L_2(\Omega)},\end{aligned}$$

and we are left with the summation argument to form the L_2 norm of the error over Ω. But in general we cannot use the localization property of the right-hand side of (11), since we are stuck with

$$|f(x) - s_f(x)| \leq P(x)|f - s_f|_\Phi$$

for all $x \in \Omega$. This only yields

$$\|f - s_f\|_{L_2(\Omega)} \leq \|P\|_{L_2(\Omega)} |f - s_f|_\Phi$$

and is off by a factor $\sqrt{h}$ if written down in the cubic spline setting, ignoring the additional information. Still, this technique allows to prove error bounds of the form

$$\begin{aligned}|f(x) - s_f(x)| &\leq P(x)\|P\|_{L_2(\Omega)} \|L^*Lf\|_{L_2(\Omega)} \\ \|f - s_f\|_{L_2(\Omega)} &\leq \|P\|_{L_2(\Omega)}^2 \|L^*Lf\|_{L_2(\Omega)}\end{aligned}$$

that roughly double the previously available orders, but still are off from the optimal orders in the cubic case by 1/2 or 1, respectively, because P, being optimally bounded as in (11), has only $\mathcal{O}(h^{3/2})$ behavior. It it reasonable to conjecture that the above inequalities are off from optimality by orders $d/2$ and d in a d-variate setting, respectively. We hope that this presentation helps to clarify the arising problems and to encourage further research in this direction.

Acknowledgements. Help in proofreading was provided by H. Wendland.

References

1. Ahlberg J. H., Nilson E. N., and Walsh J. L., *The Theory of Splines and Their Applications*, Academic Press, 1967
2. Buhmann M. D., Multivariable Interpolation using Radial Basis Functions, University of Cambridge, 1989
3. Jetter K. and Stöckler J., A Generalization of de Boor's stability result and symmetric preconditioning , Advances in Comp. Math. **3** (1995) 353–367
4. Light W. A. and Wayne H., Some remarks on power functions and error estimates for radial basis function approximation, J. Approx. Theory, to appear
5. Madych W. R. and Nelson S. A., Multivariate interpolation and conditionally positive definite functions, Approx. Theory Appl. **4** (1988) 77–89

6. Madych W. R. and Nelson S. A., Multivariate interpolation and conditionally positive definite functions II, Math. Comp. **54** (1990) 211–230
7. Schaback R., Konstruktion und algebraische Eigenschaften von M–Spline–Interpolierenden, Numer. Math. **21** (1973) 166–180
8. Schaback R., Optimal recovery in translation–invariant spaces of functions, Annals of Numerical Mathematics **4** (1997) 547–555
9. Schaback R., Improved error bounds for scattered data interpolation by radial basis functions, 1997, submitted
10. Wendland H., Sobolev–type error estimates for interpolation by radial basis functions, in: *Curves and Surfaces in Geometric Design*, A. Le Méhauté and C. Rabut and L.L. Schumaker (eds.), Vanderbilt University Press, Nashville, TN, 1997
11. Wu Z. and Schaback R., Local error estimates for radial basis function interpolation of scattered data, IMA J. Numer. Anal. **13** (1993) 13–27

Robert Schaback
Institut für Numerische und Angewandte Mathematik
Universität Göttingen
Lotzestraße 16–18
D-37073 Göttingen
e-mail: schaback@math.uni-goettingen.de
Web: http://www.num.math.uni-goettingen.de/schaback

Wavelet Modelling of High Resolution Radar Imaging and Clinical Magnetic Resonance Tomography

Walter Schempp

Abstract. The speed with which clinical magnetic resonance imaging (MRI) systems spread throughout the world was phenomenal. Coherent wavelets allow for a unified model of the multichannel perfect reconstruction analysis–synthesis filter bank of high resolution radar imaging and MRI. The geometric quantization construction of matched bank filters depends upon the Kepplerian spatiotemporal strategy which succeeds in the synchronous and stroboscopic summation over phase histories in local frequency encoding channels. The Kepplerian planetary clockwork of quantum holography is implemented in symplectic affine planes by Fourier analysis of the Heisenberg nilpotent Lie group G, and the associated reconstructing symbolic calculus on the selected energetic stratum of the unitary dual $\hat{G}$ or the quantized calculus of the $C^{\star}$–algebra of G. The neural network performed by quantum holograms allows for localization of cortical activations of the human brain.

A radar system employs a directional antenna that radiates energy within a narrow beam in a known direction. One unique feature of the synthetic aperture radar (SAR) imaging modality is that its spatial resolution capability is independent of the platform altitude over the subplatform nadir track. This is a result of the fact that the SAR image is formed by simultaneously storing the phase histories and the differential time delays in local frequency encoding subbands of wideband radar, none of which is a function of the range from the radar sensor to the scene. It is this unique capability which allows the acquisition of high resolution images from satellite altitude as long as the received echo response has sufficient strength above the noise level.

The Kepplerian spatiotemporal strategy of physical astronomy succeeds in the synchronous and stroboscopic summation over phase histories in local frequency

Multivariate Approximation and Splines
G. Nürnberger, J. W. Schmidt, and G. Walz (eds.), pp. 259–274.

ISBN 3-7643-5654-5.

encoding channels, and suggests the implementation of a matched filter bank by orbit stratification in a symplectic affine plane. Application of this procedure leads to the landmark observation of the earliest SAR pioneer, Carl A. Wiley, that motion is the solution of the high resolution radar imagery and phased array antenna problem of holographic recording. Whereas the Kepplerian spatiotemporal strategy transferred to quantum holography is realized in SAR imaging by the range Doppler principle [2], [12], it is the solvable affine Lauterbur encoding principle which takes place in clinical MRI [4], [17]. At the background of both high resolution imaging techniques lies the construction of a multichannel coherent wavelet perfect reconstruction analysis–synthesis filter bank of matched filter type [3]. Beyond these applications to local frequency encoding subbands, the Kepplerian spatiotemporal strategy leads to the concept of Feynman path integral or summation over phase histories.

As approved by quantum electrodynamics and photonics [8], [19], geometric quantization allows for a semi–classical approach to the interference pattern of quantum holography [16], [17]. Indeed, the unitary dual $\hat{G}$ of the Heisenberg nilpotent Lie group G consisting of the equivalence classes of irreducible unitary linear representations of G allows for a coadjoint orbit fibration by symplectic affine planes $\mathcal{O}_\nu$, $(\nu \neq 0)$ spatially located in tomographic slices inside the vector space dual $\mathrm{Lie}(G)^\star$ of the real Heisenberg Lie algebra $\mathrm{Lie}(G)$ [13]. This is a consequence of the Kirillov homeomorphism

$$\hat{G} \longrightarrow \mathrm{Lie}(G)^\star/\mathrm{CoAd}_G(G)$$

which is at the basis of the geometric quantization in terms of the foliation of the projective space $\mathbf{P}(\mathbf{R} \times \mathrm{Lie}(G)^\star)$ generated by the coadjoint action of G. In terms of standard coordinates, the Heisenberg group G consists of the set of unipotent matrices

$$\left\{ \begin{pmatrix} 1 & x & z \\ 0 & 1 & y \\ 0 & 0 & 1 \end{pmatrix} \,\middle|\, x, y, z \in \mathbf{R} \right\}$$

under the matrix multiplication law. If the nilpotent matrices $\{P, Q, I\}$ denote the canonical basis of the three–dimensional real vector space $\mathrm{Lie}(G)$, where

$$\exp_G P = \begin{pmatrix} 1 & 1 & 0 \\ 0 & 1 & 0 \\ 0 & 0 & 1 \end{pmatrix}, \quad \exp_G Q = \begin{pmatrix} 1 & 0 & 0 \\ 0 & 1 & 1 \\ 0 & 0 & 1 \end{pmatrix}, \quad \exp_G I = \begin{pmatrix} 1 & 0 & 1 \\ 0 & 1 & 0 \\ 0 & 0 & 1 \end{pmatrix}$$

holds under the matrix exponential diffeomorphism $\exp_G : \mathrm{Lie}(G) \longrightarrow G$, the linear commutator provides the canonical commutation relations of quantum mechanics

$$[P, I] = [Q, I] = 0, \quad [P, Q] = I.$$

Of course, $\mathrm{Lie}(G)$ fails to be an associative algebra, but the Jacobi identity

$$\Big[P, [Q, I]\Big] + \Big[Q, [I, P]\Big] + \Big[I, [P, Q]\Big] = 0$$

is trivially satisfied. As a consequence, quantum mechanics is steeped in symplecticism, owing to the anti–symmetry of the Lie bracket operation occurring in the canonical commutation relations and which characterize Lie(G).

- The center of Lie(G) which is given by the one–dimensional commutator ideal

$$[\mathrm{Lie}(G), \mathrm{Lie}(G)] = \mathrm{Lie}(G)^{[2]} = \mathbf{R}I$$

  is aligned with the direction of the external static magnetic field along the longitudinal axis of the magnet bore.
- The subset $\{P, Q\}$ of the set of nilpotent matrices $\{P, Q, I\}$ forms a symplectic basis of the quotient $\mathrm{Lie}(G)/\mathrm{Lie}(G)^{[2]}$.

The Lie bracket $[.,.]$ which denotes the linear commutator acts as the Poisson bracket $\{.,.\}$ of classical Hamiltonian mechanics on the real vector space $\mathcal{C}^{\infty}_{\mathbf{R}}(\mathbf{R} \oplus \mathbf{R})$ of smooth real–valued functions on the projectively immersed, symplectic affine plane $\mathbf{R} \oplus \mathbf{R}$. For proton densities $p, q \in \mathcal{C}^{\infty}_{\mathbf{R}}(\mathbf{R} \oplus \mathbf{R})$, the Poisson bracket reads

$$\{p, q\} = \frac{\partial p}{\partial y}\frac{\partial q}{\partial x} - \frac{\partial p}{\partial x}\frac{\partial q}{\partial y}.$$

The Poisson bracket is compatible with the Leibniz rule of the derivation $\{r, .\}$. It is given by

$$\{r, pq\} = \{r, p\}\, q + p\, \{r, q\}$$

for the pointwise multiplication of functions $r, p, q \in \mathcal{C}^{\infty}_{\mathbf{R}}(\mathbf{R} \oplus \mathbf{R})$. Moreover, it forms a symplectic affine invariant of the plane $\mathbf{R} \oplus \mathbf{R}$. The density differentials

$$\mathrm{d}p = \frac{\partial p}{\partial x}\,\mathrm{d}x + \frac{\partial p}{\partial y}\,\mathrm{d}y, \quad \mathrm{d}q = \frac{\partial q}{\partial x}\,\mathrm{d}x + \frac{\partial q}{\partial y}\,\mathrm{d}y$$

act as symplectic gradients according to the identity

$$\mathrm{d}p \wedge \mathrm{d}q = -\{p, q\} \,.\, \omega_1$$

for $p, q \in \mathcal{C}^{\infty}_{\mathbf{R}}(\mathbf{R} \oplus \mathbf{R})$, and give rise via contraction to the rotational curvature form

$$\omega_1 = \mathrm{d}x \wedge \mathrm{d}y$$

of the generic coadjoint orbit $\mathcal{O}_1$ of G. Then the integrability condition of the Poincaré lemma reads

$$\mathrm{d}\{p, q\} = \{\mathrm{d}p, \mathrm{d}q\}.$$

The action of the symplectic gradients establishes that the bracket transition

$$\Phi : \{.,.\} \mapsto [.,.]$$

from Poisson to Lie is at the basis of the solvable affine Lauterbur subband encoding technique of spatial localization by linear magnetic field gradients operating on the L^2–sections of a homogeneous hologram line bundle. The Lie algebra morphism Φ from $\mathcal{C}^\infty_{\mathbf{R}}(\mathbf{R} \oplus \mathbf{R})$ endowed with the Poisson bracket of classical Hamiltonian mechanics, to the real vector space of vector fields on the symplectic affine plane $\mathbf{R} \oplus \mathbf{R}$ with respect to their natural Lie bracket, is given by the symplectically transposed, or alternatively twisted exterior differentiation

$$\Phi = \mathrm{d}^t.$$

It takes the first coordinate function $x = (x, 0)$ of the symplectic affine plane $\mathbf{R} \oplus \mathbf{R}$ to $-Q \in \mathrm{Lie}(G)$, and assigns to the second coordinate function $y = \binom{y}{0}$ of the symplectic affine plane $\mathbf{R} \oplus \mathbf{R}$ the matrix $P \in \mathrm{Lie}(G)$. Thus the canonical commutation relations of quantum mechanics read in Poisson bracket notation

$$\{x, x\} = \{y, y\} = 0, \quad \{y, x\} = 1.$$

It follows from the identities

$$\Phi(y) = P, \qquad \Phi(x) = -Q,$$

which read explicitly on the Lie algebra level

$$\Phi\left(\begin{pmatrix} y \\ 0 \end{pmatrix}\right) = \begin{pmatrix} 0 & 1 & 0 \\ 0 & 0 & 0 \\ 0 & 0 & 0 \end{pmatrix}, \quad \Phi((x,0)) = \begin{pmatrix} 0 & 0 & 0 \\ 0 & 0 & -1 \\ 0 & 0 & 0 \end{pmatrix},$$

that the linear differential forms $\mathrm{d}y$ and $-\mathrm{d}x$ are the coordinates with respect to the basis $\{P^\star, Q^\star\}$, dual to the canonical basis $\{P, Q\}$ of the symplectic affine plane $\mathrm{Lie}(G)/\mathrm{Lie}(G)^{[2]}$.

Let $\mathrm{Z}^1(\mathcal{O}_\nu)$ denote the real Lie algebra of the de Rham cocycles of degee 1 of the affine symplectic plane $\mathcal{O}_\nu$. It follows from the exact sequence

$$\{0\} \longrightarrow \mathbf{R} \longrightarrow \mathcal{C}^\infty_{\mathbf{R}}(\mathrm{Lie}(G)/\mathrm{Lie}(G)^{[2]}) \xrightarrow{\Phi} \mathrm{Z}^1(\mathcal{O}_\nu) \longrightarrow \{0\}$$

that $\mathcal{C}^\infty_{\mathbf{R}}(\mathrm{Lie}(G)/\mathrm{Lie}(G)^{[2]})$ is a one–dimensional central extension of the Lie algebra of vector fields $\mathrm{Z}^1(\mathcal{O}_\nu)$. The exact sequence can be embedded in a de Rham cohomology sequence

$$\{0\} \to \mathrm{H}^0(\mathrm{Lie}(G)/\mathrm{Lie}(G)^{[2]}) \to \mathcal{C}^\infty_{\mathbf{R}}(\mathrm{Lie}(G)/\mathrm{Lie}(G)^{[2]}) \xrightarrow{\Phi} \mathrm{Z}^1(\mathcal{O}_\nu) \to \mathrm{H}^1(\mathcal{O}_\nu) \to \{0\}.$$

The affine group

$$\mathbf{GA}(\mathbf{R}) \hookrightarrow \mathbf{GL}(2, \mathbf{R})$$

of the real line $\mathbf{R}$, the "$at + b$" group of affine linear transformations of the line, forms a two–dimensional solvable Lie group which is given by the matrices

$$\mathbf{GA}(\mathbf{R}) = \left\{ \begin{pmatrix} a & b \\ 0 & 1 \end{pmatrix} \,\middle|\, a \neq 0, \, b \in \mathbf{R} \right\}$$

of dilation $a \in \mathbf{R}, a \neq 0$, shift $b \in \mathbf{R}$, and off–set term 1. Its operation on the Hamiltonian action of Lie(G) by symplectomorphisms on $\mathrm{Z}^1(\mathcal{O}_\nu)$ allows to embed the solvable affine Lauterbur subband encoding technique of wideband signal wavelets for spatial localization into the real vector space dual Lie$(G)^\star$.
In order to do this, complete the basis $\{P^\star, Q^\star\}$ to the basis $\{P^\star, Q^\star, I^\star\}$ of the real vector space dual Lie$(G)^\star$ such that it is dual to the canonical basis $\{P, Q, I\}$ of Lie(G). Different from the saddle–shaped, non–uniform magnetic flux density, chosen to achieve spatial localization by Damadian, the other visionary MRI pioneer [10], the solvable affine Lauterbur magnetic field gradients are represented by linear differential forms

$$\alpha = \mathrm{d}y, \quad \beta = -\mathrm{d}x$$

imposed on the homogeneous symplectic affine plane spanned by its symplectic basis $\{P^\star, Q^\star\}$ in Lie$(G)^\star$. Then the exterior product mapping $\wedge^2\Phi$ provides the distinguished closed exterior differential 2–form

$$\omega_\nu = \nu \,.\, \alpha \wedge \beta = \nu \,.\, \mathrm{d}x \wedge \mathrm{d}y \quad (\nu \neq 0)$$

on $\mathcal{O}_\nu \hookrightarrow$ Lie$(G)^\star$.

- The bandwidth of the transmitted radiofrequences controls the width of the selected tomographic slice.
- After tomographic slice selection by resonance with the center frequency ν, the operation of the solvable affine Lie group $\mathbf{GA}(\mathbf{R})$ on the spatial linear magnetic field gradients $-\mathrm{d}y$ and $\mathrm{d}x$, respectively, inside the symplectic affine planes of incidence $\mathcal{O}_\nu$ $(\nu \neq 0)$, is switched in sequence instead of simultaneously.
- Pulsing instead of continuous ramp changes by the action of the solvable affine Lie group $\mathbf{GA}(\mathbf{R})$ allows for sequencing the spatial linear magnetic field gradients.

When the spatial linear magnetic field gradients are removed, the nuclear spins return to the original local frequency of Larmor precession, but retain the phase shift caused by the phase encoding linear gradient. Most clinical scanner systems use 128, 192, 256, or even 512 incremental phase encoding steps to acquire the signal wavelet data of a scan.
Due to the relaxation phenomena [17], MRI is basically a slow scan modality. Conventional SE methods require long data acquisition times since only one hologram line is collected for each excitation pulse. Fast spin echo (FSE) imaging, however, markedly reduces scan times. In FSE imaging, the phase encoding linear gradient for each echo in the pulse train is changed so that multiple channel data are acquired within a given repetition time. The excitation profiles can be performed by use of multiwavelets. Each echo in the train contributes both spatial resolution and image contrast. Although FSE methods have been employed in neurofunctional MRI and have even shown advantages with respect to selectivity

for capillary blood oxygenation level dependent (BOLD) effect, the signal wavelet increases are significantly smaller than for gradient recalled echoes (GREs) under the condition of good macroscopic magnetic flux density homogeneity.
Recall that for a non–trivial transvection a basis of the underlying vector space may be chosen such that with respect to this basis the associated matrix has all entries on its main diagonal equal 1, and all the other matrix entries, with exactly one exception, equal zero [17]. The coordinate functions of the transvections which give rise to the Hamiltonian action of G by symplectomorphisms on $Z^1(\mathcal{O}_\nu)$ now read on the Lie group level

$$\begin{pmatrix} y \\ \nu \end{pmatrix} = \begin{pmatrix} 1 & 1 & 0 \\ 0 & 1 & 0 \\ 0 & 0 & 1 \end{pmatrix}, \qquad (x, \nu) = \begin{pmatrix} 1 & 0 & 0 \\ 0 & 1 & -1 \\ 0 & 0 & 1 \end{pmatrix}.$$

The homogeneous hologram line bundle splitting on which the solvable affine Lauterbur subband encoding is based is then performed by multiplication on the left and right of the column and row vectors, respectively, by the gradient matrices

$$\begin{pmatrix} a & 1 \\ 0 & 1 \end{pmatrix} \in \mathbf{GA}(\mathbf{R}), \quad \begin{pmatrix} a & 0 \\ 1 & 1 \end{pmatrix} \in \mathbf{GA}(\mathbf{R})^t \qquad (a \neq 0)$$

of polarity

$$\operatorname{sign} \det \begin{pmatrix} a & 1 \\ 0 & 1 \end{pmatrix} = \operatorname{sign} a,$$

and normalized shift $b = 1$. In gradient recalled echo (GRE) imaging, GREs are formed following a single radiofrequency pulse with the echo generated by refocusing spins using rewinder gradient matrices of opposite polarity

$$\begin{pmatrix} -a & 1 \\ 0 & 1 \end{pmatrix} \in \mathbf{GA}(\mathbf{R}), \quad \begin{pmatrix} -a & 0 \\ 1 & 1 \end{pmatrix} \in \mathbf{GA}(\mathbf{R})^t \qquad (a \neq 0),$$

respectively. Indeed, the transition to the frequency domain is performed by the affine wavelet transform on the positive half–line $\mathbf{R}_+$. By structure transportation to the complex Hilbert space $L^2_{\mathbf{C}}(\mathbf{R}_+)$ of transversal sections of square integrable functions of positive energy, the affine wavelet transform is deduced from the spin labelled holomorphic discrete series representation of the special linear group $\mathbf{SL}(2, \mathbf{R})$ over $\mathbf{R}$ onto which the metaplectic group $\mathbf{Mp}(2, \mathbf{R})$ projects with kernel $\mathbf{Z}/2\mathbf{Z}$ as the group of dynamical symmetries [17].
In order to perform the structure transport onto the group of dynamical symmetries, apply the Gram–Schmidt orthonormalization procedure to $\mathbf{R} \oplus \mathbf{R}$ in order to generate the Iwasawa topological decomposition of the Lie group $\mathbf{SL}(2, \mathbf{R})$. Denote the decomposition by

$$K.\underbrace{A_0.N_0}_{},$$

where

$$N_0 = \big[\mathbf{GA}(\mathbf{R}), \mathbf{GA}(\mathbf{R})\big]$$

is the commutator subgroup of the solvable affine Lie group $\mathbf{GA}(\mathbf{R})$. It acts smoothly and transitively by bilinear maps on the projective real line $\mathbf{P}_1(\mathbf{R})$, hence biholomorphically and transitively by means of Möbius fractional linear transformations on the open upper complex half–plane

$$\mathcal{O}_+ = \{w \in \mathbf{C} \mid \Im w > 0\}.$$

The exterior differential 2–form on $\mathcal{O}_+$ given as

$$\frac{1}{y^2} \,.\, \mathrm{d}x \wedge \mathrm{d}y = \frac{i}{2y^2} \,.\, \mathrm{d}w \wedge \mathrm{d}\bar{w}$$

is invariant under the action of the group $\mathbf{SL}(2,\mathbf{R})$ by automorphisms of $\mathcal{O}_+$, so that the measure

$$\frac{1}{y^2} \,.\, \mathrm{d}x \otimes \mathrm{d}y$$

associated to the Poincaré metric of $\mathcal{O}_+$ is invariant under the biholomorphic and transitive action of $\mathbf{SL}(2,\mathbf{R})$. The maximal compact, connected component K of the Iwasawa decomposition is isomorphic to the Lie group $\mathbf{SO}(2,\mathbf{R})$ of planar rotations, hence to the reference unit circle $\mathbf{T} \hookrightarrow \mathbf{C}$ of the eccentric Kepplerian clockwork dynamics. Because the stabilizer of the point $w = i \in \mathcal{O}_+$ is given by $\mathbf{SO}(2,\mathbf{R})$, it follows the homogeneous manifold representation

$$\mathcal{O}_+ \cong \mathbf{SL}(2,\mathbf{R})/\mathbf{SO}(2,\mathbf{R}).$$

The special importance of the open upper complex half–plane $\mathcal{O}_+$ for linear magnetic field actions stems from this homogeneous manifold representation which is a special case of the general principle:

Semi–simple real Lie group/maximal compact subgroup $\cong$ symmetric manifold.

The orbit of the point $w = ir \in \mathcal{O}_+$ where $r \in]0,1]$ is formed by the circle of center $i\xi \in \mathcal{O}_+$, where the positive real number ξ is given by the Joukowski transform

$$\xi = \frac{1}{2}\left(r + \frac{1}{r}\right)$$

of external magnetic flux density correction, and has the two points $\{ir, \frac{i}{r}\}$ located on the imaginary axis as diametrically opposed points.

- The points $\{ir, \frac{i}{r}\}$ serve as off–center shimming control points of the magnetic flux density in an azimuthal polar plot transverse to the longitudinal axis of the magnet bore.

The solvable component of the Iwasawa decomposition, the underbraced closed subgroup

$$A_0.N_0 = \left\{ \begin{pmatrix} a & b \\ 0 & \frac{1}{a} \end{pmatrix} \;\middle|\; a \neq 0,\, b \in \mathbf{R} \right\}$$

of $\mathbf{SL}(2,\mathbf{R})$, forms the stabilizer of the point $\infty \in \mathbf{P}_1(\mathbf{R})$. Let $A_0.N_0 \hookrightarrow \mathbf{SL}(2,\mathbf{R})$ act on the point $w = i \in \mathcal{O}_+$ as follows:

$$\begin{pmatrix} a & b \\ 0 & \frac{1}{a} \end{pmatrix}.i = \frac{ai+b}{\frac{1}{a}} = a^2 i + ab, \qquad (a \neq 0).$$

Then the solvable affine Lie group $\mathbf{GA}(2,\mathbf{R})$ acts transitively on the planar coadjoint orbit $\mathcal{O}_+$, and $\mathbf{GA}(2,\mathbf{R})$ inherits its Haar measure from one half of the measure of $\mathcal{O}_+$ associated to the Poincaré metric. For this action which admits the 1–cocycle

$$\begin{pmatrix} a & b \\ 0 & \frac{1}{a} \end{pmatrix} \mapsto \frac{1}{a^{2m+1}}, \qquad (a \neq 0),$$

the functions

$$\mathbf{R}_+ \ni t \mapsto e^{\pi i w t} t^m \qquad \left(w \in \mathcal{O}_+\right)$$

give rise to a total set in the Hilbert space $L^2_{\mathbf{C}}(\mathbf{R}_+)$ of square integrable, complex–valued functions of positive energy, provided $2m \in \mathbf{N}$, $m > 0$.

- The affine wavelet transform is generated by unitary induction from the Pontryagin dual $\hat{K}$ of the maximal compact subgroup K of $\mathbf{SL}(2,\mathbf{R})$.

By inducing the unitary character of order $2m+1$ of the subgroup $K \hookrightarrow \mathbf{SL}(2,\mathbf{R})$, the kernel function attached to the affine wavelet transform of gradient matrix $\begin{pmatrix} a & 1 \\ 0 & 1 \end{pmatrix} \in \mathbf{GA}(\mathbf{R})$ and the spin label of half–integral or integral value

$$m \in \frac{1}{2}\mathbf{N}, \ m > 0$$

takes for the choice $w = i \in \mathcal{O}_+$ the following form in the transversal line bundle section $L^2_{\mathbf{C}}(\mathbf{R}_+)$:

$$\mathbf{R}_+ \ni t \mapsto a^{2m+1} e^{2\pi i a t} e^{-2\pi a^2 t} t^m.$$

In order to trivialize in biomedical magnetic resonance spectroscopic imaging the cocycle of the projectivized representation of spin label $2m \in \frac{1}{2}\mathbf{N}$, $m > 0$, the special linear group $\mathbf{SL}(2,\mathbf{R})$ has to replaced by the group extension $\mathbf{Mp}(2,\mathbf{R})$ which is pointwise attached to the one–dimensional center C of G as the group of dynamical symmetries.

The affine wavelet transform includes an attenuation as well as a dephasing factor. The wavelets which are diffracted by the linear gradient switching for spatial localization purposes die out due to the attenuation factor. The linear gradient reflections

$$\begin{pmatrix} a & 1 \\ 0 & 1 \end{pmatrix} \mapsto \begin{pmatrix} -a & 1 \\ 0 & 1 \end{pmatrix}, \quad \begin{pmatrix} a & 0 \\ 1 & 1 \end{pmatrix} \mapsto \begin{pmatrix} -a & 0 \\ 1 & 1 \end{pmatrix}$$

are able to rewind those phases which are caused by linear gradient switching, hence to eliminate the dispersion spectrum. The spectral leakage effect due to phase dispersion can be avoided by a repeating the data acquisition experiment with linear magnetic field gradients of opposite polarity. The close analogy to the dephasing effect originating from relaxation, and the technique of SE rephasing should be observed. Of course, the linear rewinding procedure has to be in synergy with the successive radiofrequency pulse excitations of the spin isochromats in order to avoid that dispersive wavelets creep into the quantum hologram and then decrease the signal–to–noise ratio.

- Due to the strategy of untwisting by means of reversing the polarity of the linear magnetic field gradient for a time half the length of the $\frac{\pi}{2}$ pulse, or applying a π pulse, the lift out of the tomographic slice $\mathcal{O}_\nu$ is brought back and frozen in the rotating coordinate frame of quadrature reference at center frequency $\nu \neq 0$.
- Stepwise incrementation of the linear phase encoding gradient in conjunction with polarity reversal or phase conjugation flip allows for recording the quantum hologram.

In the presence of the frequency gradient of gradient matrix $\begin{pmatrix} -1 & 0 \\ 1 & 1 \end{pmatrix} \in \mathbf{GA}(\mathbf{R})^t$ of polarity -1 which acts via transport by the morphism

$$\Phi : \mathcal{C}^\infty_{\mathbf{R}}\big(\mathrm{Lie}(G)/\mathrm{Lie}(G)^{[2]}\big) \longrightarrow \mathrm{Z}^1(\mathcal{O}_\nu)$$

on the linear differential form

$$\nu(-y\,.\,\mathrm{d}x + x\,.\,\mathrm{d}y) \in \mathrm{Z}^1(\mathcal{O}_\nu),$$

the action of the stepping phase encoding gradient matrices $\begin{pmatrix} a & 1 \\ 0 & 1 \end{pmatrix} \in \mathbf{GA}(\mathbf{R})$ via Φ takes place. Then the quantum hologram of the solvable affine Lauterbur subband encoding technique of spatial localization in the affine symplectic plane $\mathcal{O}_\nu$ is generated by $\wedge^2\Phi$ and takes the form

$$K^\nu(x,y)\,.\,\omega_\nu \qquad (\nu \neq 0)$$

with density $K^\nu \in L^2_{\mathbf{C}}\big(\mathbf{R} \oplus \mathbf{R}\big)$ and non–zero $\nu \in \mathbf{R}$.

- During the presence of a readout gradient of opposite polarity $+1$, the quantum hologram $K^\nu(x,y)\,.\,\omega_\nu$ collects in the affine symplectic plane $\mathcal{O}_\nu$ the readout data and stepped phase data of the mode spectra at center frequency $\nu \neq 0$.

It will be established later on that the kernel function K^ν admits Hermitian symmetry. Harmonic analysis on the Heisenberg nilpotent group G allows for a detailed analysis of the kernel K^ν in terms of the proton density function [15], [17].

The coadjoint action of G on $\mathrm{Lie}(G)^\star$ is given by

$$\mathrm{CoAd}_G\begin{pmatrix}1 & x & z\\ 0 & 1 & y\\ 0 & 0 & 1\end{pmatrix} = \begin{pmatrix}1 & 0 & -y\\ 0 & 1 & x\\ 0 & 0 & 1\end{pmatrix}.$$

Therefore the action CoAd_G reads in terms of the coordinates $\{\alpha, \beta, \nu\}$ with respect to the dual basis $\{P^\star, Q^\star, I^\star\}$ of the real vector space dual $\mathrm{Lie}(G)^\star$ as follows:

$$\mathrm{CoAd}_G\begin{pmatrix}1 & x & z\\ 0 & 1 & y\\ 0 & 0 & 1\end{pmatrix}(\alpha P^\star + \beta Q^\star + \nu I^\star) = (\alpha - \nu y)P^\star + (\beta + \nu x)Q^\star + \nu I^\star.$$

The linear varieties

$$\mathcal{O}_\nu = \mathrm{CoAd}_G(G)(\nu I^\star) = \mathbf{R}P^\star + \mathbf{R}Q^\star + \nu I^\star \quad (\nu \neq 0)$$

actually are symplectic affine planes in the sense that they are in the natural way compatibly endowed with both the structure of an affine plane and a symplectic structure. Their rotational curvature forms are exactly the standard symplectic form of $\mathbf{R} \oplus \mathbf{R}$, dilated by the frequency ν. The corresponding equivalence classes of irreducible unitary linear representations U^ν of G acting on the standard complex Hilbert space $L^2_{\mathbf{C}}(\mathbf{R})$ of square integrable wave functions on the bi–infinite stratigraphic time line $\mathbf{R}$ are infinite dimensional and can be realized as Hilbert–Schmidt integral operators with kernels K^ν in $L^2_{\mathbf{C}}(\mathbf{R} \oplus \mathbf{R})$ [13]. Despite the strong non–linearity of dissipative nuclear spin systems, symplectic linear response theory holds exactly in MRI processing. Therefore the symplectic affine planes $\mathcal{O}_\nu$ $(\nu \neq 0)$ in $\mathrm{Lie}(G)^\star$ are contiguous, adjacently decoupled energetic strata, predestinate to implement the Kepplerian spatiotemporal strategy, and to carry quantum holograms [14].

- In radar imaging, $\nu \neq 0$ denotes the center frequency of the transmitted pulse train, whereas in clinical MRI ν is the frequency of the rotating coordinate frame defined by tomographic slice selection of the energetic stratum.

The confocal plane $\nu = 0$ in $\mathrm{Lie}(G)^\star$ consists of the single point orbits

$$\mathcal{O}_\infty = \{\varepsilon_{(\alpha,\beta)} \mid (\alpha, \beta) \in \mathbf{R} \oplus \mathbf{R}\},$$

corresponding to the one–dimensional representations of G, or unitary characters of G/C. As the reconstruction plane $\mathbf{P}(\mathbf{R} \times \mathcal{O}_\infty)$, it plays a fundamental role in the coherent optical processing of radar data [2], morphological MRI (Figure 1), and neurofunctional MRI (Figure 2) detection for the recording of brain activities [17]. It
follows from this classification of the coadjoint orbits of G in $\mathrm{Lie}(G)^\star$ the highly remarkable fact that there exists no finite dimensional irreducible unitary linear

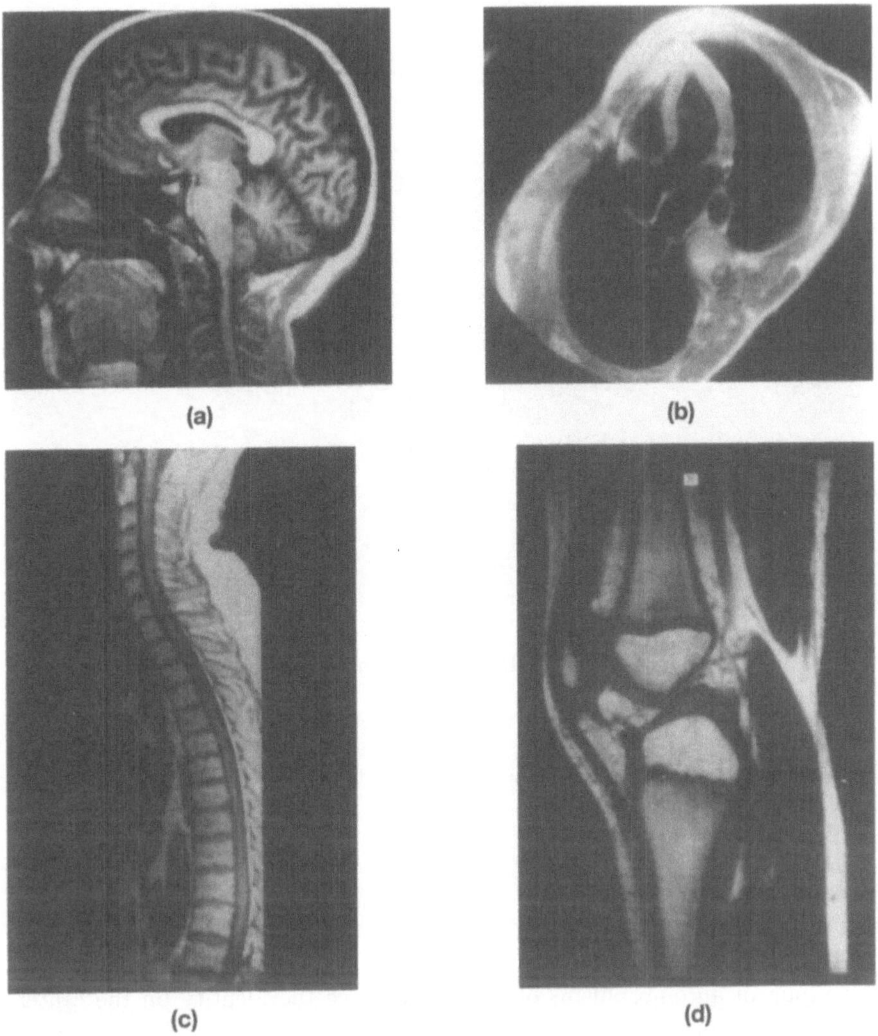

Fig. 1. Clinico–morphological anatomy displayed by MRI.

representation of G having dimension > 1. Hence the irreducible unitary linear representations of G which are not characters are infinite dimensional. Their coefficient functions define the holographic transforms.

Let C denote the one–dimensional center of G transversal to the plane carrying the spin excitation profiles or quantum holograms (Figure 3). Then

$$C = [G, G] = \mathbf{R}.\exp_G I$$

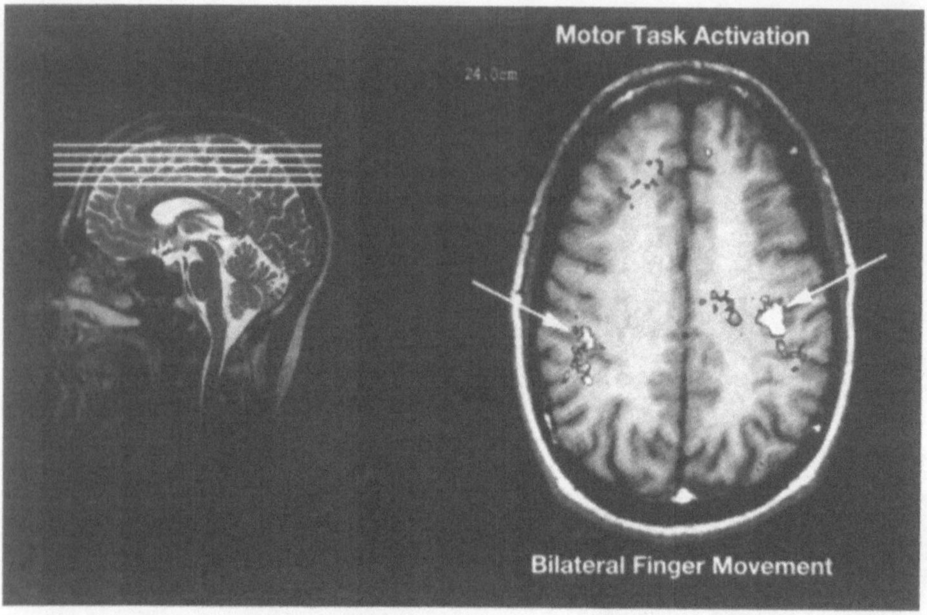

Fig. 2. Neurofunctional MRI displaying neuronal activation.

is spanned by the central transvection $\exp_G I$. In coordinate–free terms, G forms the non–split central group extension

$$C \triangleleft G \longrightarrow G/C$$

where G/C is transversal to the line C, and bicontinuously isomorphic to the symplectic affine plane $\mathbf{R} \oplus \mathbf{R}$.

- The group of automorphisms of G which induce the identity on the center C of G is isomorphic to the semi–direct product of the special linear group $\mathbf{SL}(2, \mathbf{R})$ and $\mathbf{R} \oplus \mathbf{R}$.

The irreducible unitary linear representations of G associated to the coadjoint orbit $\mathcal{O}_\nu$ are square integrable mod C. Indeed, it is well known that a coadjoint orbit is a linear variety if and only if one (and hence all) of the corresponding irreducible unitary linear representations is square integrable modulo its kernel. This forces the dimension of G to be *odd*. It is reasonable to regard square integrability as an essential part of the Stone–von Neumann theorem of quantum mechanics, because a representation of a nilpotent Lie group is determined by its central character χ_ν if and only if it is square integrable modulo center. Thus χ_ν allows for selection of the tomographic slice $\mathcal{O}_\nu$ with coordinate frame rotating at frequency $\nu \neq 0$.
Inspection of the spectrum of G reveals a continuous open projection onto C, taking a copy of $\mathbf{R} \oplus \mathbf{R}$ onto $\{0\}$ and a single point onto each non–zero and a

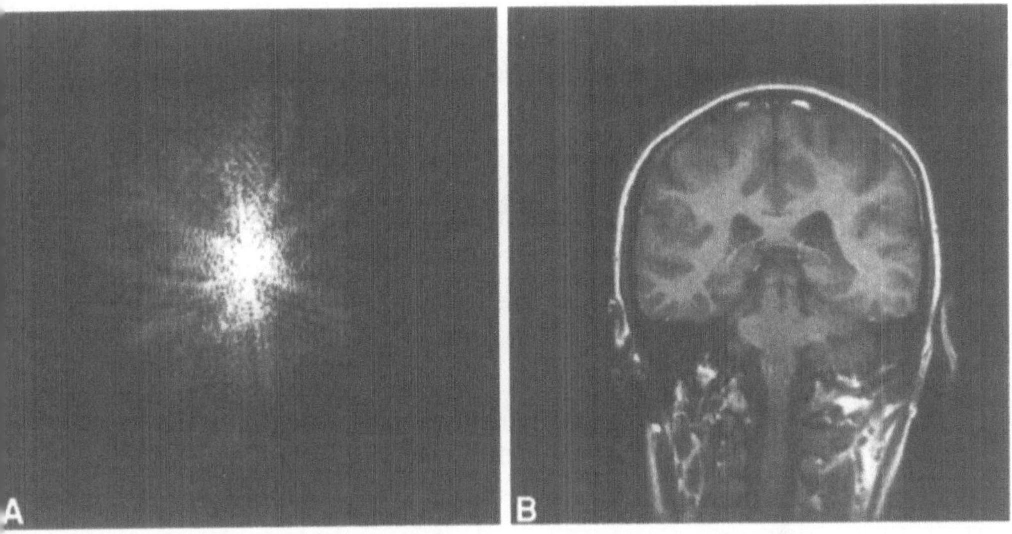

Fig. 3. Quantum hologram (A) and its reconstruction (B).

single point onto each non–zero $\nu \in C$. In terms of quantized calculus, the $C^\star$–algebra of the Heisenberg group G is the algebra of continuous sections vanishing at infinity of a continuous field, or bundle, of $C^\star$–algebras $(\mathcal{A})_{\nu \in C}$, with $\mathcal{A}_\nu$ the quotient corresponding to the closed subset mapping into $\{\nu\}$. Since all points of the copy of $\mathbf{R} \oplus \mathbf{R}$ which maps into $\{0\}$ are one–dimensional representations of G, it follows

$$\mathcal{A}_0 \cong \mathcal{C}_0(\mathbf{R} \oplus \mathbf{R})$$

For each $\nu \neq 0$, since the spectrum of $\mathcal{A}_\nu$ is a single point, and the irreducible unitary linear representation corresponding to the coadjoint orbit $\mathcal{O}_\nu$ is infinite–dimensional, it follows that $\mathcal{A}_\nu$ is isomorphic to the simple $C^\star$–algebra $\mathcal{K}^\nu(L^2_{\mathbf{C}}(\mathbf{R}))$ of compact operators acting on the standard complex Hilbert space $L^2_{\mathbf{C}}(\mathbf{R})$. Clearly

$$\mathcal{A}_\nu \cong \mathcal{K}^\nu(L^2_{\mathbf{C}}(\mathbf{R})) \qquad (\nu \neq 0)$$

includes the Hilbert–Schmidt operators on $L^2_{\mathbf{C}}(\mathbf{R})$ as a norm dense ideal.
In analogy to the Kepplerian conchoid construction [18], the $C^\star$–algebra of the Heisenberg group G may be naturally viewed as the crossed product of $\mathcal{C}_0(\mathbf{R} \oplus \mathbf{R})$ by the action of the central factor C which in each line $\{\nu\} \times \mathbf{R}$ of the bundle of lines parallel to the second axis of $\mathbf{R} \oplus \mathbf{R}$ consists of translation, but scaled by the first coordinate ν of this line, so that each point of the second coordinate axis itself is left fixed under the scaling procedure.

- The kernel function $K^\nu \in L^2_{\mathbf{C}}(\mathbf{R} \oplus \mathbf{R})$ associated to the central character χ_ν represents a planar spin excitation profile or quantum hologram. It defines a multichannel coherent wavelet perfect reconstruction analysis–synthesis filter bank of matched filter type. The reconstruction of the phase histories in local frequency encoding subbands of K^ν is performed by the symplectically reformatted two–dimensional Fourier transform.

The kernel functions admit the Hermitian symmetry

$$K^\nu(x,y) = \bar{K}^\nu(y,x) \quad \big((x,y) \in \mathbf{R} \oplus \mathbf{R}\big)$$

so that the associated Hilbert–Schmidt integral operators in $\mathcal{K}^\nu\big(L^2_{\mathbf{C}}(\mathbf{R})\big)$ are self–adjoint. It is this kernel symmetry which reflects the symmetry inherent to the geometric quantization approach.

- The Hermitian symmetry inherent to the kernel function allows for an application of the one–half Fourier, or alternatively, $\frac{1}{2}$–NEX reconstruction technique.

The $\frac{1}{2}$–NEX reconstruction method reduces the data acquisition time by 50 % and retains the spatial resolution. However, the signal–to–noise ratio is reduced by a factor of $\frac{1}{\sqrt{2}}$.
The compact self–adjoint integral operator defined by the Hermitian kernel $K^\nu \in L^2_{\mathbf{C}}(\mathbf{R} \oplus \mathbf{R})$ has a singular value decomposition associated to to its discrete spectrum as a least square fit. The Karhunen–Loève basis is known to form the minimizer of non–negligible amplitudes in the computational realization of the planar spin excitation profiles or quantum holograms (Figure 1) attached to U^ν. The Karhunen–Loève transform (KLT) can be applied in order to reparametrize the spatial coordinates of the quantum hologram via the normalizing action of the metaplectic group $\mathbf{Mp}(2,\mathbf{R})$ which projects onto $\mathbf{SL}(2,\mathbf{R})$ with kernel $\mathbf{Z}/2\mathbf{Z}$.
The KLT both decorrelates the input and optimizes the redistribution of the wavelet energy in the L^2–sense. As an adaptive image transform algorithm [22] it allows for extraction of the prominent features of the final image with the fewest number of measurements. Such a block compression design performed by the best zonal sampler is of particular importance for use in hospital picture and archiving systems (PACS).
The geometric quantization approach leads to non–locality phenomenon of quantum mechanics [14], and to major application areas of pulsed signal recovery methods, the corner turn algorithm in the digital processing of high resolution SAR data [21], the spin–warp procedure in clinical MRI [7], [15], and finally to the variants of the ultra–high–speed echo–planar imaging (EPI) technique of neurofunctional MRI [17], and MRI–guided interventions [6]. The capacity of MRI to obtain both structural and functional information promises to dramatically alter the way of studying the human brain *in vivo*. MRI–guided surgical management recently allowed for the first time to perform brain operations without craniotomy. Combined with multi–slice acquisition, it is the spin–warp version of Fourier transform MRI which is used almost exclusively in current routine clinical examinations [5], [9]

and now forms a general basis for the majority of MRI studies and preoperative assessment [4], [17].
There have been dramatic advances in the methods used to display MRI scans, which are able now to display finer details and more subtle lesions. The non–invasive detection of multiple sclerosis plaques [1], [20], provided one of the early success stories of clinical MRI. It is worthwile to observe that, for the benefit of humankind, the theoretical and clinical applications of MRI are still proceeding at a rapid pace. In conclusion, it can only be assumed that the inexorable progress of MRI will continue, and that there are many more improvements and discoveries of new clinical applications around the corner.

References

1. Barkovich A. J., Pediatric Neuroimaging. Second edition, Raven Press, New York 1995
2. Cutrona L. J., Leith E. M., Porcello L. J., and Vivian W. E., On the application of coherent optical processing techniques to synthetic–aperture radar. Proc. IEEE 54, 1026–1032 (1966)
3. Davies E.R., Electronics, Noise and Signal Recovery. Academic Press, London, San Diego, New York 1993
4. Gadian D. G., NMR and its Applications to Living Systems. Second edition, Oxford University Press, Oxford, New York, Tokyo 1995
5. Gillies R. J. (Editor), NMR in Physiology and Biomedicine. Academic Press, San Diego, New York, Boston 1994
6. Jolesz F. A., MRI–guided interventions. The Coolidge Sci. Rev. 2, 1–25 (1994)
7. Keller P. J., How big is k–space? Int. J. Neuroradiology 2, 274–289 (1996)
8. Mallick S. and Rajbenbach H., Photorefractive nonlinear optics and optical computing. In: Optical Phase Conjugation, M. Gower, D. Proch, editors, pp. 342–363, Springer–Verlag, Berlin, Heidelberg, New York 1994
9. Marseille G. J., MRI scan time reduction through nonuniform sampling. Ph.D. Thesis, Delft University of Technology, Ponsen & Looijen, Wageningen 1997
10. Mattson J. and Simon M., The Pioneers of NMR and Magnetic Resonance in Medicine: The Story of MRI. Bar–Ilan University Press, Ramat Gan 1996
11. Müller W. K., Ziegler R., Bauer A., and Soldner E. H., Virtual reality in surgical arthroscopic training. J. Imag. Guid. Surg. 1, 288–294 (1995)
12. Rihaczek A. W. and Hershkowitz S. J., Radar Resolution and Complex–Image Analysis. Artech House, Boston, London 1996
13. Schempp W., Harmonic Analysis on the Heisenberg Nilpotent Lie Group, with Applications to Signal Theory. Pitman Research Notes in Mathematics Series, Vol. 147, Longman Scientific and Technical, London 1986
14. Schempp W., Geometric analysis: The double–slit interference experiment and magnetic resonance imaging. Cybernetics and Systems '96, Vol. 1, pp. 179–183, Austrian Society for Cybernetic Studies, Vienna 1996

15. Schempp W., Wavelets in high resolution radar imaging and clinical magnetic resonance imaging. Proc. IWISP '96, Manchester, United Kingdom: Third International Workshop on Image and Signal Processing on the Theme of Advances in Computational Intelligence, B.G. Mertzios, P. Liatsis, editors, pp. 73–80, Elsevier, Amsterdam, Lausanne, New York 1996
16. Schempp W., Wavelets and image processing: A geometric analysis approach to magnetic resonance tomography. Proc. First International Workshop on Advanced Signal Processing for Medical MRI–MRS, B. G. Mertzios, editor, pp. 80–83, Democritus University of Thrace, Xanthi, Hellas 1997
17. Schempp W., Magnetic Resonance Imaging: Mathematical Foundations and Applications. John Wiley & Sons, New York, Chichester, Brisbane (in print)
18. Stephenson B., Kepler's Physical Astronomy. Princeton University Press, Princeton, NJ 1994
19. Tschudi T., Phase conjugation in optical signal processing. In: Optical Phase Conjugation, M. Gower, D. Proch, editors, pp. 364–380, Springer–Verlag, Berlin, Heidelberg, New York 1994
20. van der Knaap M. S. and Valk J., Magnetic Resonance of Myelin, Myelination, and Myelin Disorders. Second edition, Springer–Verlag, Berlin, Heidelberg, New York 1995
21. Wehner D. R., High Resolution Radar. Artech House, Norwood, MA and London 1987
22. Wickerhauser M. V., Custom wavelet packet image compression design. Proc. IWISP '96, Manchester, United Kingdom: Third International Workshop on Image and Signal Processing on the Theme of Advances in Computational Intelligence, B.G. Mertzios, P. Liatsis, editors, pp. 47–52, Elsevier, Amsterdam, Lausanne, New York 1996

Walter Schempp
Lehrstuhl für Mathematik I
Universität Siegen
D–57068 Siegen
Germany
`schempp@mathematik.uni-siegen.d400.de`

A New Interpretation of the Sampling Theorem and Its Extensions

Gerhard Schmeisser and Jürgen J. Voss

Abstract. We start with the classical sampling theorem for bandlimited signals and comment on its various extensions. In Section 2 we discuss the problem of sampling harmonic functions. In the subsequent section we consider non-uniform sampling of bandlimited signals and entire functions of exponential type. We present a new sampling theorem which can be deduced from a recent result by one of us. As the main result of the whole paper, we show in Section 4 that there is an equivalence between sampling of signals and sampling of entire harmonic functions. This is applied to the theorem in Section 3. As a consequence, we get a new uniqueness theorem for entire harmonic functions of exponential type.

1. Introduction

The classical sampling theorem as it may be attributed to Whittaker (1915), Kotel'nikov (1933), Shannon (1949) and many others (see [4] for numerous references) is a statement on the validity of the interpolation formula

$$f(t) \;=\; \sum_{n=-\infty}^{\infty} f\left(\frac{n\pi}{\sigma}\right) \frac{\sin(\sigma t - n\pi)}{\sigma t - n\pi} \;=:\; C_\sigma[f](t). \tag{1}$$

In the language of electrical engineers it reads as follows.

Theorem A. *A signal f of finite energy which is defined on the whole real line and is bandlimited to the interval $[-\sigma,\sigma]$ can be reconstructed from its samples $f(n\pi/\sigma)$ $(n \in \mathbb{Z})$ by the formula (1). The series converges absolutely at each $t \in \mathbb{R}$ and uniformly on any compact subset of $\mathbb{R}$.*

Multivariate Approximation and Splines
G. Nürnberger, J. W. Schmidt, and G. Walz (eds.), pp. 275–288.

ISBN 3-7643-5654-5.

Some of the terms appearing in this statement may need a mathematical formulation. A signal f is nothing but a real-valued continuous function. Finite energy means that the integral $\int_{-\infty}^{\infty} |f(t)|^2\,dt$ exists, or equivalently, $f \in L^2(\mathbb{R})$. Such a function is known to have a Fourier transform

$$\widehat{f}(x) = \int_{-\infty}^{\infty} f(t)e^{-ixt}\,dt$$

in the L^2 sense. Finally, bandlimited to $[-\sigma,\sigma]$ means that $\widehat{f}(x) = 0$ almost everywhere outside $[-\sigma,\sigma]$. Here the properties "bandlimited" and "finite energy" are linked. We may separate them as follows. From the Paley–Wiener theorem [2, Theorem 6.8.1] we get that a function $f \in L^2(\mathbb{R})$ which is bandlimited to $[-\sigma,\sigma]$ has an extension into the whole complex plane to an entire function f such that

$$\limsup_{r\to\infty} \frac{1}{r} \max_{|z|=r} \log|f(z)| \leq \sigma. \tag{2}$$

Any entire function f satisfying (2) is said to be of exponential type σ. Hence the restriction to $\mathbb{R}$ of any entire function of exponential type σ may be seen as a function which is bandlimited to $[-\sigma,\sigma]$ no matter whether it really belongs to $L^2(\mathbb{R})$ or not. If in addition it is real-valued on $\mathbb{R}$, then we may conceive it as a signal that is bandlimited to $[-\sigma,\sigma]$ in a generalized sense. Under these agreements

$$f(t) = \cosh t = \frac{e^t + e^{-t}}{2} \tag{3}$$

will be bandlimited to $[-1,1]$.

Theorem A has a very extensive literature (for surveys see [4; 10; 12; 17]). It has been generalized in various ways. Let us briefly reconsider its diverse assumptions and indicate how they have been relaxed.

(a) Real-valued function f

There is absolutely no problem in extending Theorem A to complex-valued functions f provided that all the other assumptions hold. In fact,

$$\varphi := \frac{1}{2}(f+\overline{f}) \quad\text{and}\quad \psi := \frac{1}{2i}(f-\overline{f})$$

are real-valued functions satisfying the hypothesis of Theorem A. Therefore

$$f(t) = \varphi(t) + i\psi(t) = C_\sigma[\varphi](t) + i\,C_\sigma[\psi](t) = C_\sigma[f](t).$$

(b) Finite energy or $f \in L^2(\mathbb{R})$

If this condition is violated, then the convergence of the series in (1) may fail. However, there is a multiplier technique which restores convergence. Let φ be an entire function of exponential type ε which attains the value 1 at the origin and is real-valued on the real line. Suppose that for every $x \in \mathbb{R}$ the function $f\,\varphi(x-\cdot)$ belongs to $L^2(\mathbb{R})$. If f is bandlimited to $[-\sigma, \sigma]$, then Theorem A applies to $f\,\varphi(x-\cdot)$ with σ replaced by $\tau := \sigma + \varepsilon$. Therefore

$$f(t)\,\varphi(x-t) = \sum_{n=-\infty}^{\infty} f\left(\frac{n\pi}{\tau}\right)\varphi\left(x - \frac{n\pi}{\tau}\right)\frac{\sin(\tau t - n\pi)}{\tau t - n\pi}.$$

In particular, setting $x = t$, we arrive at

$$f(t) = \sum_{n=-\infty}^{\infty} f\left(\frac{n\pi}{\tau}\right)\Lambda(\tau t - n\pi), \tag{4}$$

where $\Lambda(z) := \varphi(z/\tau)\cdot(\sin z)/z$. This technique has been thoroughly studied in [13]. There exist multipliers φ which admit functions f that grow faster than any polynomial on the real line. However, there is no formula (4) that applies to the function (3).

(c) Bandlimited to $[-\sigma, \sigma]$

If f is not bandlimited but satisfies all the other conditions of Theorem A, then $f(t) - C_\sigma[f](t) \not\equiv 0$. This difference is called the *aliasing error*. Being vague, we may consider a non-bandlimited signal f as one which is bandlimited to $(-\infty, \infty)$ and hope that

$$f(t) = \lim_{\sigma\to\infty} C_\sigma[f](t).$$

Under appropriate assumptions on the decrease of $\widehat{f}$, this equation does indeed hold [5]. Its statement is referred to as the *generalized sampling theorem.*

(d) Uniform nodes

In Theorem A, the samples are taken on the set of uniform points $n\pi/\sigma$ ($n = 0, \pm1, \pm2, \ldots$). Sometimes irregular nodes are of interest. For instance, it may be that we aim at uniform nodes but errors cause some irregularity. This is what the engineers call *jitter*. Besides, there are problems in theory [7] and applications (*e.g.*, in laser-Doppler or geophysical signal processing) where irregularity comes in a natural way.

If we have nodes s_n ($n \in \mathbb{Z}$) which are slight displacements of the points $n\pi/\sigma$ ($n \in \mathbb{Z}$), then

$$G(z) := (z - s_0)\prod_{n=1}^{\infty}\left(1 - \frac{z}{s_n}\right)\left(1 - \frac{z}{s_{-n}}\right)$$

will define an entire function of exponential type σ. Now we may look for the validity of the Lagrange interpolation formula

$$f(t) = \sum_{n=-\infty}^{\infty} f(s_n) \frac{G(t)}{(t-s_n)G'(s_n)}.$$

Clearly, it reduces to (1) for $s_n = n\pi/\sigma$ $(n \in \mathbb{Z})$. Such formulae and variants in the spirit of the modification under (b) have become a very active field of research. We shall present a precise result of this type in Section 3 (see Theorem 1 below).

(e) Reconstruction on $\mathbb{R}$

Theorem A provides a reconstruction of f on $\mathbb{R}$. However, under the assumptions of Theorem A the series in (1) converges for complex t as well and the convergence is uniform on compact subsets of $\mathbb{C}$. Therefore, by a theorem of Weierstrass and the uniqueness theorem for holomorphic functions, we can conclude that

$$f(z) = \sum_{n=-\infty}^{\infty} f\left(\frac{n\pi}{\sigma}\right) \frac{\sin(\sigma z - n\pi)}{\sigma z - n\pi} \quad \text{for} \quad z \in \mathbb{C}, \tag{5}$$

where on the left hand side f stands for the extension of the signal f to an entire function of exponential type as given by the Paley–Wiener theorem. We may also take the real parts on both sides of (5). Introducing $u(x,y) := \operatorname{Re} f(x+iy)$,

$$A(x,y) := \operatorname{Re} \frac{\sin(x+iy)}{x+iy} = \frac{x \sin x \cosh y + y \cos x \sinh y}{x^2+y^2} \tag{6}$$

and assuming f to be real-valued on $\mathbb{R}$, we deduce that

$$u(x,y) = \sum_{n=-\infty}^{\infty} u\left(\frac{n\pi}{\sigma}, 0\right) A(\sigma x - n\pi, \sigma y). \tag{7}$$

This is a reconstruction formula for a certain class of harmonic functions. Unfortunately, it fails for such a simple harmonic function as $u(x,y) \equiv y$.

2. Sampling of Entire Harmonic Functions

Definition 1. *An entire harmonic function of exponential type* σ *is a function* $u : \mathbb{R}^2 \to \mathbb{R}$ *which is harmonic on* $\mathbb{R}^2$ *and satisfies the growth condition*

$$\limsup_{r\to\infty} \frac{1}{r} \max\{\log|u(x,y)| \;:\; x^2+y^2=r^2\} \le \sigma.$$

It is known [3] that u is an entire harmonic function of exponential type σ if and only if it is the real part of an entire function f of exponential type σ. Clearly, along with u the imaginary part v of f is also an entire harmonic function of exponential type σ called a *conjugate* of u. Using this term, we can say that the reconstruction formula (7) holds for all those entire harmonic functions u of exponential type σ which belong to L^2 on $\mathbb{R}$ and have a conjugate v that vanishes along $\mathbb{R}$. In the absence of the last condition, formula (7) fails as the example

$$u(x,y) \equiv y \tag{8}$$

shows. A revised form of (7), which does not need a condition on v, is the formula [15, Corollary 1.6]

$$\frac{u(x,y)+u(x,-y)}{2} = \sum_{n=-\infty}^{\infty} u\left(\frac{n\pi}{\sigma},0\right) A(\sigma x - n\pi, \sigma y).$$

Unfortunately, it does not yield a reconstruction of u itself but only of a symmetrization of u.

While an entire function f of exponential type less than σ is uniquely determined by its values on the points $n\pi/\sigma$ $(n \in \mathbb{Z})$ as a consequence of Carlson's theorem [2, Theorem 9.2.1], the example (8) shows that even the restriction of u to the real line is not enough information for determining u. As an analogue of Carlson's theorem in the case of harmonic functions, Boas [3] came up with the following result.

Theorem B. *Let u be an entire harmonic function of exponential type less than π such that $u(n,0) = u(n,1) = 0$ for $n = 0, \pm 1, \pm 2, \dots$. Then u is identically zero.*

Formulae for the reconstruction of u from its values on the lattice points

$$\{(n,0),\,(n,1)\;:\;n \in \mathbb{Z}\}$$

were first given by Ching and Chui [6] and Andersen [1]. More powerful results were later established by Rahman and Schmeisser [14].

The example (8) clearly shows that a reconstruction of u by an interpolation formula with nodes only on the real line is not possible even if we involve in addition derivatives of u on $\mathbb{R}$. However, the situation changes favourably if we consider directional derivatives perpendicular to the real line. Using the notation

$$D_1 u := \frac{\partial}{\partial x} u, \qquad D_2 u := \frac{\partial}{\partial y} u \tag{9}$$

and introducing

$$B(x,y) := \int_0^y A(x,\eta)\, d\eta \tag{10}$$

with the help of (6), we may state the following theorem [15, p. 144, Theorem 3.3].

Theorem C. *Let u be an entire harmonic function of exponential type $\sigma > 0$. Let $u(\cdot, 0)$ and $D_2u(\cdot, 0)$ belong to $L^2(\mathbb{R})$. Then at each point $(x, y) \in \mathbb{R}^2$*

$$u(x,y) = \sum_{n=-\infty}^{\infty} \left(u\left(\frac{n\pi}{\sigma}, 0\right) A(\sigma x - n\pi, \sigma y) + \frac{1}{\sigma} D_2 u\left(\frac{n\pi}{\sigma}, 0\right) B(\sigma x - n\pi, \sigma y) \right)$$

with A given by (6) and B by (10). The series converges absolutely and uniformly on any compact subset of $\mathbb{R}^2$.

The proof of Theorem C as given in [15] needs two lemmas. One of them is the following uniqueness theorem.

Theorem D. *Let u be an entire harmonic function of exponential type $\sigma > 0$ and let $u(\cdot, 0)$ and $D_2u(\cdot, 0)$ belong to $L^2(\mathbb{R})$. Suppose that*

$$u\left(\frac{n\pi}{\sigma}, 0\right) = D_2 u\left(\frac{n\pi}{\sigma}, 0\right) = 0$$

for all integers n. Then u is identically zero.

We shall see in Section 4 that Theorem C may be directly deduced from Theorem A as an equivalent form of the latter. Moreover, Theorem D comes out as a corollary.

3. Non-Uniform Sampling of Entire Functions

Non-uniform sampling has gained a lot of attention. As some relevant contributions we mention Yen [22], Higgins [9], Seip [16], and Hinsen [11]. Further progress was recently made by Voss [18; 19; 20]. Although in his theorems the nodes may be non-real, we shall consider only real nodes in the present paper. They may be introduced as follows.

Definition 2. *Let $\delta \in (0, 1]$ and L be a non-negative integer. Then $\mathcal{N}(\delta, L)$ denotes the set of all sequences $(s_n)_{n\in\mathbb{Z}}$ with the following properties. For every $n \in \mathbb{Z}$:*

(i) $s_n \in \mathbb{R}$;

(ii) $s_n \neq 0$ *if* $n \neq 0$;

(iii) $s_{n+1} - s_n > \delta$;

(iv) $|\, s_n - n \,| \leq L$.

We may consider the sequences in $\mathcal{N}(\delta, L)$ as displaced integers which are separated by δ. An interesting example of such a sequence, which is of significance in applications [7], arises from the zeros of the Bessel function J_α of the first kind of order α. If $\alpha > -1$, then $J_\alpha(z)/z^\alpha$ is an entire function of exponential type 1 with infinitely many zeros, which are all real and simple (see [21] for this and the following statements). They may be denoted by $j_n(\alpha)$ $(n = \pm 1, \pm 2, \ldots)$ and ordered such that

$$j_{-n}(\alpha) = -j_n(\alpha) \quad \text{and} \quad 0 < j_1(\alpha) < j_2(\alpha) < \cdots.$$

Moreover,

$$j_n(\alpha) = \left(n + \frac{\alpha}{2} - \frac{1}{4}\right)\pi + O\left(\frac{1}{n}\right) \quad \text{as} \quad n \to \infty.$$

Hence, setting

$$s_n := \begin{cases} j_n(\alpha)/\pi & \text{for } n = -1, -2, \ldots, \\ j_{n+1}(\alpha)/\pi & \text{for } n = 0, 1, \ldots, \end{cases}$$

we get a sequence which belongs to $\mathcal{N}(\delta, L)$ for some appropriate δ and L.

For a sequence $\boldsymbol{s} = (s_n)_{n\in\mathbb{Z}} \in \mathcal{N}(\delta, L)$ the canonical product

$$G(\boldsymbol{s}, z) := (z - s_0) \prod_{n=1}^{\infty} \left(1 - \frac{z}{s_n}\right)\left(1 - \frac{z}{s_{-n}}\right) \tag{11}$$

is an entire function of exponential type π with polynomial growth on the real line [19, Lemma 2.28]. The quotient

$$\frac{G(\boldsymbol{s}, z)}{(z - s_n)\, G'(\boldsymbol{s}, s_n)}$$

is a fundamental function for Lagrange interpolation with respect to the nodes $(s_n)_{n\in\mathbb{Z}}$ since it attains the value 1 at $z = s_n$ and vanishes at all the other nodes.

Next we want to be more precise about the multipliers we shall consider.

Definition 3. *Let ε be a positive real number and ψ a positive continuous function defined on $[0, \infty)$. Then $\mathcal{M}(\varepsilon, \psi)$ denotes the class of all entire functions φ with the following properties:*

(i) *φ is of exponential type ε;*

(ii) *$\varphi(x) \in \mathbb{R}$ for $x \in \mathbb{R}$;*

(iii) *$\varphi(0) = 1$;*

(iv) *for every compact subset $\mathcal{K}$ of $\mathbb{C}$, there exists a number $c > 0$ such that*

$$\left|\varphi(z - t)\right| \le c\psi\left(|t|\right) \quad \text{for} \quad z \in \mathcal{K},\ t \in \mathbb{R}.$$

Condition (iv) shows that the function ψ is essentially a prescribed majorant. A rapidly decreasing ψ will be of special interest.

Examples

For $\psi(t) := (1+t^k)^{-1}$ with $k \in \mathbb{N}$, the function

$$\varphi(z) := \left(\frac{\sin(\varepsilon z/k)}{\varepsilon z/k}\right)^k$$

belongs to $\mathcal{M}(\varepsilon, \psi)$. In [8] the authors have considered a majorant ψ with

$$\psi(t) = O\left(\exp\left(\frac{-t}{(\log t)^\alpha}\right)\right) \quad \text{as} \quad t \to \infty, \tag{12}$$

where $\alpha > 1$, and constructed a function φ belonging to $\mathcal{M}(\varepsilon, \psi)$. This is nearly the best one can do. Indeed, it follows from the discussion in [8] that the class $\mathcal{M}(\varepsilon, \psi)$ is empty if

$$\int_1^\infty \frac{|\min\{0, \log \psi(t)\}|}{t^2}\, dt = \infty.$$

Thus $\mathcal{M}(\varepsilon, \psi)$ is empty if $\psi(t) = O(e^{-t})$ as $t \to \infty$ and also if (12) holds with $\alpha = 1$.

Theorem 1. *With reference to Definitions 2–3, let* $\boldsymbol{s} = (s_n)_{n\in\mathbb{Z}} \in \mathcal{N}(\delta, L)$ *and* $\varphi \in \mathcal{M}(\varepsilon, \psi)$, *where* $\varepsilon \in (0, \pi)$. *Let* f *be an entire function of exponential type* $\sigma < \pi - \varepsilon$ *satisfying*

$$f(s_n) = O\left(\frac{1}{|s_n|^{4L}\ \psi(|s_n|)}\right) \quad \text{as} \quad n \to \pm\infty. \tag{13}$$

Then

$$f(z) = \sum_{n=-\infty}^{\infty} f(s_n)\Lambda_n(\boldsymbol{s}, z),$$

where

$$\Lambda_n(\boldsymbol{s}, z) := \varphi(z - s_n)\, \frac{G(\boldsymbol{s}, z)}{(z - s_n)\, G'(\boldsymbol{s}, s_n)} \tag{14}$$

with G *defined by (11). The series converges uniformly on every compact subset of* $\mathbb{C}$.

Proof: Let $\mathcal{K}$ be a compact subset of $\mathbb{C}$. For $\zeta \in \mathcal{K}$ define $F_\zeta(z) := f(z)\varphi(z-\zeta)z^{4L}$. Then F_ζ is an entire function of exponential type less than π. Moreover, (13) and property (iv) in Definition 3 imply that $F_\zeta(s_n) = O(1)$ as $n \to \pm\infty$ uniformly for $\zeta \in \mathcal{K}$. By a result of Duffin and Schaeffer [2, Theorem 10.5.3], it follows that F_ζ is bounded on $\mathbb{R}$ uniformly for $\zeta \in \mathcal{K}$. Hence there exists a positive C such that

$$|f(x)\varphi(x-\zeta)| \le C\left(|x|+1\right)^{-4L} \quad \text{for} \quad x \in \mathbb{R},\ \zeta \in \mathcal{K}.$$

Now in case $L \ge 1$ a result proved in [20, Theorem 1.1] applies and provides the conclusion of Theorem 1. In case $L = 0$, where the nodes are equidistant, Theorem 1 can be settled by the approach used in [13]. □

4. An Equivalent Form of Sampling Theorems

We shall now show that any two (not necessarily different) sampling theorems for bandlimited signals imply a sampling theorem for entire harmonic functions of exponential type. Conversely, from such a sampling theorem for entire harmonic functions of exponential type, two sampling theorems for bandlimited signals can be deduced.

Consider two sampling theorems for signals f bandlimited to $[-\sigma, \sigma]$ with the reconstruction formulae

$$f(t) = \sum_{n=-\infty}^{\infty} f(s_n) L_n(t) =: C_{1,\sigma}[f](t) \tag{15}$$

and

$$f(t) = \sum_{n=-\infty}^{\infty} f(t_n) M_n(t) =: C_{2,\sigma}[f](t), \tag{16}$$

respectively. We suppose that both series converge uniformly with respect to t on compact subsets of $\mathbb{C}$ so that the formulae extend to entire functions of exponential type σ as we have explained in Section 1 under (e).

The hypothesis of the following theorem may need an explanation. Let u be an entire harmonic function of exponential type σ. Then there exists an entire function g of exponential type σ which coincides with u on the real line. This will be seen in the first part of the proof below. Hence the restriction of u to $\mathbb{R}$ may be understood as a signal bandlimited to $[-\sigma, \sigma]$ in a generalized sense as explained in the introduction. Therefore the restriction of u to $\mathbb{R}$ is a possible candidate for applying (15) or (16). The same considerations hold for $D_2 u$.

Main Theorem. *Let u be an entire harmonic function of exponential type σ. Suppose that the restrictions of u and $D_2 u$ to $\mathbb{R}$ are such that (15) and (16), respectively, are applicable. Then*

$$u(x, y) = \sum_{n=-\infty}^{\infty} \left(u(s_n, 0) \operatorname{Re} L_n(x + iy) + D_2 u(t_n, 0) \int_0^y \operatorname{Re} M_n(x + i\eta)\, d\eta \right) \tag{17}$$

for $(x, y) \in \mathbb{R}^2$. Conversely, let

$$u(x, y) = \sum_{n=-\infty}^{\infty} \left(u(s_n, 0) A_n(x, y) + D_2 u(t_n, 0) B_n(x, y) \right) \tag{18}$$

be a reconstruction formula for entire harmonic functions of exponential type σ and suppose that the series

$$\sum_{n=-\infty}^{\infty} D_2 u(t_n, 0) D_2 B_n(x, y) \tag{19}$$

converges uniformly on compact subsets of $\mathbb{R}^2$. *Then*

$$f(t) = \sum_{n=-\infty}^{\infty} f(s_n) A_n(t,0) \qquad (t \in \mathbb{R}) \tag{20}$$

and

$$g(t) = \sum_{n=-\infty}^{\infty} g(t_n) D_2 B_n(t,0) \qquad (t \in \mathbb{R}) \tag{21}$$

are reconstruction formulae for signals f *and* g *which are bandlimited to* $[-\sigma, \sigma]$ *and satisfy the hypothesis on* $u(\cdot,0)$ *and* $D_2 u(\cdot, 0)$, *respectively.*

Proof: Let u be an entire harmonic function of exponential type σ as specified in the theorem. Then there exists a conjugate harmonic function v so that

$$f(x+iy) := u(x,y) + iv(x,y)$$

is an entire function of exponential type σ (see Section 2). The functions

$$g(z) := \frac{f(z) + \overline{f(\overline{z})}}{2} \quad \text{and} \quad h(z) := \frac{f(z) - \overline{f(\overline{z})}}{2i}$$

are also entire functions of exponential type σ. Moreover,

$$g(x) = u(x,0), \quad \text{and} \quad h(x) = v(x,0) \quad \text{for} \quad x \in \mathbb{R}.$$

By the Cauchy–Riemann equations, we have $D_1 v = -D_2 u$. Hence using our assumptions on u and $D_2 u$, and recalling that the formulae (15) and (16) extend to entire functions of exponential type σ, we conclude that

$$g(z) = C_{1,\sigma}[u](z) \quad \text{and} \quad h'(z) = C_{2,\sigma}[D_1 v](z) = -C_{2,\sigma}[D_2 u](z).$$

In order to get $h(z)$, we integrate h' along the line-segment connecting $\operatorname{Re} z$ and z. In the representation of h' by the sampling series (16), the integration may be performed inside the summation sign since we have assumed uniform convergence. Writing z as $x + iy$ with x, $y \in \mathbb{R}$, we get

$$\begin{aligned} h(z) &= h(x) + i \int_0^y h'(x+i\eta)\, d\eta \\ &= v(x,0) - i \int_0^y C_{2,\sigma}[D_2 u](x+i\eta)\, d\eta \\ &= v(x,0) - i \sum_{n=-\infty}^{\infty} D_2 u(t_n, 0) \int_0^y M_n(x+i\eta)\, d\eta. \end{aligned}$$

Since

$$f(z) = g(z) + ih(z) \quad \text{and} \quad u(x,y) = \operatorname{Re} f(x+iy),$$

we conclude that

$$\begin{aligned} u(x,y) &= \operatorname{Re} g(z) \; - \; \operatorname{Im} h(z) \\ &= \operatorname{Re} C_{1,\sigma}[u](z) \; + \; \operatorname{Re} \int_0^y C_{2,\sigma}[D_2 u](x+i\eta)\, d\eta \\ &= \sum_{n=-\infty}^{\infty} u(s_n,0) \operatorname{Re} L_n(z) \; + \sum_{n=-\infty}^{\infty} D_2 u(t_n,0) \int_0^y \operatorname{Re} M_n(x+i\eta)\, d\eta. \end{aligned}$$

This completes the proof of (17).

Now we turn to the converse statement. The signal f is the restriction to $\mathbb{R}$ of an entire function f of exponential type σ which is real-valued on the real line. Therefore $u(x,y) := \operatorname{Re} f(x+iy)$ is an entire harmonic function of exponential type σ such that

$$u(t,0) \; = \; f(t) \quad \text{and} \quad D_2 u(t,0) \; \equiv \; 0 \quad \text{for} \quad t \in \mathbb{R}.$$

Now, applying (18) to this particular function u, we obtain (20).

The signal g has also an extension to an entire function, denoted again by g, which is of exponential type σ and real-valued on the real line and so is

$$G(z) \; := \; \int_0^z g(\zeta)\, d\zeta.$$

Therefore

$$u(x,y) \; := \; \operatorname{Re} i\, G(x+iy) \quad \text{and} \quad v(x,y) \; := \; \operatorname{Im} i\, G(x+iy)$$

are a pair of conjugate entire harmonic functions of exponential type σ with the properties

$$u(t,0) \equiv 0 \quad \text{and} \quad D_2 u(t,0) = -D_1 v(t,0) = -G'(t) = -g(t) \quad \text{for} \quad t \in \mathbb{R}. \tag{22}$$

Hence employing (18), we get

$$u(x,y) = - \sum_{n=-\infty}^{\infty} g(t_n) B_n(x,y). \tag{23}$$

Now we differentiate both sides with respect to the second variable. Because of the assumption on the series (19), we may interchange differentiation and summation on the right hand side of (23). Therefore (22) readily implies (21). □

With the main theorem, we easily see that Theorem C is an equivalent form of Theorem A. The corresponding equivalent form of Theorem 1 reads as follows.

Theorem 2. *With reference to Definitions 2–3, let*

$$\boldsymbol{s} = (s_n)_{n\in\mathbb{Z}} \in \mathcal{N}(\delta, L), \qquad \boldsymbol{t} = (t_n)_{n\in\mathbb{Z}} \in \mathcal{N}(\delta, L),$$

and $\varphi \in \mathcal{M}(\varepsilon, \psi)$, *where* $\varepsilon \in (0, \pi)$. *Let* u *be an entire harmonic function of exponential type* $\sigma < \pi - \varepsilon$ *satisfying*

$$\left.\begin{aligned} u(s_n, 0) &= O\left(\frac{1}{|\,s_n\,|^{4L}\ \psi(|\,s_n\,|)}\right) \\ D_2u(t_n, 0) &= O\left(\frac{1}{|\,t_n\,|^{4L}\ \psi(|\,t_n\,|)}\right) \end{aligned}\right\} \quad \text{as} \quad n \to \pm\infty.$$

Then

$$u(x,y) = \sum_{n=-\infty}^{\infty} \left(u(s_n, 0)\mathrm{Re}\,\Lambda_n(\boldsymbol{s}, x+iy) + D_2u(t_n, 0) \int_0^y \mathrm{Re}\,\Lambda_n(\boldsymbol{t}, x+i\eta)\, d\eta \right),$$

where Λ_n *is defined by (14). The series converges uniformly on every compact subset of* $\mathbb{R}^2$.

The following uniqueness theorem is an obvious consequence.

Corollary 1. *Let* $(s_n)_{n\in\mathbb{Z}}$ *and* $(t_n)_{n\in\mathbb{Z}}$ *be two sequences as in Theorem 2. Let* u *be an entire harmonic function of exponential type less than* π *such that*

$$u(s_n, 0) = D_2u(t_n, 0) = 0$$

for all integers n. *Then* u *is identically zero.*

Comparing Theorem D for $\sigma = \pi$ with Corollary 1, we note that the former has an additional condition on the functions $u(\cdot, 0)$ and $D_2u(\cdot, 0)$ but allows the exponential type to be equal to π. The example

$$u(x,y) := \mathrm{Re}\,\sin(\pi(x+iy))$$

shows that in Theorem D the additional condition cannot be cancelled.

References

1. Andersen K. F., On the representation of harmonic functions by their values on lattice points, J. Math. Anal. Appl. **49** (1975), 692–695; Corrigendum **58** (1977), 437.
2. Boas R. P., *Entire Functions*, Academic Press, New York, 1954.
3. Boas R. P., A uniqueness theorem for harmonic functions, J. Approx. Theory **5** (1972), 425–427.
4. Butzer P. L., Splettstößer W., and Stens R. L., The sampling theorem and linear prediction in signal analysis, Jahresber. Deutsch. Math. Verein. **90** (1988), 1–70.
5. Butzer P. L. and Stens R. L., Sampling theory for not necessarily band-limited functions; a historical overview, SIAM Review **34** (1992), 40–53.
6. Ching C.-H. and Chui C. K., A representation formula for harmonic functions, Proc. Amer. Math. Soc. **39** (1973), 349–352.
7. Frappier C. and Olivier P., A quadrature formula involving zeros of Bessel functions, Math. Comp. **60** (1993), 303-316.
8. Gervais R., Rahman Q. I., and Schmeisser G., A bandlimited function simulating a duration-limited one, in *Anniversary Volume on Approximation Theory and Functional Analysis*, Butzer P. L. et al. (eds.), Birkhäuser, Basel, 1984, 355–362.
9. Higgins J. R., A sampling theorem for irregularly spaced sample points, IEEE Trans. Inform. Theory **IT-22** (1976), 621–622.
10. Higgins J. R., Five short stories about the cardinal series, Bull. Amer. Math. Soc. **12** (1985), 45–89.
11. Hinsen G., Irregular sampling of bandlimited L^p-functions, J. Approx. Theory **72** (1993), 346–364.
12. Jerri A. J., The Shannon sampling theorem – its various extensions and applications: a tutorial review, Proc. IEEE **65** (1977), 1565–1596.
13. Rahman Q. I. and Schmeisser G., Reconstruction and approximation of functions from samples, in *Delay Equations, Approximation and Application*, Meinardus G. and Nürnberger G. (eds.), Birkhäuser, Basel, 1985, 213–233.
14. Rahman Q. I. and Schmeisser G., Representation of entire harmonic functions by given values, J. Math. Anal. Appl. **115** (1986), 461–469.
15. Schmeisser G., Sampling theorems for entire harmonic functions of exponential type, in *1995 Workshop on Sampling Theory and Applications*, Bilinskis I. et al. (eds.), Institute of Electronics and Computer Science, Riga, 1995, 140–145.
16. Seip K., An irregular sampling theorem for functions bandlimited in a generalized sense, SIAM J. Appl. Math. **47** (1987), 1112–1116.

17. Stenger F., Numerical methods based on Whittaker cardinal, or sinc functions, SIAM Review **23** (1981), 165–224.
18. Voss J., A contribution to nonuniform sampling with complex nodes, in *1995 Workshop on Sampling Theory and Applications*, Bilinskis I. et al. (eds.), Institute of Electronics and Computer Science, Riga, 1995, 59–64.
19. Voss J. J., *Irreguläres Abtasten von ganzen und harmonischen Funktionen von exponentiellem Typ,* Diplomarbeit, Erlangen, 1995.
20. Voss J. J., A sampling theorem with nonuniform complex nodes, J. Approx. Theory (to appear).
21. Watson G. N., *A treatise on the theory of Bessel functions,* 2nd ed., Cambridge University Press, Cambridge, 1952.
22. Yen J. L., On nonuniform sampling of bandwidth-limited signals, IRE Trans. Circuit Theory **CT-3** (1956), 251–257.

Gerhard Schmeisser
Mathematisches Institut
Universität Erlangen-Nürnberg
D-91054 Erlangen
`schmeisser@mi.uni-erlangen.de`

Jürgen J. Voss
Mathematisches Institut
Universität Erlangen-Nürnberg
D-91054 Erlangen
`voss@mi.uni-erlangen.de`

Gridded Data Interpolation with Restrictions on the First Order Derivatives

Jochen W. Schmidt and Marion Walther

Abstract. This paper is concerned with restricted interpolation of data sets given on rectangular grids using biquadratic C^1 and biquartic C^2 splines on refined grids. In extension of monotonicity preserving constraints, we consider piecewise constant lower and upper bounds for the first partial derivatives. Utilizing the corresponding univariate results as well as the tensor product structure, sufficient conditions for fulfilling the considered restrictions are constructed. Furthermore, the solvability of the arising inequalities can always be assured for strictly compatible data when placing the additional knots suitably.

AMS (MOS) classification: 65D07, 41A15, 41A29.

Keywords: Splines on refined grids, interpolation in a first order derivative strip, tensor product splines, constructive existence results, placements of the additional knots.

1. Introduction

Suppose a rectangular grid

$$\Delta \times \Sigma = \{x_0 < x_1 < \cdots < x_n\} \times \{y_0 < y_1 < \cdots < y_m\} \tag{1.1}$$

and corresponding data values $z_{i,j}$, $i = 0, \dots, n$, $j = 0, \dots, m$, are given. We are interested in a smooth function s defined on the domain $\Omega = [x_0, x_n] \times [y_0, y_m]$ interpolating at the data sites,

$$s(x_i, y_j) = z_{i,j}, \qquad i = 0, \dots, n, \ j = 0, \dots, m. \tag{1.2}$$

Depending on physical or aesthetic reasons, it may be desirable or even essential that the function s preserves certain shape properties of the data such as nonnegativity, monotonicity, or convexity. More general, range restrictions on the function values or on the derivatives may occur. Recently, gridded data interpolation sub-

Multivariate Approximation and Splines
G. Nürnberger, J. W. Schmidt, and G. Walz (eds.), pp. 289–305.

ISBN 3-7643-5654-5.

ject to restrictions on the function values was investigated in [11] and [13] using biquadratic and biquartic tensor product splines on refined rectangular grids

$$\Delta_1 \times \Sigma_1 = \{x_0 < \xi_1 < x_1 < \cdots < \xi_n < x_n\} \times \{y_0 < \eta_1 < y_1 < \cdots < \eta_m < y_m\}. \tag{1.3}$$

In addition, we mention the paper [8] where range restricted interpolation of scattered data is considered. Monotonicity preserving interpolatior ,applying biquadratic splines on $\Delta_1 \times \Sigma_1$ was treated in the paper [1]. Earlier papers in monotone interpolation using other types of splines are [2,3,5]; further we refer to the review paper [4]. In the present paper we are concerned with the construction of a smooth interpolating function s which satisfies the two-sided restrictions on the first order partial derivatives

$$L_{i,j} \leq \partial_1 s(x,y) \leq U_{i,j}, \tag{1.4}$$

$$K_{i,j} \leq \partial_2 s(x,y) \leq V_{i,j} \quad \forall (x,y) \in [x_{i-1}, x_i] \times [y_{j-1}, y_j], \tag{1.5}$$

$i = 1, \ldots, n$, $j = 1, \ldots, m$. The bounds $L_{i,j}$, $U_{i,j}$, $K_{i,j}$,$V_{i,j}$ are assumed to be given. Obviously, for $L_{i,j} = K_{i,j} = 0$, $U_{i,j} = V_{i,j} = \infty$ we are led to the monotonicity preserving interpolation. Also the special monotonicity problems treated in [5] can be included. We are in the position to offer a computational method which is always successful provided the data and the bounds are strictly compatible in a suitable sense. In contrast to [14], where rational splines are used in monotone interpolation, no heuristical search procedure is needed in order to fix the occurring parameters. In the forthcoming paper [15], the univariate problem that corresponds to (1.2), (1.4), (1.5) has been solved from a numerical point of view. This was done by applying quadratic C^1 and quartic C^2 splines on refined grids. In that paper also some applications are described which lead to the mathematical model of interpolation subject to two-sided restrictions on the first order derivative.

For handling the present interpolation problem (1.2), (1.4), (1.5) we select classes of splines for which no implicit systems of equations for satisfying the interpolation as well as the smoothness conditions occur. Sufficient conditions for the restrictions (1.4), (1.5) are derived using the corresponding univariate results as well as a nonnegativity lemma on tensor product splines. The result is a set of nonlinear inequalities. This system can shown to be solvable if the additional knots are suitably placed. Thus, the main point in solving the present interpolation problem is the placement of the additional knots introduced in refining the original grid. Finally, results of our numerical experiments are presented.

2. Univariate Case

2.1 Nonnegative and Strip Interpolation with Quadratic C^1 Splines

Suppose a grid $\Delta : x_0 < x_1 < \cdots < x_n$ of data sites is given, and the step sizes are denoted by $h_i = x_i - x_{i-1}$. We consider a quadratic C^1 spline with respect to the refinement Δ_1 of the grid Δ defined by adding one knot

$$\xi_i = \alpha_i x_i + \beta_i x_{i-1}, \; \alpha_i, \, \beta_i > 0, \; \alpha_i + \beta_i = 1 \tag{2.1}$$

in each interval $[x_{i-1}, x_i]$. As usual, the space of quadratic C^1 splines on Δ_1 is denoted by $S_2^1(\Delta_1)$. A spline $s \in S_2^1(\Delta_1)$ can be described in Bernstein-Bézier representation, namely

$$\begin{aligned} s(x) &= z_{i-1}v^2 + 2a_i vu + \zeta_i u^2 \quad \text{for } x \in [x_{i-1}, \xi_i], \\ s(x) &= \zeta_i v^2 + 2A_i vu + z_i u^2 \quad \text{for } x \in [\xi_i, x_i], \end{aligned} \tag{2.2}$$

$i = 1, \dots, n$, where u and v are the barycentric coordinates of x with respect to the considered interval, i.e.,

$$\begin{aligned} u &= (x - x_{i-1})/(\alpha_i h_i),\ v = (\xi_i - x)/(\alpha_i h_i) \text{ for } x \in [x_{i-1}, \xi_i], \\ u &= (x - \xi_i)/(\beta_i h_i),\ v = (x_i - x)/(\beta_i h_i) \quad \text{for } x \in [\xi_i, x_i]. \end{aligned} \tag{2.3}$$

In this way the continuity of s and the interpolation conditions

$$s(x_i) = z_i,\ i = 0, \dots, n, \tag{2.4}$$

are assured automatically. Requiring that the first derivative of s is continuous on $[x_0, x_n]$, we straightforwardly obtain the following sufficient and necessary conditions

$$\begin{aligned} a_i &= z_{i-1} + \frac{\alpha_i h_i}{2} p_{i-1},\ A_i = z_i - \frac{\beta_i h_i}{2} p_i, \\ \zeta_i &= \alpha_i A_i + \beta_i a_i,\ i = 1, \dots, n, \end{aligned} \tag{2.5}$$

where the parameters

$$p_i = s'(x_i),\ i = 0, \dots, n, \tag{2.6}$$

are introduced. Consequently, a quadratic spline on a fixed refined grid Δ_1 is uniquely determined by the function values (2.4) and the first order derivatives (2.6) at the original knots $x_0, \dots, x_n$. In convex and monotone interpolation these splines are exposed in [16].

It follows immediately from the Bernstein-Bézier representation of polynomials that the nonnegativity of all B-ordinates on the interval $[x_{i-1}, x_i]$ implies the nonnegativity of s on this interval. Since the B-ordinate ζ_i is a positive linear combination of a_i and A_i, we get

Lemma 1. *Assume $i \in \{1, \dots, n\}$ is fixed and $s \in S_2^1(\Delta_1)$. If s satisfies the conditions*

$$z_{i-1}, a_i, A_i, z_i \geq 0, \tag{2.7}$$

then s is nonnegative on $[x_{i-1}, x_i]$.

Obviously, if the ratios α_i, β_i are considered as fixed the conditions (2.7) depend linearly on the parameters z_{i-1}, z_i, p_{i-1}, p_i. By applying the lemma to each subinterval, a sufficient condition for the nonnegativity of s on $[x_0, x_n]$ reads $z_0, a_1, A_1, z_1, \dots, z_{n-1}, a_n, A_n, z_n \geq 0$. For nonnegative data values z_i this system of inequalities is solved by $p_0 = p_1 = \dots = p_n = 0$, i.e., nonnegative quadratic spline interpolants always exist in $S_2^1(\Delta_1)$; compare with [9].

Remark. Due to the C^1 conditions the data values z_i for $i = 1,\ldots,n-1$ are positive linear combinations of the neighbouring B-ordinates A_i and a_{i+1}. Thus, in order to assure the nonnegativity of s on $[x_0, x_n]$ in the above set of conditions $z_1,\ldots,z_{n-1} \geq 0$ can be dropped.

Next, in view of the above restrictions (1.4), (1.5), we are interested in interpolants $s \in S_2^1(\Delta_1)$ that satisfy the following strip constraints on the first order derivative,

$$L_i \leq s'(x) \leq U_i, \qquad x \in [x_{i-1}, x_i], \tag{2.8}$$

$i = 1,\ldots,n$. The bounds L_i and U_i are assumed to be strictly compatible with the slopes $\tau_i = (z_i - z_{i-1})/h_i$. This means in the present case

$$L_i < \tau_{i-1}, \tau_i < U_i, \qquad i = 1,\ldots,n, \tag{2.9}$$

where τ_0 is defined, *e.g.*, by $\tau_0 = \tau_1$.

Since the first order derivative of s is piecewise polygonal with respect to the grid Δ_1, we have

Lemma 2. *For fixed $i \in \{1,\ldots,n\}$ a spline $s \in S_2^1(\Delta_1)$ is monotone (increasing) on $[x_{i-1}, x_i]$ if and only if*

$$p_{i-1} \geq 0,\ s'(\xi_i) = 2\tau_i - \alpha_i p_{i-1} - \beta_i p_i \geq 0,\ p_i \geq 0. \tag{2.10}$$

Considering the differences $s' - L_i$, $U_i - s'$ the conditions

$$L_i \leq p_{i-1},\ p_i,\ 2\tau_i - \alpha_i p_{i-1} - \beta_i p_i \leq U_i, \quad i = 1,\ldots,n, \tag{2.11}$$

are seen to be equivalent to the strip constraints (2.8). Dealing with the solvability of the problem (2.11) we are led to the following proposition proved earlier in [15].

Proposition 1. *Assume the data and bounds are strictly compatible in the sense of (2.9). If the ratios $\alpha_i \in (0,1)$ are chosen according to*

$$\frac{1}{\alpha_i} \geq \max\left\{\frac{\tau_i - \tau_{i-1}}{U_i - \tau_i}, \frac{\tau_i - \tau_{i-1}}{L_i - \tau_i}\right\},\ i = 1,\ldots,n, \tag{2.12}$$

then interpolation subject to the strip conditions (2.8) is always successful in $S_2^1(\Delta_1)$.

Proof: Using the feasible particular choices

$$p_i = \tau_i, \qquad i = 0,\ldots,n, \tag{2.13}$$

the first and second set of inequalities formulated in (2.11) are consequences of the compatibility assumptions (2.9). On the other hand, this step motivates the definition (2.9) of compatibility. The requirement (2.12) implies

$$L_i - \tau_i \leq \alpha_i(\tau_i - \tau_{i-1}) \leq U_i - \tau_i,$$

being equivalent to

$$L_i \leq 2\tau_i - \alpha_i \tau_{i-1} - \beta_i \tau_i \leq U_i. \qquad \square$$

2.2 Nonnegative and Strip Interpolation with Quartic C^2 Splines

In this section we consider the space $S_4^2(\Delta_1)$ of quartic C^2 splines on refined grids Δ_1. Again, $s \in S_4^2(\Delta_1)$ is represented expediently in Bernstein-Bézier form,

$$
\begin{aligned}
s(x) &= z_{i-1}v^4 + 4a_i v^3 u + 6b_i v^2 u^2 + 4c_i v u^3 + \zeta_i u^4, \; x \in [x_{i-1}, \xi_i], \\
s(x) &= \zeta_i v^4 + 4C_i v^3 u + 6B_i v^2 u^2 + 4A_i v u^3 + z_i u^4, \; x \in [\xi_i, x_i],
\end{aligned} \tag{2.14}
$$

$i = 1, \ldots, n$. It can be verified or is given, e.g., in [12] that $s \in S_4^2(\Delta_1)$ if and only if

$$
\begin{aligned}
a_i &= z_{i-1} + \frac{\alpha_i h_i}{4} p_{i-1}, \; b_i = z_{i-1} + \frac{\alpha_i h_i}{2} p_{i-1} + \frac{\alpha_i^2 h_i^2}{12} P_{i-1}, \\
c_i &= \frac{1+\beta_i}{2} b_i + \frac{\alpha_i}{2} B_i - \frac{\alpha_i h_i^2}{24} \Pi_i, \\
A_i &= z_i - \frac{\beta_i h_i}{4} p_i, \; B_i = z_i - \frac{\beta_i h_i}{4} p_i + \frac{\beta_i^2 h_i^2}{2} P_i, \\
C_i &= \frac{\beta_i}{2} b_i + \frac{1+\alpha_i}{2} B_i - \frac{\beta_i h_i^2}{24} \Pi_i, \\
\zeta_i &= \alpha_i C_i + \beta_i c_i, \quad i = 1, \ldots, n,
\end{aligned} \tag{2.15}
$$

holds, where the parameters

$$
p_i = s'(x_i), \; P_i = s''(x_i), \; i = 0, \ldots, n, \; \Pi_i = s''(\xi_i), \; i = 1, \ldots, n, \tag{2.16}
$$

are used. Obviously, for fixed $i \in \{1, \ldots, n\}$ the conditions

$$
z_{i-1}, \; a_i, \; b_i, \; c_i, \; C_i, \; B_i, \; A_i, \; z_i \geq 0,
$$

are sufficient for the nonnegativity of $s \in S_4^2(\Delta_1)$ on $[x_{i-1}, x_i]$. If we consider the subspace $\tilde{S}^2(\Delta_1)$ of $S_4^2(\Delta_1)$ defined by choosing the parameters $P_i = 0$, $i = 0, \ldots, n$, and $\Pi_i = 0$, $i = 1, \ldots, n$, these conditions reduce to the requirements (2.7).

It can by easily established that the conditions

$$
p_{i-1}, \; \frac{4(b_i - a_i)}{\alpha_i h_i}, \; \frac{4(c_i - b_i)}{\alpha_i h_i}, p_i, \; \frac{4(B_i - C_i)}{\beta_i h_i}, \; \frac{4(A_i - B_i)}{\beta_i h_i} \geq 0, \tag{2.17}
$$

imply the monotonicity of $s \in S_4^2(\Delta_1)$ on $[x_{i-1}, x_i]$. In the subspace $\tilde{S}^2(\Delta_1)$ these conditions are identical with the requirements (2.10).

The strip constraints (2.8) are fulfilled if the system of inequalities

$$
\begin{aligned}
L_i &\leq p_{i-1}, \; \frac{4(b_i - a_i)}{\alpha_i h_i}, \; \frac{4(c_i - b_i)}{\alpha_i h_i} \leq U_i, \\
L_i &\leq p_i, \; \frac{4(B_i - C_i)}{\beta_i h_i}, \; \frac{4(A_i - B_i)}{\beta_i h_i} \leq U_i, \quad i = 1, \ldots, n,
\end{aligned} \tag{2.18}
$$

holds; cf. [15]. In $\tilde{S}^2(\Delta_1)$ these conditions reduce to (2.11).

2.3 Nonnegative and Monotone Interpolation with Quintic C^3 Splines

In order to achieve C^3 continuity without any implicit condition, quintic splines on refined grids Δ_1 may be used. The Bernstein-Bézier form of these splines $s \in S_5^3(\Delta_1)$ reads

$$s(x) = z_{i-1}v^5 + 5a_iv^4u + 10b_iv^3u^2 + 10c_iv^2u^3 + 5d_ivu^4 + \zeta_iu^5,\ x \in [x_{i-1}, \xi_i],$$
$$s(x) = \zeta_iv^5 + 5D_iv^4u + 10C_iv^3u^2 + 10B_iv^2u^3 + 5A_ivu^4 + z_iu^5,\ x \in [\xi_i, x_i],$$

$i = 1, \ldots, n$. Clearly, the interpolation condition is satisfied, and the following choice of the B-ordinates assures the C^3 property,

$$\begin{aligned}
a_i &= z_{i-1} + \frac{\alpha_i h_i}{5}p_{i-1},\ b_i = z_{i-1} + \frac{2\alpha_i h_i}{5}p_{i-1} + \frac{\alpha_i^2 h_i^2}{20}P_{i-1},\\
c_i &= z_{i-1} + \frac{3\alpha_i h_i}{5}p_{i-1} + \frac{3\alpha_i^2 h_i^2}{20}P_{i-1} + \frac{\alpha_i^3 h_i^3}{60}G_{i-1},\\
A_i &= z_i - \frac{\beta_i h_i}{5}p_i,\ B_i = z_i - \frac{2\beta_i h_i}{5}p_i + \frac{\beta_i^2 h_i^2}{20}P_i,\\
C_i &= z_i - \frac{3\beta_i h_i}{5}p_i + \frac{3\beta_i^2 h_i^2}{20}P_i - \frac{\beta_i^3 h_i^3}{60}G_i,\\
d_i &= -\beta_i^2 b_i + 2\beta_i c_i - \frac{\alpha_i^3}{\beta_i}B_i + \frac{\alpha_i^2}{\beta_i}C_i,\\
D_i &= -\alpha_i^2 B_i + 2\alpha_i C_i - \frac{\beta_i^3}{\alpha_i}b_i + \frac{\beta_i^2}{\alpha_i}c_i,\\
\zeta_i &= \beta_i d_i + \alpha_i D_i, \quad i = 1, \ldots, n.
\end{aligned} \tag{2.19}$$

The parameters are now the derivatives

$$p_i = s'(x_i),\ P_i = s''(x_i),\ G_i = s'''(x_i),\ i = 0, \ldots, n. \tag{2.20}$$

Sufficient conditions for $s \in S_5^3(\Delta_1)$ to be nonnegative or monotone increasing on $[x_0, x_n]$ are obtained when requiring the nonnegativity of all B-ordinates of s or s', respectively. It can be verified that both sets of inequalities are satisfied if $p_i = P_i = G_i = 0$, $i = 0, \ldots, n$, provided $z_i \geq 0$, $i = 0, \ldots, n$, or $\tau_i \geq 0$, $i = 1, \ldots, n$, respectively; see [15].

3. Nonnegativity Lemma for Tensor Product Splines

In this section sufficient nonnegativity conditions for tensor product splines $s \in S \otimes T$ defined on the rectangle $[a, b] \times [c, d]$ are derived. The univariate spline spaces S and T are assumed to be finite dimensional and linear. Further we suppose $S \subset C^0[a, b]$ and $T \subset C^0[c, d]$.

The justification of the following tensor product constructions is explained, e.g., in [6] and [7]. The space $S \otimes T$ is formed by all finite linear combinations of functions described by

$$(s \otimes t)(x,y) = s(x) \cdot t(y), \quad (x,y) \in [a,b] \times [c,d],\ s \in S,\ t \in T. \tag{3.1}$$

The tensor product functional $\lambda \otimes \mu$ on $S \otimes T$ of two linear functionals λ on S and μ on T is defined by linear extension from the rule

$$(\lambda \otimes \mu)(s \otimes t) = \lambda(s) \cdot \mu(t), \quad s \in S,\ t \in T. \tag{3.2}$$

The following lemma allows the construction of sufficient nonnegativity conditions in the tensor product space $S \otimes T$ by means of such conditions in S and T.

Lemma 3. *Let λ_i, $i = 1, \dots, N$, be linear functionals on S such that for $s \in S$*

$$\lambda_i(s) \geq 0,\ i = 1, \dots, N \Longrightarrow s(x) \geq 0\ \forall x \in [a,b] \tag{3.3}$$

and μ_j, $j = 1, \dots, M$, be linear functionals on T such that for $s \in T$

$$\mu_j(s) \geq 0,\ j = 1, \dots, M \Longrightarrow s(y) \geq 0\ \forall y \in [c,d]. \tag{3.4}$$

Then for $s \in S \otimes T$ the conditions

$$(\lambda_i \otimes \mu_j)(s) \geq 0,\ i = 1, \dots, N,\ j = 1, \dots, M, \tag{3.5}$$

are sufficient for

$$s(x,y) \geq 0 \quad \forall (x,y) \in [a,b] \times [c,d].$$

It is easily seen that $N \geq \dim S$ and $M \geq \dim T$. Under the special assumption $N = \dim S$ and $M = \dim T$, a proof of Lemma 3 is given in [11], while the general case is treated rigorously in the paper [10]; see also [17] for a more direct proof using the bipolar theorem.

Of course, Lemma 3 also applies if S and T are spaces of derivatives of the considered splines. Therefore the lemma is suitable for deriving sufficient conditions for the requirements (1.4), (1.5), and for others on partial derivatives.

4. Gridded Data Interpolation with Restrictions on the Derivatives

4.1 Biquadratic C^1 splines

In this section we introduce the tensor product space $S_2^1(\Delta_1) \otimes S_2^1(\Sigma_1)$. Here $S_2^1(\Delta_1)$ denotes the set of all quadratic splines with respect to the refined grid

$$\Delta_1 : \{x_0 < \xi_1 < x_1 < \cdots < \xi_n < x_n\},$$

which are continuously differentiable on the interval $[x_0, x_n]$.

The additional knots

$$\xi_i = \alpha_i x_i + \beta_i x_{i-1}, \quad i = 1, \dots, n,$$

are determined by the ratios $\alpha_i, \beta_i > 0$ with $\alpha_i + \beta_i = 1$, $i = 1, \dots, n$. Analogously, the notation $S_2^1(\Sigma_1)$ for the space of all quadratic C^1 splines with respect to the grid

$$\Sigma_1 : \{y_0 < \eta_1 < y_1 < \dots < \eta_m < y_m\}$$

is used, and the additional knots $\eta_j = \gamma_j y_j + \delta_j y_{j-1}$ are determined by the ratios $\gamma_j, \delta_j > 0$ with $\gamma_j + \delta_j = 1$, $j = 1, \dots, m$.

The tensor product space $S_2^1(\Delta_1) \otimes S_2^1(\Sigma_1)$ consists of all biquadratic splines on the refined grid $\Delta_1 \times \Sigma_1$ that are continuous and possess continuous first partial derivatives. It follows that also the mixed second partial derivative is continuous.

A general result formulated, e.g., in [6] implies that a biquadratic spline on a fixed refined grid is uniquely determined by the function values, the first partial and the second mixed partial derivatives at the original data sites, *i.e.*, the interpolation problem of finding a spline $s \in S_2^1(\Delta_1) \otimes S_2^1(\Sigma_1)$ that satisfies, with given $z_{i,j}$, $p_{i,j}$, $q_{i,j}$, and $r_{i,j}$, the conditions

$$\begin{aligned} s(x_i, y_j) = z_{i,j},\ \partial_1 s(x_i, y_j) = p_{i,j},\ \partial_2 s(x_i, y_j) = q_{i,j},\ \partial_1\partial_2 s(x_i, y_j) = r_{i,j} \\ i = 0, \dots, n,\ j = 0, \dots, m, \end{aligned} \tag{4.1}$$

is uniquely solvable.

4.2 Sufficient Conditions for Preserving the Constraints (1.4), (1.5)

At first sufficient conditions for the nonnegativity of the first order partial derivative in x-direction of a biquadratic spline on a subrectangle $[x_{i-1}, x_i] \times [y_{j-1}, y_j] \subset \Omega$ are derived, where $i \in \{1, \dots, n\}$ and $j \in \{1, \dots, m\}$ are considered as fixed. Further the refined grids Δ_1 and Σ_1 are assumed to be fixed. The linear functionals λ_1, λ_2, λ_3 introduced below result from the monotonicity conditions (2.10) while the linear functionals μ_1, μ_2, μ_3, μ_4 arise from the nonnegativity conditions (2.5), (2.7). Then in view of Lemma 3 we get

Proposition 2. *Suppose the linear functionals λ_k, $k = 1, 2, 3$, for $s \in S_2^1(\Delta_1)$ are defined by*

$$\begin{aligned} &\lambda_1(s) := s'(x_{i-1}),\ \lambda_2(s) := s'(x_i), \\ &\lambda_3(s) := \frac{2}{h_i}\left(s(x_i) - s(x_{i-1})\right) - \alpha_i s'(x_{i-1}) - \beta_i s'(x_i), \end{aligned} \tag{4.2}$$

where $h_i = x_i - x_{i-1}$ denotes the step size. Further for $s \in S_2^1(\Sigma_1)$ the functionals are given by

$$\begin{aligned} &\mu_1(s) := s(y_{j-1}),\ \mu_2(s) := s(y_j), \\ &\mu_3(s) := s(y_{j-1}) + \frac{1}{2}\gamma_j k_j s'(y_{j-1}),\ \mu_4(s) := s(y_j) - \frac{1}{2}\delta_j k_j s'(y_j), \end{aligned} \tag{4.3}$$

where $k_j = y_j - y_{j-1}$. *If* $s \in S_2^1(\Delta_1) \otimes S_2^1(\Sigma_1)$ *fulfils the conditions*

$$(\lambda_k \otimes \mu_l)(s) \geq 0, \; k = 1,2,3, \; l = 1,2,3,4, \tag{4.4}$$

then we have

$$\partial_1 s(x,y) \geq 0 \quad \forall (x,y) \in [x_{i-1}, x_i] \times [y_{j-1}, y_j] \,.$$

Next, we derive sufficient conditions for the strip requirement (1.4) for a biquadratic spline $s \in S_2^1(\Delta_1) \otimes S_2^1(\Sigma_1)$ restricted to a subrectangle $[x_{i-1}, x_i] \times [y_{j-1}, y_j]$ with fixed $i \in \{1, \dots, n\}$, $j \in \{1, \dots, m\}$.

Proposition 3. *The requirements*

$$L_{i,j} \leq (\lambda_k \otimes \mu_l)(s) \leq U_{i,j} \qquad k = 1,2,3, \; l = 1,2,3,4, \tag{4.5}$$

imply

$$L_{i,j} \leq \partial_1 s(x,y) \leq U_{i,j} \qquad \forall (x,y) \in [x_{i-1}, x_i] \times [y_{j-1}, y_j] \,.$$

Proof: For $u = U_{i,j} x$ and $s \in S_2^1(\Delta_1) \otimes S_2^1(\Sigma_1)$ the difference $u - s$ also belongs to the space $S_2^1(\Delta_1) \otimes S_2^1(\Sigma_1)$. Applying Proposition 2 we obtain that the conditions

$$(\lambda_k \otimes \mu_l)(u - s) \geq 0, \qquad k = 1,2,3, \; l = 1,2,3,4, \tag{4.6}$$

are sufficient for $\partial_1(u - s) \geq 0$, i.e., for

$$\partial_1 s(x,y) \leq U_{i,j} \qquad \forall (x,y) \in [x_{i-1}, x_i] \times [y_{j-1}, y_j] \,.$$

Further, it can be easily verified that the relations

$$(\lambda_k \otimes \mu_l)(u) = U_{i,j}, \qquad k = 1,2,3, \; l = 1,2,3,4,$$

hold true. For example, we have

$$(\lambda_3 \otimes \mu_1)(u) = 2U_{i,j} h_i / (x_i - x_{i-1}) - \alpha_i U_{i,j} - \beta_i U_{i,j} = U_{i,j} \,.$$

Thus, the conditions (4.6) are equivalent to

$$(\lambda_k \otimes \mu_l)(s) \leq U_{i,j}, \qquad k = 1,2,3, \; l = 1,2,3,4.$$

Analogously, the inequalities for the lower bounds are proved . □

Considering the sufficient conditions (4.5) on each subrectangle, sufficient requirements for the considered constraints (1.4) are obtained. Analogously, the strip constraints for the first partial derivative with respect to the y-direction (1.5) can be handled. Sufficient for (1.5) are the requirements

$$K_{i,j} \leq (\mu_l \otimes \lambda_k)(s) \leq V_{i,j}, \quad l = 1,2,3,4, \; k = 1,2,3; \tag{4.7}$$

but now the functionals μ_l are defined by (4.3) with α_i substituted for γ_j, β_i for δ_j, h_i for k_j, and x_i for y_j, while the functionals λ_k are given by (4.2) with the reverse substitution.

4.3 Existence and Construction of Constrained C^1 Interpolants

We now consider the systems of inequalities (4.5) and (4.7) being sufficient for the strip conditions (1.4) and (1.5) on the first order derivatives. The parameters $p_{i,j}$, $q_{i,j}$, $r_{i,j}$, $i = 0,\dots,n$, $j = 0,\dots,m$, and the ratios α_i, $i = 1,\dots,n$, and γ_j, $j = 1,\dots,m$, are at our disposal to fulfil these systems of inequalities. In order to assure the solvability we have to assume that the data and the bounds in (1.4) and (1.5) are strictly compatible. In extending the definition (2.9) from the univariate case, strict compatibility is now introduced by

$$L_{i,j} < \tau_{i-1,j-1}, \tau_{i-1,j}, \tau_{i,j-1}, \tau_{i,j} < U_{i,j} \quad i = 1,\dots,n,\ j = 1,\dots,m, \tag{4.8}$$

and

$$K_{i,j} < \sigma_{i-1,j-1}, \sigma_{i-1,j}, \sigma_{i,j-1}, \sigma_{i,j} < V_{i,j} \quad i = 1,\dots,n,\ j = 1,\dots,m. \tag{4.9}$$

Here $\tau_{i,j}$ and $\sigma_{i,j}$ are the partial slopes,

$$\tau_{i,j} = (z_{i,j} - z_{i-1,j})/h_i, \quad i = 1,\dots,n,\ j = 0,\dots,m, \tag{4.10}$$

and

$$\sigma_{i,j} = (z_{i,j} - z_{i,j-1})/k_j, \quad i = 0,\dots,n,\ j = 1,\dots,m. \tag{4.11}$$

In addition, we may choose, e.g.,

$$\tau_{0,j} = \tau_{1,j},\ j = 0,\dots,m, \quad \sigma_{i,0} = \sigma_{i,1},\ i = 0,\dots,n, \tag{4.12}$$

and

$$\begin{aligned} &\varrho_{i,j} = (z_{i,j} - z_{i-1,j} - z_{i,j-1} + z_{i-1,j-1})/(h_i k_j), \\ &\varrho_{0,0} = \varrho_{1,1},\ \varrho_{0,j} = \varrho_{1,j},\ \varrho_{i,0} = \varrho_{i,1},\ i = 1,\dots,n,\ j = 1,\dots,m. \end{aligned} \tag{4.13}$$

In order to find a solution of the sufficient conditions the choices

$$p_{i,j} = \tau_{i,j},\ q_{i,j} = \sigma_{i,j},\ r_{i,j} = 0 \quad i = 0,\dots,n,\ j = 0,\dots,m, \tag{4.14}$$

turn out to be suitable. We get the following main result of this paper.

Theorem 1. *Suppose the data and the bounds on the first order derivatives are strictly compatible, i.e., (4.8) and (4.9) are assumed to be valid. The ratios are chosen according to*

$$\alpha_i \in (0,1),\ 1/\alpha_i \ge \max\{M_i, X_i\},\ i = 1,\dots,n, \tag{4.15}$$

and

$$\gamma_j \in (0,1),\ 1/\gamma_j \ge \max\{N_j, Y_j\},\ j = 1,\dots,m, \tag{4.16}$$

with

$$\begin{aligned} M_i = \max_{j=1,\dots,m} \max \Big\{ &\frac{\tau_{i,j} - \tau_{i-1,j}}{U_{i,j} - \tau_{i,j-1}}, \frac{\tau_{i,j} - \tau_{i-1,j}}{L_{i,j} - \tau_{i,j-1}}, 2\frac{\tau_{i,j-1} - \tau_{i-1,j-1}}{U_{i,j} - \tau_{i,j-1}}, \\ &2\frac{\tau_{i,j-1} - \tau_{i-1,j-1}}{L_{i,j} - \tau_{i,j-1}}, \frac{\tau_{i,j} - \tau_{i-1,j}}{U_{i,j} - \tau_{i,j}}, \frac{\tau_{i,j} - \tau_{i-1,j}}{L_{i,j} - \tau_{i,j}} \Big\}, \\ X_i = 2h_i \max_{j=1,\dots,m} \max \Big\{ &\frac{\varrho_{i-1,j}}{V_{i,j} - \sigma_{i-1,j}}, \frac{\varrho_{i-1,j}}{K_{i,j} - \sigma_{i-1,j}} \Big\}, \end{aligned}$$

$$
\begin{aligned}
N_j = \max_{i=1,\ldots,n} \max \Big\{ & \frac{\sigma_{i,j} - \sigma_{i,j-1}}{V_{i,j} - \sigma_{i-1,j}}, \frac{\sigma_{i,j} - \sigma_{i,j-1}}{K_{i,j} - \sigma_{i-1,j}}, 2\frac{\sigma_{i-1,j} - \sigma_{i-1,j-1}}{V_{i,j} - \sigma_{i-1,j}}, \\
& 2\frac{\sigma_{i-1,j} - \sigma_{i-1,j-1}}{K_{i,j} - \sigma_{i-1,j}}, \frac{\sigma_{i,j} - \sigma_{i,j-1}}{V_{i,j} - \sigma_{i,j}}, \frac{\sigma_{i,j} - \sigma_{i,j-1}}{K_{i,j} - \sigma_{i,j}} \Big\}, \\
Y_j = & 2k_j \max_{i=1,\ldots,n} \max \Big\{ \frac{\varrho_{i,j-1}}{U_{i,j} - \tau_{i,j-1}}, \frac{\varrho_{i,j-1}}{L_{i,j} - \tau_{i,j-1}} \Big\}.
\end{aligned}
$$

Then there always exist biquadratic C^1 spline interpolants s on the grid $\Delta_1 \times \Sigma_1$ satisfying the restrictions

$$
L_{i,j} \le \partial_1 s(x,y) \le U_{i,j},\ K_{i,j} \le \partial_2 s(x,y) \le V_{i,j} \qquad i = 1,\ldots,n,\ j = 1,\ldots,m.
$$

One of these interpolants is given by (4.14).

Proof: It is shown that the system (4.5) is solved on all subrectangles by the values (4.14) and the ratios α_i, γ_j satisfying

$$
1/\alpha_i \ge M_i,\ i = 1,\ldots,n, \qquad 1/\gamma_j \ge Y_j,\ j = 1,\ldots,m.
$$

Let $i \in \{1,\ldots,n\}$ and $j \in \{1,\ldots,m\}$ be fixed. Utilizing (4.14), for $k = 1,2$ and $l = 1,2,3,4$ the conditions (4.5) turn out to be consequences of the compatibility assumptions (4.8). Analogous to the univariate case, the requirements

$$
1/\alpha_i \ge \max_{j=1,\ldots,m} \max \left\{ \frac{\tau_{i,j-1} - \tau_{i-1,j-1}}{U_{i,j} - \tau_{i,j-1}}, \frac{\tau_{i,j-1} - \tau_{i-1,j-1}}{L_{i,j} - \tau_{i,j-1}} \right\}
$$

and

$$
1/\alpha_i \ge \max_{j=1,\ldots,m} \max \left\{ \frac{\tau_{i,j} - \tau_{i-1,j}}{U_{i,j} - \tau_{i,j}}, \frac{\tau_{i,j} - \tau_{i-1,j}}{L_{i,j} - \tau_{i,j}} \right\}
$$

imply (4.5) for $k = 3$ and $l = 1,2$. Of course, these inequalities are valid if the conditions (4.15) hold true. Next, using (4.14) we get

$$
\begin{aligned}
(\lambda_3 \otimes \mu_3)(s) = & \frac{2}{h_i}(z_{i,j-1} - z_{i-1,j-1}) - \alpha_i p_{i-1,j-1} - \beta_i p_{i,j-1} \\
& + \frac{\gamma_j k_j}{h_i}(q_{i,j-1} - q_{i-1,j-1}) - \frac{\gamma_j k_j}{2}(\alpha_i r_{i-1,j-1} + \beta_i r_{i,j-1}) \\
= & 2\tau_{i,j-1} - \alpha_i \tau_{i-1,j-1} - \beta_i \tau_{i,j-1} + \gamma_j k_j \varrho_{i,j-1} \\
= & \alpha_i(\tau_{i,j-1} - \tau_{i-1,j-1}) + \tau_{i,j-1} + \gamma_j k_j \varrho_{i,j-1}.
\end{aligned}
$$

If the conditions

$$
1/\alpha_i \ge 2 \max_{j=1,\ldots,m} \max \left\{ \frac{\tau_{i,j-1} - \tau_{i-1,j-1}}{U_{i,j} - \tau_{i,j-1}}, \frac{\tau_{i,j-1} - \tau_{i-1,j-1}}{L_{i,j} - \tau_{i,j-1}} \right\}
$$

and

$$1/\gamma_j \geq 2k_j \max_{i=1,\dots,n} \max\left\{\frac{\varrho_{i,j-1}}{U_{i,j}-\tau_{i,j-1}}, \frac{\varrho_{i,j-1}}{L_{i,j}-\tau_{i,j-1}}\right\}$$

which are consequences of (4.15), (4.16) are required, the inequalities

$$\begin{aligned}(L_{i,j}-\tau_{i,j-1})/2 &\leq \alpha_i(\tau_{i,j-1}-\tau_{i-1,j-1}) \leq (U_{i,j}-\tau_{i,j-1})/2,\\ (L_{i,j}-\tau_{i,j-1})/2 &\leq \gamma_j k_j \varrho_{i,j-1} \leq (U_{i,j}-\tau_{i,j-1})/2\end{aligned}$$

hold. Adding these inequalities the desired relation $L_{i,j} \leq (\lambda_3 \otimes \mu_3)(s) \leq U_{i,j}$ is obtained. Further, in view of

$$\begin{aligned}(\lambda_3 \otimes \mu_4)(s) &= \alpha_i(\tau_{i,j}-\tau_{i-1,j}) + \gamma_j(\tau_{i,j}-\tau_{i,j-1}) + \tau_{i,j-1}\\ &= \gamma_j[\tau_{i,j}+\alpha_i(\tau_{i,j}-\tau_{i,j-1})] + (1-\gamma_j)[\tau_{i,j-1}+\alpha_i(\tau_{i,j}-\tau_{i,j-1})],\end{aligned}$$

the estimates $L_{i,j} \leq (\lambda_3 \otimes \mu_4)(s) \leq U_{i,j}$ are established for all $\gamma_j \in (0,1)$ if

$$1/\alpha_i \geq \max\left\{\frac{\tau_{i,j}-\tau_{i-1,j}}{U_{i,j}-\tau_{i,j}}, \frac{\tau_{i,j}-\tau_{i-1,j}}{L_{i,j}-\tau_{i,j}}, \frac{\tau_{i,j}-\tau_{i-1,j}}{U_{i,j}-\tau_{i,j-1}}, \frac{\tau_{i,j}-\tau_{i-1,j}}{L_{i,j}-\tau_{i,j-1}}\right\}.$$

These requirements are incorporated in (4.15). Consequently, the inequalities (4.5) are solved by the above parameters and ratios. It can be verified analogously that the set of inequalities (4.7) is fulfilled by the parameters (4.14) and ratios if they are chosen according to

$$1/\alpha_i \geq X_i,\ i=1,\dots,n, \qquad 1/\gamma_j \geq N_j,\ j=1,\dots,m. \qquad \square$$

4.4 Constrained C^2 and C^3 Interpolation

The problem of strip interpolation (1.2), (1.4), (1.5) is likewise solvable in the tensor product space $S_4^2(\Delta_1) \otimes S_4^2(\Sigma_1)$, i.e., the smoothness C^2 can be achieved. Indeed, in the subspaces $\tilde{S}^2(\Delta_1) \subset S_4^2(\Delta_1)$ and $\tilde{S}^2(\Sigma_1) \subset S_4^2(\Sigma_1)$ the sufficient nonnegativity and monotonicity conditions are identical with the ones in $S_2^1(\Delta_1)$ and $S_2^1(\Sigma_1)$, respectively. Therefore, in $\tilde{S}^2(\Delta_1) \otimes \tilde{S}^2(\Sigma_1)$ again we are led to the requirements (4.5) and (4.7), and the choices (4.15) and (4.16) of the ratios α_i and γ_j are suitable also in $S_4^2(\Delta_1) \otimes S_4^2(\Sigma_1)$.

Monotonicity preserving C^3 interpolation is always successful in the tensor product space $S_5^3(\Delta_1) \otimes S_5^3(\Sigma_1)$ since this holds true in the univariate case. In addition, now the ratios can be fixed a priori. However, it must left as an open problem whether the strip interpolation (1.2), (1.4), (1.5) has solutions in this space of C^3 splines.

4.5 Choice Functionals

The main result of the preceding considerations says that the constrained interpolation problem (1.2), (1.4), (1.5) is always solvable in $S_2^1(\Delta_1) \otimes S_2^1(\Sigma_1)$. One solution is given by the parameters (4.14) and the ratios

$$\begin{aligned} 1/\alpha_i &= \max\{2, M_i, X_i\}, \quad i = 1, \ldots, n, \\ 1/\gamma_j &= \max\{2, N_j, Y_j\}, \quad j = 1, \ldots, m. \end{aligned} \tag{4.17}$$

However, this is not the only spline solution. In general, it is possible to find improved splines. To this end, usually an objective function is introduced which is minimized subject to the shape constraints, in the present case to (4.5), (4.7). For simplicity the ratios are fixed a priori, e.g., according to (4.17). There are several choice functionals in use. We mention the bivariate Holladay and the thin plate functionals. For practical reasons we prefer a fit-and-modify-method. This method consists of a modification of an initial approximation s^0 in order to fulfil the shape conditions. Here we have selected the special functional

$$\sum_{i=1}^{n} \sum_{j=1}^{m} \sum_{k,l \in \{1,2,3,4\}} \left[\left(\mu_k^i \otimes \mu_l^j \right) (s - s^0) \right]^2 \tag{4.18}$$

where the functionals $\mu_l = \mu_l^j$ and $\mu_k = \mu_k^i$ are defined by (4.3) directly or with the substitution mentioned subsequently to (4.7).

5. Numerical Experiments

For testing purposes programs have been implemented in $MATLAB^{TM}$ utilizing the Spline and Optimization Toolboxes. In all tests the ratios are taken according to the formula (4.17) while the derivative parameters $p_{i,j}$, $q_{i,j}$, $r_{i,j}$ are computed in different ways. We have plotted the spline EXIST; see Figure 2. Here the parameters (4.14) are used which are suitable in proving the existence (Theorem 1). Next these parameters are determined by minimizing the functional (4.18) without any constraints. The resulting spline in Figure 3 is denoted by FMUNC. The spline FMOPT is obtained by minimizing (4.18) but now subject to the strip constraints (4.5) and (4.7). For computing the spline FMOPT a quadratic non-separable optimization problem in $3(n+1)(m+1)$ unknowns has to be solved numerically. In Figures 2–4 the partial derivative $\partial_2 s$ is plotted, too.

We have tested several monotonicity examples from [2,3,5], but now supplied with strictly compatible lower and upper bounds. For graphical representation the following data values $z_{i,j}$ visualized in Figure 1 are used.

$x_i \backslash y_j$	-0.07	0.33	0.55	0.69	0.84	0.93	0.98	1.02	1.08	1.13
-2.30	-34.54	-34.54	-34.54	-34.54	-34.54	-34.54	-3.06	-2.86	-2.37	-1.89
-1.61	-13.82	-13.80	-13.82	-13.82	-13.82	-2.68	-2.28	-1.92	-1.60	-1.30
-0.92	-10.10	-10.10	-10.10	-10.10	-2.52	-1.88	-1.63	-1.39	-1.17	-0.95
-0.51	-7.26	-7.26	-7.26	-4.82	-2.22	-1.56	-1.32	-1.10	-0.90	-0.71
-0.22	-5.66	-5.66	-4.88	-3.34	-1.98	-1.41	-1.15	-0.92	-0.72	-0.54
0.00	-4.53	-4.13	-3.35	-2.73	-1.78	-1.28	-1.05	-0.81	-0.60	-0.41

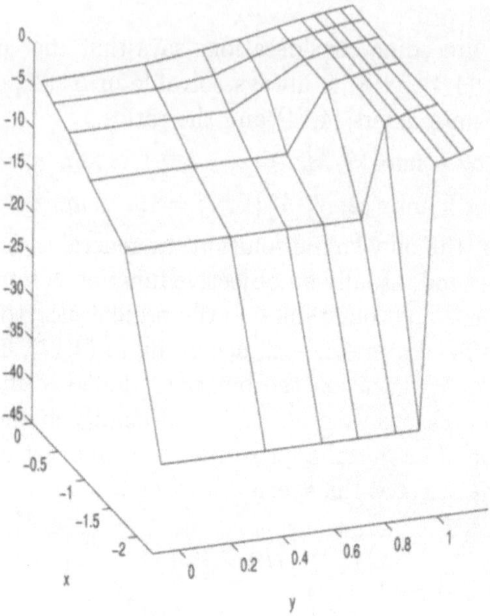

Fig. 1. Data points.

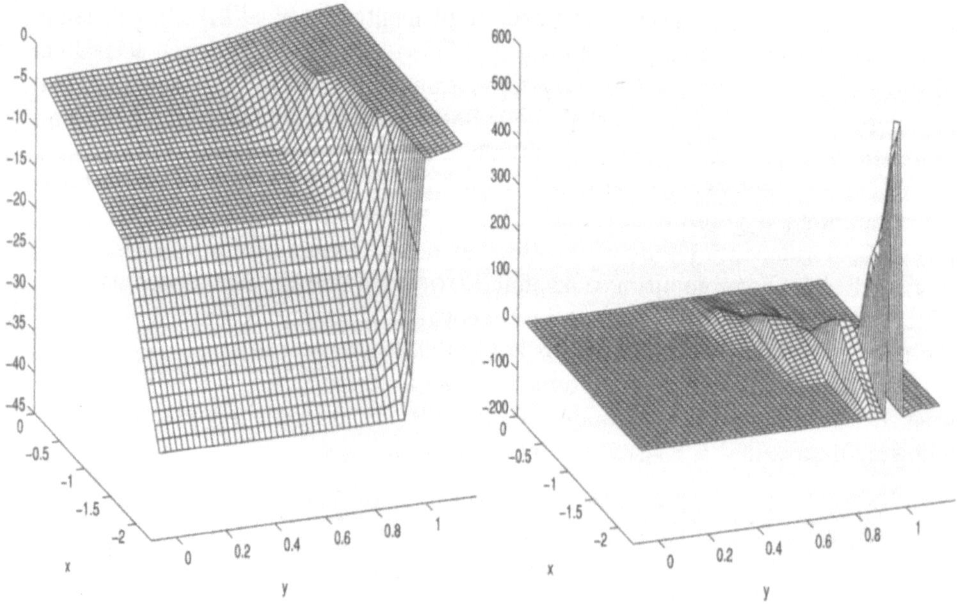

Fig. 2. Spline EXIST (derivatives (4.14)).

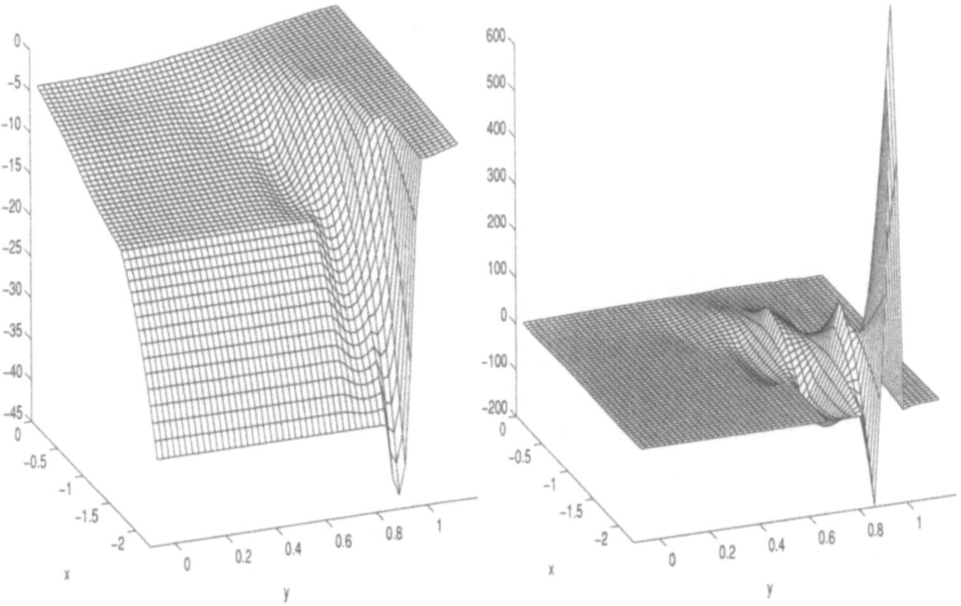

Fig. 3. Spline FMUNC (unconstrained fit and modify).

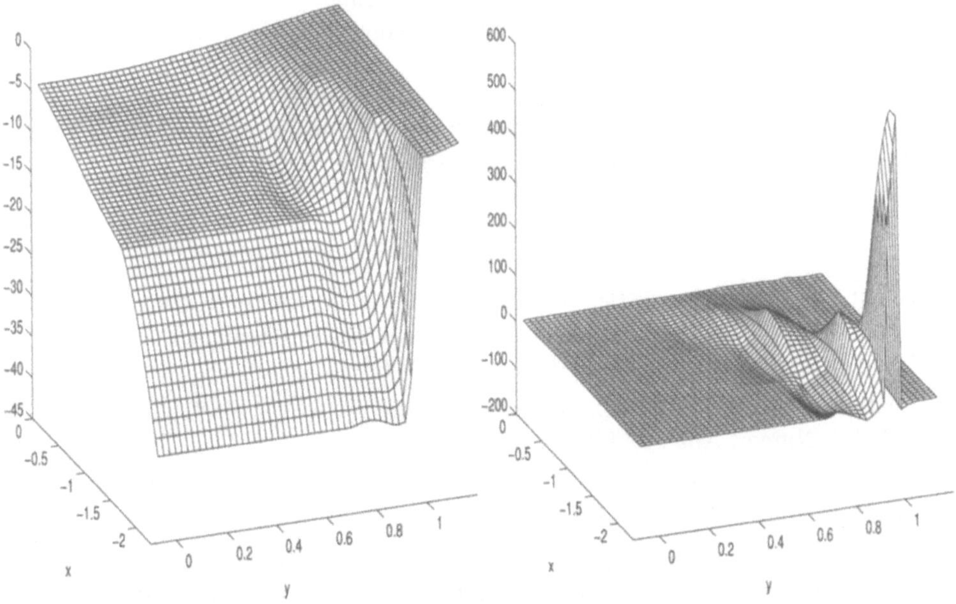

Fig. 4. Spline FMOPT (constrained fit and modify).

The lower and upper bounds on the first order partial derivatives are prescribed as follows

$$\begin{aligned} L_{i,j} &= \min\{\tau_{i-1,j-1}, \tau_{i-1,j}, \tau_{i,j-1}, \tau_{i,j}\} - \varepsilon, \\ K_{i,j} &= \min\{\sigma_{i-1,j-1}, \sigma_{i-1,j}, \sigma_{i,j-1}, \sigma_{i,j}\} - \varepsilon, \\ U_{i,j} &= \max\{\tau_{i-1,j-1}, \tau_{i-1,j}, \tau_{i,j-1}, \tau_{i,j}\} + \varepsilon, \\ V_{i,j} &= \max\{\sigma_{i-1,j-1}, \sigma_{i-1,j}, \sigma_{i,j-1}, \sigma_{i,j}\} + \varepsilon \quad i = 1, \ldots, 5,\ j = 1, \ldots, 9, \end{aligned} \tag{5.1}$$

where $\varepsilon = 20$. Applying formula (4.17) we obtain

$$\begin{aligned} &\alpha_1 = 0.5, \alpha_2 = 0.0185, \alpha_3 = 0.1443, \alpha_4 = 0.5, \alpha_5 = 0.5, \\ &\gamma_1 = \gamma_2 = \gamma_3 = 0.5, \gamma_4 = 0.1979, \gamma_5 = 0.0808, \gamma_6 = 0.0159, \\ &\gamma_7 = 0.0160, \gamma_8 = \gamma_9 = 0.5. \end{aligned}$$

The initial spline s^0 needed in the objective functions (4.18) is computed by means of divided difference approximations for the partial derivatives $\partial_1 s^0$, $\partial_2 s^0$, and $\partial_1\partial_2 s^0$ at the grid points.

Comparing the Figures 2–4, the spline FMOPT seems to be most pleasant; the spline EXIST in this case is acceptable, too.

Finally we remark that parameters $\varepsilon > 0$ smaller than $\varepsilon = 20$ are tempting in the bounds (5.1). But, if ε tends to zero so also the ratios (4.17). Therefore, it is natural that numerical problems may arise for small $\varepsilon > 0$. In addition, the resulting interpolants though theoretical from C^1 look then like being continuous only.

Acknowledgements. The work of the second author was supported by the Deutsche Forschungsgemeinschaft under grant Schm 968/2-2.

References

1. Asaturyan S. and Unsworth K., A C^1 monotonicity preserving surface interpolation scheme, in *Mathematics of Surfaces III*, Handscomb D. C. (ed.), Oxford University Press, Oxford, 1989, 243–266.
2. Beatson R. K. and Ziegler Z., Monotonicity preserving surface interpolation, SIAM J. Numer. Anal. **22** (1985), 401–411.
3. Carlson R. E. and Fritsch F. N., Monotone piecewise bicubic interpolation, SIAM J. Numer. Anal. **22** (1985), 386–400.
4. Costantini P., Shape-preserving interpolation with variable degree polynomial splines, in *Advanced Course on Fairshape*, Hoschek J. and Kaklis P. (eds.), Teubner, Stuttgart, 1996, 87–114.
5. Costantini P. and Fontanella F., Shape-preserving bivariate interpolation, SIAM J. Numer. Anal. **27** (1990), 488–506.

6. de Boor C., *A Practical Guide to Splines*, Springer, New York, 1978.
7. Greub W. H., *Multilinear Algebra*, Springer, Berlin-Heidelberg-New York, 1967.
8. Herrmann M., Mulansky B., and Schmidt J. W., Scattered data interpolation subject to piecewise quadratic range restrictions, in *Scattered Data Fitting*, Le Méhauté A., Schumaker L. L., Traversoni L. (guest eds.), J. Comput. Appl. Math. **73** (1996), 209–223.
9. Lahtinen A., Positive Hermite interpolation by quadratic splines, SIAM J. Math. Anal. **24** (1993), 223–233.
10. Mulansky B., Tensor products of convex cones, this proceedings.
11. Mulansky B. and Schmidt J. W., Nonnegative interpolation by biquadratic splines on refined rectangular grids, in *Wavelets, Images and Surface Fitting*, Laurent P.-J., Le Méhauté A., and Schumaker L. L. (eds.), AK Peters, Wellesley, 1994, 379–386.
12. Mulansky B. and Schmidt J. W., Constructive methods in convex C^2 interpolation using quartic splines, Numer. Algorithms **12** (1996), 111–124.
13. Mulansky B., Schmidt J. W., and Walther M., Tensor product spline interpolation subject to piecewise bilinear lower and upper bounds, in *Advanced Course on Fairshape*, Hoschek J. and Kaklis P. (eds.), Teubner, Stuttgart, 1996, 201–216.
14. Schmidt J. W., Positive, monotone, and S-convex C^1-interpolation on rectangular grids, Computing **48** (1992), 363–371.
15. Schmidt J. W., Interpolation in a derivative strip, Computing **58** (1997).
16. Schumaker L. L., On shape preserving quadratic spline interpolation, SIAM J. Numer. Anal. **20** (1983), 854–864.
17. Walther M., Interpolation of gridded data by tensor product splines subject to piecewise constant lower and upper bounds for the first partial derivatives, Preprint MATH–NM–16–1996, TU Dresden, 1996.

Jochen W. Schmidt
Institute of Numerical Mathematics
Technical University of Dresden
D-01062 Dresden, Germany
`jschmidt@math.tu-dresden.de`

Marion Walther
Institute of Numerical Mathematics
Technical University of Dresden
D-01062 Dresden, Germany
`walther@math.tu-dresden.de`

Affine Frames and Multiresolution

Joachim Stöckler

Abstract. We use generalized Laurent operators for an extended study of multivariate affine frames. Our main concern are frames which are generated by multiresolution with a single scaling function. No stability constraints are posed on the translates of the scaling function. The connection to the transfer operator is worked out for these families. Moreover, we give a new representation of the lifting scheme.. This provides more insight into the problem of controlling the stability of the wavelet bases during the process of lifting. The relation to generalized Toeplitz operators is discussed.

1. Introduction

The interest in wavelet bases over the past two decades has naturally extended to a special type of frames, which are called *wavelet frames* or *affine frames*. Their definition by dilation and translation is similar to wavelet bases, but they allow redundancy and are therefore more flexible with respect to certain applications. Various aspects of such frames are discussed in [1, 4, 5, 9, 10, 11, 17, 18, 19, 20].

Our general setup is as follows. We consider families in $L_2(\mathbb{R}^d)$, which are generated from a finite subset $\Psi = \{\psi_i;\ i \in I\}$ of $L_2(\mathbb{R}^d)$ by means of dilation and translation. These operations are defined by two real and invertible $d \times d$ matrices M and L. M is supposed to be expansive, i.e. all eigenvalues of M have absolute value larger than 1, and L defines a lattice $\mathcal{L} = L\mathbb{Z}^d$. Then we let

$$X = (\Psi, M, \mathcal{L}) := \{|\det M|^{j/2}\psi_i(M^j \cdot - Lk);\ j \in \mathbb{Z},\ k \in \mathbb{Z}^d,\ i \in I\}. \tag{1}$$

Without further assumptions we refer to X as an *affine family*. Our treatment is based on two operators

$$\mathcal{T}_X : L_2(\mathbb{R}^d) \longrightarrow \ell_2(I \times \mathbb{Z} \times \mathbb{Z}^d), \qquad f \mapsto \left(\langle f, \eta\rangle_{L_2(\mathbb{R}^d)}\right)_{\eta \in X}, \tag{2}$$

Multivariate Approximation and Splines
G. Nürnberger, J. W. Schmidt, and G. Walz (eds.), pp. 307–320.

ISBN 3-7643-5654-5.

which will be called the *analysis operator*, and its adjoint, the *synthesis operator*

$$T_X^* : \ell_2(I \times \mathbb{Z} \times \mathbb{Z}^d) \longrightarrow L_2(\mathbb{R}^d), \qquad \mathbf{d} = (d_\eta)_{\eta \in X} \mapsto \sum_{\eta \in X} d_\eta \, \eta \, . \tag{3}$$

This terminology is used since the coefficients of $T_X f$ contain localized information of f and its Fourier transform $\widehat{f}$. The affine family X is an *affine frame*, if the operator T_X is bounded from above and below, so there exist two constants $A, B > 0$ such that

$$A\|f\|_{L_2(\mathbb{R}^d)} \leq \|T_X f\|_{\ell_2(I\times\mathbb{Z}\times\mathbb{Z}^d)} \leq B\|f\|_{L_2(\mathbb{R}^d)} \tag{4}$$

holds. Hence affine frames are total in $L_2(\mathbb{R}^d)$, but can be linearly dependent. As a consequence of the lower estimate in (4), they allow stable representations of any function $f \in L_2(\mathbb{R}^d)$. Finding frame decompositions of a given function f can be viewed as an infinite least squares problem: there always exists a unique solution $\mathbf{d} \in \ell_2(I \times \mathbb{Z} \times \mathbb{Z}^d)$ which has minimal ℓ_2-norm and gives $f = T_X^* \mathbf{d}$. We call X a *Bessel family* if the upper bound in (4) holds.

In this paper we continue our work in [19] and describe several aspects of affine frames. The central technique, which is described in section 2, is a combination of analysis and synthesis operators of two affine frames in such a way that a generalized Laurent operator is obtained. Several new results have been found in this way, see [19,20,21]. In section 3 we deal with the special case, where the family Ψ is constructed by multiresolution. This issue appears in the work of [1,5,17,18,21]. We show that fast pyramidal algorithms can be used for the frame operators instead of numerical integration. Furthermore, the case of compactly supported generators deserves special attention. The connection to the transfer operator, which is one of the main tools for regularity estimates of scaling functions [8,10,12,14,23], is worked out here. The norm estimate of Theorem 3 shows that all generalized Laurent operators which come up in this way have a nice structure. A new representation of the lifting scheme [3,22] is presented in section 4. It amounts to a lower triangular operator matrix whose diagonal is the identity operator. This gives some new insight into the yet unsolved problem under which conditions the lifting scheme produces a new pair of biorthogonal wavelet bases. Although we cannot give simple sufficient conditions, we believe that further research based on techniques from operator algebra might lead to interesting results in this area.

Let us shortly fix some of our notations. M^* denotes the transpose of M. We identify sequences $\mathbf{c} \in \ell_2(\mathbb{Z}^d)$ and the corresponding Fourier series $C \in L_2(Q)$ without further notice. Here $Q = (-\pi, \pi)^d$. We also use the bracket product

$$[\, f \, , g \,] = \sum_{k \in \mathbb{Z}^d} f(\,.+2\pi k)\, \overline{g}(\,.+2\pi k)$$

which defines a function in $L_1(Q)$ whenever $f, g \in L_2(\mathbb{R}^d)$. The Fourier transform is defined as $\widehat{f} = \int_{\mathbb{R}^d} f(x) e^{-ix\cdot} \, dx$.

2. Operator Analysis

Given a scaling matrix $M \in \mathbb{R}^{d \times d}$ and two finite families $\Psi = \{\psi_i;\ i \in I\}$ and $\Theta = \{\theta_i;\ i \in I\} \subset L_2(\mathbb{R}^d)$, we define the affine families

$$X = (\Psi, M, \mathcal{L}_X) \qquad \text{and} \qquad Y = (\Theta, M, \mathcal{L}_Y) \tag{5}$$

as in (1). In most cases of interest we will choose $\mathcal{L}_Y = \mathbb{Z}^d$. If both X and Y are Bessel families, their analysis operators (2) are bounded. Our study of affine frames is based on the following principle.

Proposition 1. *If the families X and Y in (5) are Bessel families, then the combination of analysis and synthesis operators*

$$\mathcal{T}_X \mathcal{T}_Y^* : \ell_2(I \times \mathbb{Z} \times \mathbb{Z}^d) \longrightarrow \ell_2(I \times \mathbb{Z} \times \mathbb{Z}^d)$$

is a generalized Laurent operator; i.e. it is a bounded linear operator which commutes with the bilateral shift on $\ell_2(H)$ where $H = \ell_2(I \times \mathbb{Z}^d)$.

Let us recall, for the reader's convenience, the definition of the bilateral shift

$$\mathcal{U} : \ell_2(H) \longrightarrow \ell_2(H), \qquad \mathcal{U}\left((\mathbf{c}_j)_{j\in\mathbb{Z}}\right) = (\mathbf{c}_{j-1})_{j\in\mathbb{Z}} \,.$$

Here the biinfinite vector $(\mathbf{c}_j)_{j\in\mathbb{Z}}$ is an element of $\ell_2(H)$, hence its entries $\mathbf{c}_j$ belong to H and the series $\sum_{j\in\mathbb{Z}} \|\mathbf{c}_j\|_H^2$ is finite. The proof of Proposition 1 is given in [19,20] and it relies on the scaling invariance of both families X and Y with respect to the same scaling matrix M.

The choice of two different generating families Ψ and Θ introduces much more freedom for the study of affine frames than the usual approach in the literature, where only one family is considered, see [10,17]. A special form of this can be found in [23] where regularity estimates for scaling functions are investigated.

Two standard techniques from operator theory are useful which are connected to generalized Laurent operators. Their detailed description can be found in [2,13]. First we can consider $\mathcal{T}_X \mathcal{T}_Y^*$ as a biinfinite operator matrix

$$\mathcal{T}_X \mathcal{T}_Y^* = \begin{pmatrix} \ddots & \ddots & \ddots & \cdots \\ \ddots & \boxed{S_0} & S_{-1} & S_{-2} & \cdots \\ \ddots & S_1 & \boxed{S_0} & S_{-1} & \ddots \\ \cdots & S_2 & S_1 & \boxed{S_0} & \ddots \\ & \cdots & \ddots & \ddots & \ddots \end{pmatrix} \tag{6}$$

where each S_j, $j \in \mathbb{Z}$, is a bounded linear operator on H. (As usual, we put frames around the entries on the main diagonal of biinfinite matrices. The same will be done for the 0-th entry of biinfinite vectors.) The exact definition of S_j is obtained in the following way. For arbitrary $\mathbf{c} \in H$ we let

$$\bar{\mathbf{c}} := (\ldots, 0, 0, \boxed{\mathbf{c}}, 0, 0, \ldots) \in \ell_2(H).$$

Then the equation

$$\mathcal{T}_X \mathcal{T}_Y^*(\bar{\mathbf{c}}) = (\ldots, S_{-2}\mathbf{c}, S_{-1}\mathbf{c}, \boxed{S_0\mathbf{c}}, S_1\mathbf{c}, S_2\mathbf{c}, \ldots)$$

defines all operators S_j, $j \in \mathbb{Z}$. Boundedness of $\mathcal{T}_X \mathcal{T}_Y^*$ implies that $\sum_{j\in\mathbb{Z}} S_j^* S_j$ is a bounded selfadjoint operator on H. Vice versa, this condition on S_j, $j \in \mathbb{Z}$, is not sufficient for the boundedness of $\mathcal{T}_X \mathcal{T}_Y^*$, in general. Instead, the relations

$$\left\| \sum_{j\in\mathbb{Z}} S_j^* S_j \right\|_{\mathcal{B}(H)}^{1/2} \le \|\mathcal{T}_X \mathcal{T}_Y^*\|_{\mathcal{B}(\ell_2(H))} \le \sum_{j\in\mathbb{Z}} \|S_j\|_{\mathcal{B}(H)} \tag{7}$$

can be proven in the same way as for usual Laurent matrices (where $H = \mathbb{C}$). We will see in section 3, however, that the finiteness of the left and the right hand sides in (7) are equivalent for certain frame operators.

The second tool for generalized Laurent operators is to build from $\mathcal{T}_X \mathcal{T}_Y^*$ the operator-valued function

$$F(z) := \sum_{j\in\mathbb{Z}} z^j S_j, \qquad z \in \mathbb{C} \quad \text{with} \quad |z| = 1. \tag{8}$$

This defines a (weakly) measurable function, which maps the unit circle $\mathbb{T}$ to the Banach algebra $\mathcal{B}(H)$ of bounded linear operators on H. A central theorem about generalized Laurent operators [2,page 235] states that

$$\|\mathcal{T}_X \mathcal{T}_Y^*\|_{\mathcal{B}(\ell_2(H))} = \operatorname*{ess\,sup}_{|z|=1} \|F(z)\|_{\mathcal{B}(H)} , \tag{9}$$

and that the mapping of generalized Laurent operators to $L_\infty(\mathbb{T}, \mathcal{B}(H))$ is one-to-one. Note that the second inequality in (7) can be easily deduced from (9).

Finally we mention the explicit representation of S_j in (6), which is contained in [20].

Proposition 2. *Let X and Y in (5) be Bessel families where $\mathcal{L}_Y = \mathbb{Z}^d$. Then each operator S_j in (6) is a square operator matrix of size $|I|$ with entries*

$$S_j^{(m,n)}(C) = \frac{|\det M|^{j/2}}{|\det L_X|} \left[C(M^{*j} L_X^{*\,-1} \cdot)\, \widehat{\theta}_n(M^{*j} L_X^{*\,-1} \cdot),\ \widehat{\psi}_m(L_X^{*\,-1} \cdot) \right]$$

for all $m, n \in I$ and $C \in L_2(Q)$.

3. Affine Frames by Multiresolution

We further deal with a special situation where both families Ψ and Θ are generated by scaling functions in $L_2(\mathbb{R}^d)$. This issue is important for numerical applications of frames. Several papers address this topic, see [1,5,17,18,21]. Let us first give the needed specifications.

The scaling matrix M is an integer matrix. We denote by Γ_j a complete set of representers of the cosets of $M^{*j}\mathbb{Z}^d/\mathbb{Z}^d$. Moreover, we are given two functions φ, $\varrho \in L_2(\mathbb{R}^d)$ which are compactly supported and satisfy the scaling relations

$$\widehat{\varphi}(M^*\cdot) = m_0\widehat{\varphi}, \qquad \widehat{\varrho}(M^*\cdot) = \widetilde{m}_0\widehat{\varrho}$$

a.e. in $\mathbb{R}^d$ where m_0 and $\widetilde{m}_0$ are trigonometric polynomials in $\mathbb{R}^d$. In this setting we do not require stability or existence of frame bounds for the translates of φ or ϱ. This is in accordance with [17], but differs completely from the frame multiresolution analysis which was introduced in [1]. Finally, the *wavelets* are defined by their corresponding scaling relations

$$\widehat{\psi}_i(M^*\cdot) = m_i\widehat{\varphi}, \qquad \widehat{\theta}_i(M^*\cdot) = \widetilde{m}_i\widehat{\varrho}, \qquad i \in I,$$

with trigonometric polynomials m_i, $\widetilde{m}_i$. For later use we let $\psi_0 = \varphi$ and $\theta_0 = \varrho$. Note that we restrict our attention to finite masks for the scaling functions and wavelets. Further investigations are needed for more general cases, see e.g. the framework in [7]. Unlike the construction of wavelet bases from multiresolution analysis, the cardinality of I is not connected to the determinant of the scaling matrix. The affine families to be considered are $X = (\Psi, M, \mathbb{Z}^d)$ and $Y = (\Theta, M, \mathbb{Z}^d)$.

The characterization of tight affine frames X which are defined by multiresolution from a scaling function φ was recently obtained by Ron and Shen [17,Theorem 6.5] under mild assumptions on the Fourier transform of φ. In this section we present a connection of the frame operator and the *transfer operator* in [6,8,12,14]. This is the mapping T_u on $L_2(Q)$ with

$$T_u(f) = \sum_{k\in\Gamma_1} u({M^*}^{-1}(\cdot + 2\pi k))\; f({M^*}^{-1}(\cdot + 2\pi k))$$

where u is a given function in $L_\infty(Q)$. Note that its adjoint is given by

$$T_u^*(f) = |\det M|\,\overline{u}\cdot f(M^*\cdot), \qquad f \in L_2(Q),$$

which corresponds to upsampling and convolution of the corresponding sequences of Fourier coefficients. The representation of S_j in Proposition 2 can be transformed by means of the scaling relations. With $\mu = |\det M|$ and by simple algebraic manipulations we obtain

$$S_0^{(i,n)}(C)(\xi) = [\,\widehat{\theta}_n\,,\,\widehat{\psi}_i\,](\xi)\cdot C(\xi),$$

$$S_j^{(i,n)}(C)(\xi) = \mu^{j/2}\widetilde{m}_n(M^{*j-1}\xi)\prod_{\nu=0}^{j-2}\widetilde{m}_0(M^{*\nu}\xi)\;[\,\widehat{\varrho}\,,\,\widehat{\psi}_i\,](\xi)\cdot C(M^{*j}\xi)$$

for $j > 0$, and

$$S_j^{(i,n)}(C)(\xi) = \mu^{j/2} \sum_{k \in \Gamma_{|j|}} \left[\widehat{\theta}_n, \widehat{\psi}_i(M^{*|j|}\cdot)\right](M^{*j}(\xi + 2\pi k)) \cdot C(M^{*j}(\xi + 2\pi k))$$
$$= \mu^{j/2} T_{\overline{m_i}} \circ T_{\overline{m_0}}^{|j|-1} \left([\widehat{\theta}_n, \widehat{\varphi}] \cdot C\right)(\xi)$$

for $j < 0$.

3.1 Pyramidal algorithm

The above representation of the entries of the square operator matrix S_j gives rise to efficient numerical algorithms. For this purpose we recall that the functions m_i, $\widetilde{m}_i$, $i \in I' = I \cup \{0\}$, are trigonometric polynomials. Moreover, we can conclude, by means of the Poisson Summation Formula, that all functions $[\widehat{\theta}_n, \widehat{\psi}_i]$, $i, n \in I'$, are trigonometric polynomials, too. Hence S_0 represents a matrix filter which acts on a vector $\mathbf{c} \in H = \ell_2(I \times \mathbb{Z}^d)$ and has finite filters in each entry. S_j for $j > 0$ consists of FIR filters and upsampling, while S_j for $j < 0$ uses downsampling instead. The overall operator

$$\mathcal{T}_X \mathcal{T}_Y^*(\mathbf{c}_j)_{j \in \mathbb{Z}} = \left(\sum_{k \in \mathbb{Z}} S_k \, \mathbf{c}_{j-k}\right)_{j \in \mathbb{Z}}, \qquad (\mathbf{c}_j)_{j \in \mathbb{Z}} \in \ell_2(H),$$

can be decomposed into three parts. The first part is the filtering

$$(\mathbf{c}_j)_{j \in \mathbb{Z}} \mapsto (S_0 \mathbf{c}_j)_{j \in \mathbb{Z}}.$$

The second part

$$(\mathbf{c}_j)_{j \in \mathbb{Z}} \mapsto \left(\sum_{k > 0} S_k \, \mathbf{c}_{j-k}\right)_{j \in \mathbb{Z}}$$

corresponds to Mallat's pyramidal algorithm [15] with filters $\widetilde{m}_i$, $i \in I'$, which is often called the reconstruction algorithm for the wavelets $\widetilde{\psi}_i$. The only difference consists in the final filtering with FIR filters $[\widehat{\varrho}, \widehat{\psi}_i]$ in each component. The third part

$$(\mathbf{c}_j)_{j \in \mathbb{Z}} \mapsto \left(\sum_{k < 0} S_k \, \mathbf{c}_{j-k}\right)_{j \in \mathbb{Z}}$$

uses $[\widehat{\theta}_n, \widehat{\varphi}]$ as initial filters in each component $\mathbf{c}_j$, $j \in \mathbb{Z}$, and then applies the decomposition algorithm with filters $\overline{m_i}$, $i \in I'$. Hence the computational complexity for the evaluation of $\mathcal{T}_X \mathcal{T}_Y^*$ is comparable to the sum of complexities for the wavelet reconstruction and decomposition with the specified filters.

3.2 Operator norms

In order to find exact expressions for the operator norm of S_j, $j \in \mathbb{Z}$, we further build the selfadjoint operators $S_j^* S_j$ for $j > 0$ and $S_j S_j^*$ for $j < 0$. The products are chosen in this way in order to obtain pure multiplication operators; more precisely, the computation with adjoints as in [19] shows that these are square operator matrices of size $|I|$ with entries

$$(S_j^* S_j)^{(\nu,n)}(C)(\xi) = T_{\overline{\widetilde{m}_\nu} \widetilde{m}_n} \circ T^{j-1}_{|\widetilde{m}_0|^2} \Big(\sum_{i \in I} \Big| [\widehat{\varrho}, \widehat{\psi}_i] \Big|^2 \Big)(\xi) \cdot C(\xi) \tag{10}$$

for $j > 0$, and

$$(S_j S_j^*)^{(\nu,n)}(C)(\xi) = T_{\overline{m_\nu} m_n} \circ T^{|j|-1}_{|m_0|^2} \Big(\sum_{i \in I} \Big| [\widehat{\theta}_i, \widehat{\varphi}] \Big|^2 \Big)(\xi) \cdot C(\xi)$$

for $j < 0$, each of them acting on $L_2(Q)$. Note that the product of transfer operators in these formulas is applied to the functions

$$g := \sum_{i \in I} \Big| [\widehat{\varrho}, \widehat{\psi}_i] \Big|^2, \qquad h := \sum_{i \in I} \Big| [\widehat{\theta}_i, \widehat{\varphi}] \Big|^2,$$

but not to C. The same argument as in the previous section shows that g and h are trigonometric polynomials. Hence the vector spaces

$$E_g = \operatorname{Span} \{ T^j_{\widetilde{u}}(g);\ j \geq 0 \} \qquad \text{with} \qquad \widetilde{u} = |\widetilde{m}_0|^2,$$
$$E_h = \operatorname{Span} \{ T^j_u(h);\ j \geq 0 \} \qquad \text{with} \qquad u = |m_0|^2$$

are finite dimensional spaces of trigonometric polynomials, see [8,Prop. 3.1] for more details. The Perron-Frobenius theory for non-negative operators on finite dimensional space leads to the following result.

Theorem 3. *Let X and Y be Bessel families defined by multiresolution with compactly supported scaling functions and wavelets. Then the operators S_j in (6) satisfy*

$$\sum_{j \in \mathbb{Z}} \|S_j\|_{\mathcal{B}(H)} < \infty. \tag{11}$$

In particular, formal matrix calculus is allowed for multiplication and transposition of the biinfinite operator matrices in (6).

Proof: We show a stronger result, namely that (11) holds if and only if

$$\Big\| \sum_{j>0} S_j^* S_j \Big\|_{\mathcal{B}(H)} < \infty \qquad \text{and} \qquad \Big\| \sum_{j<0} S_j S_j^* \Big\|_{\mathcal{B}(H)} < \infty. \tag{12}$$

Then the assertion follows from (7), since boundedness of $\mathcal{T}_X \mathcal{T}_Y^*$ implies that the first series in (12) is finite, and boundedness of $\mathcal{T}_Y \mathcal{T}_X^*$ gives finiteness of the second series.

Let us now prove the equivalence of (11) and (12). Obviously, (11) implies (12) by the triangle inequality and since $\ell_1(\mathbb{Z}) \subset \ell_2(\mathbb{Z})$. For the converse implication we only deal with the first series in (12), since the second series is treated analogously. We also omit the trivial case where all $\widetilde{m}_n$, $n \in I$, are identically zero. Let $\nu = n \in I$ be chosen such that $\widetilde{m}_n$ is a nontrivial trigonometric polynomial. Then (10) and (12) imply that

$$\Big\| \sum_{j>0} (S_j^* S_j)^{(n,n)}(C) \Big\|_{L_2(Q)} = \Big\| \sum_{j>0} T_{|\widetilde{m}_n|^2} \circ T_{\widetilde{u}}^{j-1}(g) \cdot C \Big\|_{L_2(Q)} < \infty$$

for all $C \in L_2(Q)$ with norm 1. Since the last operator is a multiplication on $L_2(Q)$, we further conclude that

$$\Big\| \sum_{j>0} T_{|\widetilde{m}_n|^2} \circ T_{\widetilde{u}}^{j-1}(g) \Big\|_{L_\infty(Q)} < \infty.$$

Let r denote the spectral radius of the transfer operator $T_{\widetilde{u}}$ when restricted to its invariant subspace E_g. Since E_g is finite dimensional and the operator is non-negative on E_g, we know that r is an eigenvalue of $T_{\widetilde{u}}$ with a non-negative eigenfunction $f \in E_g$, $f \neq 0$. If we write f in terms of the basis of E_g, which is constituted by the Krylov sequence g, $T_{\widetilde{u}}(g)$, $T_{\widetilde{u}}^{\dim E_g - 1}(g)$, then the triangle inequality leads directly to

$$\Big(\sum_{j>0} r^{j-1} \Big) \cdot \| T_{|\widetilde{m}_n|^2}(f) \|_{L_\infty(Q)} = \Big\| \sum_{j>0} T_{|\widetilde{m}_n|^2} \circ T_{\widetilde{u}}^{j-1}(f) \Big\|_{L_\infty(Q)} < \infty.$$

Since f is non-negative, we have $T_{|\widetilde{m}_n|^2}(f) \neq 0$, and hence $r < 1$ follows. From here it is easy to prove (11). Note that by (10)

$$\|S_j\|_{\mathcal{B}(H)}^2 = \|S_j^* S_j\|_{\mathcal{B}(H)} \le \Big(\sum_{\nu, n \in I} \| T_{\overline{\widetilde{m}_\nu} \widetilde{m}_n} \circ T_{\widetilde{u}}^{j-1}(g) \|_{L_\infty(Q)}^2 \Big)^{1/2}$$

where the summation enters because $S_j^* S_j$ is a multiplication operator on vectors in $H = L_2(Q)^I$. The spectral radius formula assures that there exists a constant $K = K(\varepsilon)$ such that

$$\|T_{\widetilde{u}}^{j-1}(g)\|_{L_\infty(Q)} \le K (r+\varepsilon)^{j-1} \|g\|_{L_\infty(Q)} \qquad \text{for all} \qquad j > 0.$$

If we let $\varepsilon > 0$ be small enough such that $r + \varepsilon < 1$, then

$$\sum_{j>0} \|S_j\|_{\mathcal{B}(H)} \le \widetilde{K} \sum_{j>0} (r+\varepsilon)^{(j-1)/2} < \infty,$$

where the constant $\widetilde{K}$ includes the remaining terms $\|g\|^{1/2}_{L_\infty(Q)}$ and the sum over the operator norms of $T_{\widetilde{\widetilde{m_\nu}} \widetilde{m_n}}$. Thus we have shown that the part of the series in (11) for $j > 0$ is finite. As mentioned above, the remaining part is treated in exactly the same way. $\square$

The proof contains a result which is interesting by itself.

Corollary 4. *Under the same assumptions as in Theorem 3 the transfer operator $T_{\widetilde{u}}$ restricted to E_g has spectral radius strictly less than 1. The same is true for the transfer operator T_u restricted to E_h.*

Remark. A similar technique for the proof has been applied in several articles, see e.g. [8,12].

4. The Lifting Scheme

There is a nice application of the generalized Laurent operator in connection with the lifting scheme which was introduced in [3,22]. As in [22] we confine ourselves to the univariate case and scaling by $M = (2)$. The starting point are two biorthogonal multiresolution analyses and associated compactly supported wavelet bases. Hence with the same notation as in section 3 the functions φ and ϱ are generators of two multiresolution analyses of $L_2(\mathbb{R})$, with respect to scaling by 2 and translates by $\mathbb{Z}$, and they satisfy the biorthogonality relation

$$[\widehat{\varrho}, \widehat{\varphi}] = 1 \qquad \text{a.e. in } \mathbb{R}. \tag{13}$$

Furthermore, the functions ψ and θ are the generators of two biorthogonal wavelet bases. (We can drop the index set I, since only one wavelet is needed for each family. We still use m_0, m_1 etc. for the scaling filters.) This means that the affine families $X = (\psi, 2, \mathbb{Z})$ and $Y = (\theta, 2, \mathbb{Z})$ are biorthogonal Riesz bases of $L_2(\mathbb{R})$. The orthogonality relations can be expressed by the identities for the bracket products

$$[\widehat{\theta}, \widehat{\varphi}] = 0, \qquad [\widehat{\varrho}, \widehat{\psi}] = 0, \qquad \text{and} \qquad [\widehat{\theta}, \widehat{\psi}] = 1 \tag{14}$$

which hold a.e. in $\mathbb{R}$. The representation of the operators S_j in section 3 gives

$$\mathcal{T}_X\,\mathcal{T}_Y^* = \begin{pmatrix} \ddots & \ddots & \ddots & \ddots & & \\ \ddots & 0 & \boxed{\mathrm{id}_H} & 0 & 0 & \cdots \\ \cdots & 0 & 0 & \boxed{\mathrm{id}_H} & 0 & \ddots \\ & & \ddots & \ddots & \ddots & \ddots \end{pmatrix} = \mathrm{id}_{\ell_2(H)}\,,$$

which is another way to express the biorthogonality. Here H is simply $\ell_2(\mathbb{Z})$.

The lifting scheme tries to change the scaling filters m_i and $\widetilde{m}_i$, $i \in \{0,1\}$, in a specific way such that new biorthogonal wavelet bases are generated. Let us call the new filters p_i and $\widetilde{p}_i$. Then we define in analogy with [22]

$$\begin{aligned} p_0 &= m_0, & \widetilde{p}_0 &= \widetilde{m}_0 + \widetilde{m}_1\, \overline{t(2\cdot)}, \\ p_1 &= m_1 - m_0\, t(2\cdot), & \widetilde{p}_1 &= \widetilde{m}_1 \end{aligned} \tag{15}$$

where t is a trigonometric polynomial. This choice of new filters implies that orthogonality still holds for the filters, i.e.

$$\begin{pmatrix} p_0 & p_0(\cdot+\pi) \\ p_1 & p_1(\cdot+\pi) \end{pmatrix} \begin{pmatrix} \widetilde{p}_0 & \widetilde{p}_0(\cdot+\pi) \\ \widetilde{p}_1 & \widetilde{p}_1(\cdot+\pi) \end{pmatrix}^* = \begin{pmatrix} 1 & 0 \\ 0 & 1 \end{pmatrix}.$$

It is still unanswered under which (explicit) conditions on t the new filters give rise to biorthogonal multiresolution analyses and wavelet bases. While we were not yet able to find precise conditions on t, we still hope that the following results provide more insight into the subject.

Let us denote by $\widetilde{X} = (\eta, 2, \mathbb{Z})$ the new family with

$$\widehat{\eta}(2\cdot) = p_1\widehat{\varphi}, \qquad \text{hence} \qquad \widehat{\eta} = \widehat{\psi} - t\,\widehat{\varphi}.$$

The following is an auxiliary result which can be proved by the usual integration techniques (Parseval's identity and periodization), with applications of the scaling relations for η, θ and the orthogonality relations (13), (14).

Lemma 5. *Let $\eta_{j,k} := 2^{j/2}\eta(2^j\cdot - k)$ and $\theta_{j,k}$ be analogously defined for any $j, k \in \mathbb{Z}$. Then taking inner products in $L_2(\mathbb{R})$ we obtain*

$$\langle \theta_{\ell,\kappa}, \eta_{j,k}\rangle = \begin{cases} 0 & , \text{ if } \ell > j, \\ \delta_{\kappa,k} & , \text{ if } \ell = j, \end{cases}$$

where δ denotes the Kronecker symbol, and if $\ell < j$ we have

$$\langle \theta_{\ell,\kappa}, \eta_{j,k}\rangle = -\frac{2^{(j-\ell)/2}}{2\pi}\int_Q e^{ik\xi}\left(\overline{t(\xi)}\widetilde{m}_1(2^{j-\ell-1}\xi)\prod_{\nu=0}^{j-\ell-2}\widetilde{m}_0(2^\nu\xi)e^{-i2^{j-\ell}\kappa\xi}\right)d\xi.$$

This representation of inner products is needed in order to find the following.

Theorem 6. *Let $X = (\psi, 2, \mathbb{Z})$ and $Y = (\theta, 2, \mathbb{Z})$ be biorthogonal wavelet bases of $L_2(\mathbb{R})$, which are generated from multiresolution analyses with finite filters m_i, $\widetilde{m}_i$, $i \in \{0,1\}$. Furthermore, let t be a trigonometric polynomial defining the new filter p_1 in (15) and $\widetilde{X} = (\eta, 2, \mathbb{Z})$. Then $\widetilde{X}$ is a Bessel family if and only if $t(0) = 0$.*

Proof: Note that φ and η are compactly supported, hence their Fourier transforms are entire functions. In particular, $\widehat{\varphi}(0) \neq 0$, $m_0(0) = 1$ and $m_1(0) = 0$ are implications from the assumptions on X and Y, see [10]. The condition $t(0) = 0$ is necessary, since otherwise $\widehat{\eta}(0) \neq 0$ would contradict the general admissibility condition [4,10]. The more difficult part is to prove the sufficiency of the condition. For this we define operators S_j, $j > 0$, on $L_2(Q)$ by

$$S_j(C)(\xi) = -2^{j/2}\overline{t(\xi)}\widetilde{m}_1(2^{j-1}\xi) \prod_{\nu=0}^{j-2} \widetilde{m}_0(2^\nu \xi) \cdot C(2^j \xi). \tag{16}$$

Their connection to a generalized Laurent operator will become clear later. The same computations as in section 3 give

$$\|S_j\|^2 = \|S_j^* S_j\| = \|T_{|\widetilde{m}_1|^2} \circ T_{|\widetilde{m}_0|^2}^{j-1}(|t|^2)\|_{L_\infty(Q)}.$$

If $t(0) = 0$, then the spectral radius of the transfer operator $T_{|\widetilde{m}_0|^2}$ when restricted to the invariant subspace $E_{|t|^2}$ is less than 1 by [8,Theorem 3.1]. Here, the condition was used that the translates of ϱ define a Riesz basis of their closed linear span. Hence we obtain that

$$\sum_{j>0} \|S_j\| < \infty. \tag{17}$$

In order to show that $\widetilde{X}$ is a Bessel family, we take an arbitrary function $f \in L_2(\mathbb{R})$. Its expansion with respect to the wavelet basis Y is given by

$$f = T_Y^*(\mathbf{c}_j)_{j\in\mathbb{Z}} \qquad \text{with} \qquad (\mathbf{c}_j)_{j\in\mathbb{Z}} \in \ell_2(H).$$

The norm of the sequence $(\mathbf{c}_j)_{j\in\mathbb{Z}}$ is equivalent to the L_2-norm of f, since Y is a Riesz basis. Let us define

$$\mathbf{d}_j := (\langle f, \eta_{j,k}\rangle)_{k\in\mathbb{Z}}, \qquad j \in \mathbb{Z}.$$

If we insert $T_Y^*(\mathbf{c}_j)_{j\in\mathbb{Z}}$ for f, then a direct application of Lemma 5 gives

$$D_j = C_j + \sum_{k>0} S_k(C_{j-k}). \tag{18}$$

The boundedness of the series in (17) implies that there exist constants K and $\widetilde{K}$ with

$$\|(\mathbf{d}_j)_{j\in\mathbb{Z}}\|_{\ell_2(H)} \leq K\|(\mathbf{c}_j)_{j\in\mathbb{Z}}\|_{\ell_2(H)} \leq \widetilde{K}\|f\|_{L_2(\mathbb{R})}.$$

This proves that $\widetilde{X}$ is a Bessel family and completes the proof of the theorem. □

Note that equation (18) defines a generalized Laurent operator on $\ell_2(H)$, since shifts of the index j on the left and right hand sides of (18) are equivalent. Hence we can summarize Theorem 6 by the equation

$$\mathcal{T}_{\widetilde{X}}\mathcal{T}_Y^* = \begin{pmatrix} \ddots & \ddots & \ddots & \ddots & & \\ \ddots & S_1 & \boxed{\mathrm{id}_H} & 0 & 0 & \cdots \\ \cdots & S_2 & S_1 & \boxed{\mathrm{id}_H} & 0 & \ddots \\ & & \ddots & \ddots & \ddots & \ddots \end{pmatrix} \tag{19}$$

where the operators S_j on $L_2(Q)$ are given in (16) and satisfy (17).

Let us shortly describe some further consequences for the lifting scheme. By definition $\widetilde{X}$ is a wavelet basis if and only if $\mathcal{T}_{\widetilde{X}}$ is an invertible operator which maps $L_2(\mathbb{R})$ onto $\ell_2(\mathbb{Z}\times\mathbb{Z})$. This is equivalent to the condition that $\mathcal{T}_{\widetilde{X}}\mathcal{T}_Y^*$ in (19) is an invertible generalized Laurent operator, since Y is a wavelet basis by our starting assumptions. However, the inverse of this operator need not be of the same form as in (19). Indeed, if the inverse has this form, then the generalized Toeplitz operator

$$\mathcal{V}(C_j)_{j\geq 0} = \left(C_j + \sum_{k=0}^{j} S_k C_{j-k}\right)_{j\geq 0}$$

is also invertible, see [16,Lemma 4.2]. This is much stronger than invertibility of the generalized Laurent operator, in general. One can explain the difference with a simple example in the scalar case. Let A be the Laurent matrix

$$A = \begin{pmatrix} \ddots & \ddots & \ddots & \ddots & \ddots & & \\ \ddots & 0 & 2 & \boxed{1} & 0 & 0 & \cdots \\ \cdots & 0 & 0 & 2 & \boxed{1} & 0 & \ddots \\ & & \ddots & \ddots & \ddots & \ddots & \end{pmatrix}$$

which is lower triangular and has absolutely summable columns. Its symbol is the function $F(z) = 1+2z$, $z\in\mathbb{T}$. The operator is invertible on $\ell_2(\mathbb{Z})$ since the symbol does not vanish. The corresponding Toeplitz operator, however, is not invertible, since the winding number of $F(z)$ around 0 is 1. It turns out that the inverse of A on $\ell_2(\mathbb{Z})$ is no longer a lower triangular matrix, but the Laurent matrix with symbol $1/F(z) = \sum_{k=1}^{\infty}(-1)^{k-1}(2z)^{-k}$.

This explains how it might happen, that the new family $\widetilde{X}$ defines a wavelet basis, while the lifting scheme does not give the biorthogonal wavelet basis. On the other hand, it is likely (but only conjectured here), that the lifting scheme produces a new pair of biorthogonal wavelet bases if and only if the inverse of the operator matrix (19) is again lower triangular with finite sum of the norms in each column. This last condition can then be studied by means of the transfer operator in the following way. The adjoint of $\mathcal{T}_{\widetilde{X}} \mathcal{T}_{Y}^{*}$ defines the operator-valued function

$$F^*(z) = \mathrm{id}_H - \sum_{j>0} z^{-j} S_j^*$$

in (8). Taking adjoints in (16) leads to

$$F^*(z)(C) = \mathrm{id}_H - \sum_{j>0} 2^{-j/2} z^{-j} T_{\widetilde{\widetilde{m}}_1} \circ T_{\widetilde{\widetilde{m}}_0}^{j-1} (t \cdot C).$$

Hence the problem is transferred to finding an inverse of $F^*(z)$ on $L_2(Q)$ for all z in the closed unit disc, and the inverse must correspond to an upper triangular Laurent matrix. Since the above series is a Neumann series for $T_{\widetilde{\widetilde{m}}_0}$, we hope that further techniques from operator algebras can help to solve this problem completely.

References

1. Benedetto J. J. and Li S., Subband coding and noise reduction in multiresolution analysis frames, in: *SPIE Proceedings on Wavelet Applications in Signal and Image Processing*, San Diego, 1994, 154–165.
2. Bercovici H., *Operator Theory and Arithmetic in H^∞*, American Math. Soc., Providence, 1988.
3. Carnicer J. M., Dahmen W., and Pena J. M., Local decomposition of refinable spaces and wavelets, Appl. Comput. Harmonic Anal. **3** (1996), 127–153.
4. Chui C. K. and Shi X., Inequalities of Littlewood-Paley type for frames and wavelets, SIAM J. Math. Anal. **24** (1993), 263–277.
5. Chui C. K. and Shi X., Bessel sequences and affine frames, Appl. Comput. Harmonic Anal. **1** (1993), 29–49.
6. Cohen A. and Daubechies I., A stability criterion for biorthogonal wavelet bases and their subband coding scheme, Duke Math. J. **68** (1992), 313–335.
7. Cohen A. and Daubechies I., A new technique to estimate the regularity of refinable functions, Rev. Mat. Iberoam. **12** (1996), 527–591.
8. Cohen A., Gröchenig K., and Villemoes L., Regularity of multivariate refinable functions, preprint.
9. Daubechies I., The wavelet transform, time-frequency localization and signal analysis, IEEE Trans. Inform. Theory **36** (1990), 961–1005.
10. Daubechies I., *Ten Lectures on Wavelets*, CBMS-NSF Reg. Conf. Series in Appl. Math. 61, Soc. Industrial and Applied Math., Philadelphia, 1992.

11. Daubechies I., Grossmann A. and Meyer Y., Painless nonorthogonal expansions, J. Math. Phys. **27** (1986), 1271–1283.

12. Eirola T., Sobolev characterization of solutions of dilation equations, SIAM J. Math. Anal. **23** (1992), 1015–1030.

13. Fillmore P. A., *Notes on Operator Theory*, Van Nostrand Reinhold Co., New York, 1970.

14. Hervé L., Construction et régularité des fonctions d'échelle, SIAM J. Math. Anal. **26** (1995), 1361–1385.

15. Mallat S., Multiresolution approximations and wavelet orthonormal bases of $L_2(\mathbb{R})$, Trans. Amer. Math. Soc. **315** (1989), 69–87.

16. Rabindranathan M., On the inversion of Toeplitz operators, J. Math. and Mech. **19** (1969), 195–206.

17. Ron A. and Shen Z., Affine systems in $L_2(\mathbb{R}^d)$: the analysis of the analysis operator, to appear in J. Funct. Anal.

18. Ron A. and Shen Z., Compactly supported tight affine spline frames in $L_2(R^d)$, to appear in Math. Comp.

19. Stöckler J., *Multivariate affine Frames*, Habilitationsschrift, Universität Duisburg, 1995.

20. Stöckler J., A Laurent operator technique for multivariate frames and wavelet bases, in *Advanced Topics in Multivariate Approximation*, Fontanella F., Jetter K. and Laurent P.-J. (eds.), World Scientific, Singapore, 1996, 339–354.

21. Stöckler J., Preconditioning of the frame algorithm, preprint, December 1996.

22. Sweldens W., The lifting scheme: a custom-design construction of biorthogonal wavelets, Appl. Comput. Harmonic Anal. **3** (1996), 186–200.

23. Villemoes L. F., Wavelet analysis of refinement equations, SIAM J. Math. Anal. **25** (1994), 1433–1460.

Joachim Stöckler
Institut für Angewandte Mathematik und Statistik
Universität Hohenheim
D-70593 Stuttgart, Germany
stockler@uni-hohenheim.de

List of Participants

VLADISLAV F. BABENKO, Faculty of Mechanics and Mathematics, Dnepropetrovsk State University, pr. Gagarina 72, Dnepropetrovsk, GSP 320625, Ukraine
babenko@dsu.dp.ua

DRUMI D. BAINOV, Higher Medical Institute, P.O. Box 45, Sofia 1504, Bulgaria
dian@bgcict.acad.bg

HANS-PETER BLATT, Mathematisch-Geographische Fakultät, Katholische Universität Eichstätt, Ostenstr. 26, 85071 Eichstätt, Germany
mga009@ku-eichstaett.de

BRUNO BROSOWSKI, Fachbereich 12, Johann-Wolfgang-Goethe-Universität, Robert-Mayer-Str. 6–10, 60054 Frankfurt am Main, Germany
brosowski@math.uni-frankfurt.de

MARTIN D. BUHMANN, Mathematik Departement, ETH Zentrum, 8092 Zürich, Switzerland
mdb@math.ethz.ch

CHARLES K. CHUI, Department of Mathematics, Texas A&M University, College Station, TX 77843, USA
cchui@tamu.edu

OLEG DAVYDOV, Department of Mechanics and Mathematics, Dnepropetrovsk State University, pr. Gagarina 72, Dnepropetrovsk, GSP 320625, Ukraine
davydov@dsu.dp.ua

FRANZ-JÜRGEN DELVOS, Fachbereich Mathematik I, Universität GH Siegen, 57068 Siegen, Germany
delvos@mathematik.uni-siegen.d400.de

MARTINA FINZEL, Mathematisches Institut, Universität Erlangen-Nürnberg, Bismarckstr. 1½, 91054 Erlangen, Germay
finzel@mi.uni-erlangen.de

KLAUS FREYBURGER, Abteilung DEV.CO, SAP AG, Postfach 1461, 69185 Walldorf, Germany
klaus.freyburger@sap-ag.de

MANFRED VON GOLITSCHEK, Institut für Angewandte Mathematik und Statistik, Universität Würzburg, 97074 Würzburg, Germany
goli@mathematik.uni-wuerzburg.de

HEINZ H. GONSKA, Fachbereich Mathematik, Universität Duisburg, 47048 Duisburg, Germany
gonska@informatik.uni-duisburg.de

RENÉ GROTHMANN, Mathematisch-Geographische Fakultät, Katholische Universität Eichstätt, 85071 Eichstätt, Germany
grothm@ku-eichstaett.de

WERNER HAUSSMANN, Fachbereich Mathematik, Universität Duisburg, 47048 Duisburg, Germany
haussmann@math.uni-duisburg.de

MARGARETA HEILMANN, Fachbereich Mathematik, Bergische Universität - Gesamthochschule Wuppertal, Gaußstr. 20, 42097 Wuppertal, Germany
margareta.heilmann@math.uni-wuppertal.de

KURT JETTER, Institut für angewandte Mathematik und Statistik, Universität Hohenheim, 70593 Stuttgart, Germany
kjetter@uni-hohenheim.de

ANNA KAMONT, Mathematical Institute of Polish Academy of Sciences, ul. Abrahama 18, 81-825 Sopot, Poland
a.kamont@impan.gda.pl

BURKHARD LENZE, Fachbereich Informatik, Fachhochschule Dortmund, Postfach 105018, 44047 Dortmund, Germany
lenze@fh-dortmund.de

WU LI, Department of Mathematics and Statistics, Old Dominion University, Norfolk, VA 23529, USA
wuli@math.odu.edu

WILL A. LIGHT, Mathematics Department, University of Leicester, Leicester, LE1 7RH, United Kingdom
pwl@mcs.le.ac.uk

TOM LYCHE, Institutt for informatikk, Universitetet i Oslo, P.O. Box 1080, Blindern, 0316 Oslo, Norway
tom@ifi.uio.no

JOHN C. MASON, School of Computing and Mathematics, University of Huddersfield, Queensgate, Huddersfield, HD1 3DH, United Kingdom
j.c.mason@hud.ac.uk

BERND MULANSKY, Institut für Numerische Mathematik, Technische Universität Dresden, 01062 Dresden, Germany
mulansky@math.tu-dresden.de

PAUL G. NEVAI, Department of Mathematics, Ohio State University, Columbus, OH 43210, USA
nevai@math.ohio-state.edu

ERICH NOVAK, Mathematisches Institut, Universität Erlangen-Nürnberg, Bismarckstr. 1½, 91054 Erlangen, Germany
novak@mi.uni-erlangen.de

GÜNTHER NÜRNBERGER, Lehrstuhl für Mathematik IV, Universität Mannheim, 68131 Mannheim, Germany
nuernberger@math.uni-mannheim.de

WERNER OETTLI, Lehrstuhl für Mathematik VII, Universität Mannheim, 68131 Mannheim, Germany
oettli@math.uni-mannheim.de

ALLAN PINKUS, Department of Mathematics, Technion, Haifa, Israel
pinkus@tx.technion.ac.il

GERLIND PLONKA, Fachbereich Mathematik, Universität Rostock, 18051 Rostock, Germany
gerlind.plonka@mathematik.uni-rostock.d400.de

JÜRGEN PRESTIN, Fachbereich Mathematik, Universität Rostock, 18051 Rostock, Germany
prestin@mathematik.uni-rostock.d400.de

CHRISTOPHE RABUT, Departement de Genie Mathématique, Institut National des Sciences Appliquées, Complexe Scientifique de Rangueil, 31077 Toulouse Cedex, France
rabut@gmm.insa-tlse.fr

MANFRED REIMER, Fachbereich Mathematik, Universität Dortmund, 44221 Dortmund, Germany
reimer@euler.mathematik.uni-dortmund.de

THOMAS RIESSINGER, Fachbereich MND, Fachhochschule Frankfurt am Main, 60318 Frankfurt am Main, Germany
riessinger@rz.fh-frankfurt.d400.de

ROBERT SCHABACK, Institut für Numerische und Angewandte Mathematik, Universität Göttingen, Lotzestr. 16–18, 37083 Göttingen, Germany
schaback@math.uni-goettingen.de

WALTER SCHEMPP, Lehrstuhl für Mathematik I, Universität GH Siegen, 57068 Siegen, Germany
schempp@mathematik.uni-siegen.d400.de

KARL SCHERER, Institut für angewandte Mathematik, Rheinische Friedrich-Wilhelms-Universität Bonn, Wegelerstr. 6, 53115 Bonn
scherer@iam.uni-bonn.de

GERHARD SCHMEISSER, Mathematisches Institut, Universität Erlangen-Nürnberg, Bismarckstr. 1 1/2, 91054 Erlangen, Germany
schmeisser@mi.uni-erlangen.de

JOCHEN W. SCHMIDT, Institut für Numerische Mathematik, Technische Universität Dresden, 01062 Dresden, Germany
jschmidt@math.tu-dresden.de

MANFRED SOMMER, Mathematisch-Geographische Fakultät, Katholische Universität Eichstätt, 85071 Eichstätt, Germany
manfred.sommer@ku-eichstaett.de

GABRIELE STEIDL, Fakultät für Mathematik und Informatik, Universität Mannheim, 68131 Mannheim, Germany
steidl@kiwi.math.uni-mannheim.de

GEORG STILL, Faculty of Applied Mathematics, University of Twente, P.O. Box 217, 7500 AE Enschede, Netherlands
g.still@math.utwente.nl

JOACHIM STÖCKLER, Institut für angewandte Mathematik und Statistik, Universität Hohenheim, 70593 Stuttgart, Germany
stockler@uni-hohenheim.de

HANS STRAUSS, Institut für Angewandte Mathematik, Universität Erlangen-Nürnberg, Martensstr. 3, 91058 Erlangen, Germany
strauss@am.uni-erlangen.de

MANFRED TASCHE, Institut für Mathematik, Medizinische Universität zu Lübeck, Wallstr. 40, 18051 Lübeck, Germany
tasche@informatik.mu-luebeck.de

GUIDO WALZ, Lehrstuhl für Mathematik IV, Universität Mannheim, 68131 Mannheim, Germany
walz@math.uni-mannheim.de

FRANK ZEILFELDER, Lehrstuhl für Mathematik IV, Universität Mannheim, 68131 Mannheim, Germany
zeilfelder@math.uni-mannheim.de

International Series of Numerical Mathematics

International Series of Numerical Mathematics *is open to all aspects of numerical mathematics. Some of the topics of particular interest include free boundary value problems for differential equations, phase transitions, problems of optimal control and optimization, other nonlinear phenomena in analysis, nonlinear partial differential equations, efficient solution methods, bifurcation problems and approximation theory. When possible, the topic of each volume is discussed from three different angles, namely those of mathematical modeling, mathematical analysis, and numerical case studies.*

Titles previously published in the series

ISNM 115 **R. Bulirsch, D. Kraft (Eds):** Computational Optimal Control, 1994 (ISBN 3-7643-5015-6)

ISNM 116 **P.W. Hemker, P. Wesseling (Eds):** Multigrid Methods IV. Proceedings of the Fourth European Multigrid Conference, Amsterdam, July 6-9, 1993 (ISBN 3-7643-5030-X)

ISNM 117 **R.E. Bank, R. Bulirsch, H. Gajewski, K. Merten (Eds):** Mathematical Modelling and Simulation of Electrical Circuits and Semiconductor Devices, 1994 (ISBN 3-7643-5053-9)

ISNM 118 **W. Desch, F. Kappel, K. Kunisch (Eds):** Control and Estimation of Distributed Parameter Systems: Nonlinear Phenomena. International Conference on Control and Estimation of Distributed Parameter Systems, Vorau, July 18-24, 1993, 1994 (ISBN 3-7643-5098-9)

ISNM 119 **R. Zahar (Ed.):** Approximation and Computation: A Festschrift in Honor of Walter Gautschi, 1995 (ISBN 0-8176-3753-2)

ISNM 120 **W. Hackbusch:** Integral Equations: Theory and Numerical Treatment, 1995 (ISBN 3-7643-2871-1)

ISNM 121 **R. Jeltsch, M. Mansour (Eds)**: Stability Theory. Hurwitz Centenary Conference, Centro Stefano Franscini, Ascona, 1995, 1996 (ISBN 3-7643-5474-7)

ISNM 122 **A.M. Khludnev, J. Sokolowski:** Modelling and Control in Solid Mechanics, 1997 (ISBN 3-7643-5238-8)

ISNM 123 **C. Bandle, W.N. Everitt, L. Losonczi, W. Walter** (Eds): General Inequalities 7. 7th International Conference, Oberwolfach, November 13–18, 1995, 1997 (ISBN 3-7643-5722-3)

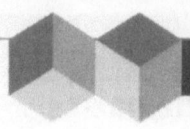

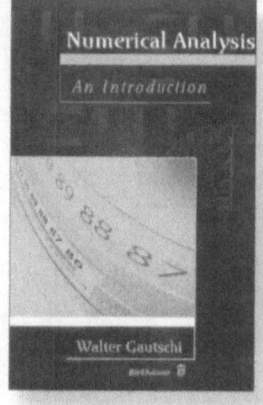